Methods in Molecular Biology

Series Editor:
John M. Walker
School of Life and Medical Sciences
University of Hertfordshire
Hatfield, Hertfordshire, United Kingdom

For further volumes:
http://www.springer.com/series/7651

Rice Grain Quality

Methods and Protocols

Editors

Nese Sreenivasulu

Grain Quality and Nutrition Center, International Rice Research Institute, Manila, Philippines

Humana Press

Editor
Nese Sreenivasulu
Grain Quality and Nutrition Center
International Rice Research Institute
Manila, Philippines

ISSN 1064-3745 ISSN 1940-6029 (electronic)
Methods in Molecular Biology
ISBN 978-1-4939-8912-6 ISBN 978-1-4939-8914-0 (eBook)
https://doi.org/10.1007/978-1-4939-8914-0

Library of Congress Control Number: 2018960445

This Humana Press imprint is published by the registered company Springer Science+Business Media, LLC part of Springer Nature.
The registered company address is: 233 Spring Street, New York, NY 10013, U.S.A.

Dedication

I dedicate this book to Dr. Bienvenido O. Juliano *(August 15, 1936–February 21, 2018), who immensely contributed significant knowledge in the area of cereal chemistry research. Dr. Juliano pioneered the research of grain quality at the International Rice Research Institute (IRRI) in 1961; he went on to spend more than 32 years at IRRI, who made significant research contributions in the area of cooking, eating, and nutritional qualities of rice grain. He also helped to build the grain-quality research capability of PhilRice, where he continued to pursue his rice research as a senior consultant/expert.*

Preface

Acceptance of new rice genotypes requires their ability to satisfy consumer preferences for premium grain quality, besides being high yielding. As rice consumers become increasingly particular about the quality of the rice, we need to ensure that modern varieties were less susceptible to breaking during milling and assure premium cooking quality with optimum texture, flavor, and aroma. Translating human health benefits by enhancing nutritional properties includes enriching micronutrients, ensuring food safety, and introducing slower digestibility factors. Recent advances made in introducing various high-throughput phenotyping methods to screen milling quality, cooking quality, and nutritional quality in the pool of breeding material are discussed in depth. This volume presents the relevant methods and detailed protocols with appropriate instructions outlined by experts to facilitate grain quality and nutritional screening in the germplasm. Detailed protocols to define seed development stages, panicle architectural traits to understand yield components, and measure physical traits such as grain dimensions using imaging techniques and chalk and head rice yield have been discussed in Chapters 1, 3, 4, 5, and 6. We need to link initial indicators of cooking quality (amylose, gel consistency, and gelatinization temperature) with texture and viscoelastic properties to capture distinct cooking quality classes. Also covered are biochemical methods that measure properties related to cooking such as starch structure and protein properties in Chapters 2, 7, 8, 9, and 10. Holistic understanding of grain quality traits by exploring metabolomics platforms to screen primary metabolites and volatiles has been described in Chapter 11. In addition, health and nutritional aspects need to be factored into rice breeding programs to contribute to the overall wellness of rice consumers. Therefore, we need to take into account breeding healthier target traits by capturing the diversity for low glycemic index and high-resistant starch and enriching nonstructural polysaccharides and micronutrients with acceptable cooking quality, texture, and palatability. Methods related to micronutrient profiling, screening heavy metals, and identifying rice cultivars with lower glycemic index have been addressed in Chapters 13–15.

We need to establish the genetic basis of the variation of cooking and eating quality traits through genome-wide association studies by studying both diverse set of lines and pre-breeding core collection. Genome-wide *-omics* analyses have provided efficient approaches to identify key genomic regions that control grain quality traits. Knowledge of these genes and the influence of specific alleles present in both domesticated and wild rice gene pools provide a robust platform for marker-assisted selection in breeding to introgress premium grain quality traits in high-yielding backgrounds. Emphasis has been given to utilizing resequencing resources for the gene discovery programs via structural genomics, exploring transcriptome, epigenetics, and genome editing technologies to unravel grain quality, and nutritional traits have been discussed in Chapters 12 and 16–18. In summary, overall emphasis has been given to holistic understanding of grain quality traits covering various phenomics technologies and links it through to gene discovery via QTL cloning and structural-functional genomics strategies.

Manila, Philippines *Nese Sreenivasulu*

Contents

Contributors

ROINAND AGUILA • *International Rice Research Institute, Los Baños, Laguna, Philippines*
CRISLINE MAE ALHAMBRA • *International Rice Research Institute, Los Baños, Laguna, Philippines*
ROSLEN ANACLETO • *International Rice Research Institute, Los Baños, Laguna, Philippines*
ANINDYA BANDYOPADHYAY • *International Rice Research Institute, Los Baños, Laguna, Philippines; Syngenta Beijing Innovation Center, Changping District, Beijing, China*
AKSHAYA BISWAL • *International Rice Research Institute, Los Baños, Laguna, Philippines*
YARIV BROTMAN • *Max-Planck-Institute of Molecular Plant Physiology, Potsdam-Golm, Germany; Department of Life Sciences, Ben-Gurion University of the Negev, Beersheba, Israel*
VITO M. BUTARDO JR. • *International Rice Research Institute, Los Baños, Laguna, Philippines; Department of Chemistry and Biotechnology, Faculty of Science, Engineering and Technology, Swinburne University of Technology, Hawthorn, VIC, Australia; Faculty of Science, Engineering and Technology, Swinburne University of Technology, Hawthorn, VIC, Australia*
RUBEN CHAVEZ • *International Rice Research Institute, Los Baños, Laguna, Philippines*
ROBERT COE • *International Rice Research Institute, Los Baños, Laguna, Philippines*
PAUL A. COUNCE • *University of Arkansas, Rice Research and Extension Center, Stuttgart, AR, USA*
ROSA PAULA O. CUEVAS • *International Rice Research Institute, Los Baños, Laguna, Philippines*
SUSHIL DHITAL • *ARC Centre of Excellence in Plant Cell Walls, Centre for Nutrition and Food Sciences, Queensland Alliance for Agriculture and Food Innovation, The University of Queensland, St Lucia, QLD, Australia*
ALISDAIR R. FERNIE • *Max-Planck-Institute of Molecular Plant Physiology, Potsdam-Golm, Germany; Center of Plant System Biology and Biotechnology, Plovdiv, Bulgaria*
AGNELO FURTADO • *Queensland Alliance for Agriculture and Food Innovation, University of Queensland, Brisbane, QLD, Australia*
ROBERT G. GILBERT • *Joint International Research Laboratory of Agriculture and Agri-Product Safety of Ministry of Education of China, Yangzhou University, Yangzhou, China; Centre for Nutrition & Food Sciences, QAAFI, University of Queensland, Brisbane, Australia*
GOPALA KRISHNAN SUBBAIYAN • *Division of Genetics, ICAR-Indian Agricultural Research Institute, New Delhi, India*
ROBERT J. HENRY • *Queensland Alliance for Agriculture and Food Innovation, University of Queensland, Brisbane, QLD, Australia*
JACKIE • *Shimadzu (Asia Pacific) Pte. Ltd., Singapore, Singapore*
ROSARIO JIMENEZ • *International Rice Research Institute, Los Baños, Laguna, Philippines*
BIENVENIDO O. JULIANO • *Philippine Rice Research Institute, Los Baños, Laguna, Philippines*

MARKUS KUHLMANN • *Leibniz Institute of Plant Genetics and Crop Plant Research, Gatersleben, Germany*
JENNINE ROSE LAPIS • *International Rice Research Institute, Los Baños, Laguna, Philippines*
CHENG LI • *Joint International Research Laboratory of Agriculture and Agri-Product Safety of Ministry of Education of China, Yangzhou University, Yangzhou, China*
HONGYAN LI • *School of Food and Chemical Engineering, Beijing Technology and Business University, Beijing, China*
CINDY LLORENTE • *International Rice Research Institute, Los Baños, Laguna, Philippines*
ARDASHIR K. MASOULEH • *Queensland Alliance for Agriculture and Food Innovation, University of Queensland, Brisbane, QLD, Australia*
KAREN A. K. MOLDENHAUER • *University of Arkansas, Rice Research and Extension Center, Stuttgart, AR, USA*
LILIA MOLINA • *International Rice Research Institute, Los Baños, Laguna, Philippines*
ERSTELLE PASION • *International Rice Research Institute, Los Baños, Laguna, Philippines*
MANDY PÜFFELD • *Leibniz Institute of Plant Genetics and Crop Plant Research, Gatersleben, Germany*
WILLIAM PAUL QUICK • *International Rice Research Institute, Los Baños, Laguna, Philippines*
MARNOL V. SANTOS • *International Rice Research Institute, Los Baños, Laguna, Philippines*
CHRISTIANE SEILER • *Leibniz Institute of Plant Genetics and Crop Plant Research, Gatersleben, Germany*
NESE SREENIVASULU • *International Rice Research Institute, Los Baños, Laguna, Philippines*
PAWAN S. TAKHAR • *Department of Food Science and Human Nutrition, University of Illinois at Urbana-Champaign, Urbana, IL, USA*
DANIEL L. E. WATERS • *Southern Cross Plant Science, Southern Cross University, Lismore, NSW, Australia; ARC ITTC for Functional Grains, Charles Sturt University, Wagga, NSW, Australia*
XIAOJIA YIN • *International Rice Research Institute, Los Baños, Laguna, Philippines*
IMAN ZAREI • *International Rice Research Institute, Los Baños, Laguna, Philippines*

Chapter 1

Improving Head Rice Yield and Milling Quality: State-of-the-Art and Future Prospects

Vito M. Butardo Jr. and Nese Sreenivasulu

Abstract

Increasing paddy yield in rice does not directly translate to enhancing food security because significant decrease in grain yield can happen during postharvest processing of the rice paddy. In parallel with enhancing paddy yield, improving the milling quality of rice is essential in ensuring food security by mitigating the impact of significant losses during the postharvest processing of rice grains. From an industrial standpoint, maximizing the milling recovery of whole grain polished rice is crucial in fetching higher revenues to rice farmers. Significant advances in rice postharvest processing technology have been achieved which are geared toward reducing the incidence of fissures and chalkiness to increase head rice yield (HRY) in rice. The genetic bases of kernel development and grain dimension are also characterized. In addition to these advancements, an integrated phenotyping suite to simultaneously characterize phenotypes related to milling quality will help in screening for breeding lines with high HRY. Toward this goal, modern imaging tools and computer algorithms are currently being developed for high-throughput characterization of rice milling quality. With the availability of more sophisticated, affordable, automated, and nondestructive phenotyping methods of milling quality, it is envisioned that significant improvement in HRY will be made possible to ensure rice food security in the future.

Key words Breeding, Chalk, Fissure, Genomics, Grain quality, Head rice yield, Milling quality

1 Introduction

Breeders traditionally focus on increasing paddy yield coupled with enhanced stress tolerance to ensure food sustainability and security for rice. However, focusing on paddy yield alone is not enough because economic losses occur as a result of quantity reduction due to poor milling quality and cooking quality deterioration can happen at any stage along the postharvest value chain [1]. It is estimated that up to 30% loss in rice grain can happen due to rice breakage as they are subjected to mechanical stress from harvesting to grain processing. Rice grain breaks when the extrinsic mechanical stresses applied to the rice grain during processing are greater than the intrinsic grain strength. This can be estimated by bending or

Nese Sreenivasulu (ed.), *Rice Grain Quality: Methods and Protocols*, Methods in Molecular Biology, vol. 1892,
https://doi.org/10.1007/978-1-4939-8914-0_1,

fracture strength, which is the maximum force a brittle material like rice grain can tolerate before it cracks due to flexural load [2]. Loss in kernel biomass is usually assessed by quantifying head rice yield (HRY), which is the proportion of rough rice that remains as head rice (intact grain) after complete milling. Head rice, on the other hand, is conventionally defined as intact grains that have ¾ of the original kernel length after complete milling. HRY is the gold standard of rice millers to quantify milling quality. Elevated proportion of broken rice grains are usually avoided because it reduces the commercial value of broken rice by half.

Usually, the milling yield in rice breeding lines varies from 70 to 79% while HRY potential is estimated at 24–74%. The significant loss in seed biomass due to kernel breakage which can otherwise feed the hungry is primarily mitigated by observing best postharvest practices as part of the rice quality management system [3]. There are currently few research and breeding efforts made to develop new cultivars with grains that have superior HRY properties. Previous studies revealed that a major way of improving HRY quality of rice is by reducing kernel weakness and susceptibility to breakage, which is usually addressed at the postharvest level by optimizing the grain drying process. The impact of dehulling and milling on breakage susceptibility of rice is well characterized [2]. This trait is associated with grain fissuring (cracking), chalkiness, maturity of rice grains, and kernel dimensions. All these factors will be discussed below after a short discussion on the best postharvest method to reduce grain loss and increase HRY.

2 Observing Best Postharvest Practices

The quality management practices in the rice postproduction sector are already comprehensively defined [3]. Based on voluminous comprehensive studies summarized in IRRI Knowledge Bank, simple harvest and postharvest practices are usually promoted to farmers and local rice processors to ensure ease of compliance. Harvest should be properly scheduled to coincide with 22–24% moisture content (MC). If at all possible, combine harvesting is encouraged to avoid delays and the harvested rice panicles should not be stacked in the field. After threshing, the harvest should be immediately dried within 24 h after harvesting. During drying, the temperature should not exceed 43 °C in a flatbed dryer or any other fixed bed batch dryer. For milling in a mixing type dryer with tempering between drying passes, temperature should not exceed 50 °C. For sun drying, the grains should have a depth between 2 and 4 cm and they should be mixed every 30 min. The aim is to slowly dry to 14% MC which ensures minimal grain breakage. After drying, the seeds should be stored safely to avoid pests, water, and re-wetting of the grains.

3 Reducing Grain Fissuring

Fissuring of rice is a major problem in the rice industry because it results in the reduction of HRY and commercial value of the grain. Rice fissures are large internal fractures that usually develop perpendicular to the length of the grain. This trait introduces weakness in the grain due to hairline cracks which usually lead to grain breakage upon dehulling and milling. Higher incidence of grain breakage leads to significant reduction in HRY. The formation of fissures is influenced by environmental conditions in the field before and during harvest, as well as during grain processing and storage. At the end of grain-filling stage, the supply of water from the vegetative tissues in rice is cut-off to initiate grain maturation. Farmers also withhold irrigated water from the farm plots to expedite grain desiccation. During this stage, MC is no longer controlled by moisture transfer within the panicle but is susceptible and varies in response to environmental fluctuations. Rain, air temperature, and humidity fluctuations are the three major environmental factors in the field that can influence the formation of fissures during and prior to harvest. At this stage, the thermal and material properties of rice grain tend to change based on relative humidity (RH), temperature (*T*), and MC. Grain fissures are usually formed in the field when the moisture MC suddenly drops below 15% [4]. The MC of individual kernels within the same panicle can vary up to 10% [2] because rice grains belonging to a single inflorescence do not mature synchronously. This variation in MC is expected to be higher among different rice plants growing on the same field. Because grains of different MC are hygroscopic, kernels with variations in MC desorb or adsorb moisture from the air until an equilibrium MC is reached. This equilibrium MC depends on both the RH and air temperature. Moisture gradient induces swelling in the grain and induces force within the endosperm, which can cause fissures. The response of rice kernels to tensile and compressive stress in a MC gradient will determine whether the grain will form fissure during grain ripening and desiccation stage [5]. Not all grains containing fissures break during milling but the role of fissure in increasing the incidence of grain breakage and reduction in HRY is well known [6, 7].

Rice grains from different varieties are usually mixed by farmers or grain processors during harvest and storage. This factor additionally contributes to the large variations in MC of grains, ranging from 16 to 26%. Suboptimal drying and storage at the postharvest stage can also lead to significant fissuring. Exposing rice kernels to extreme drying conditions usually increases the proportion of kernels with fissures. However, it is commonly observed that most rice grains do not form fissures immediately after drying [8]. Instead, the fissures emerge when the rice grains have already cooled down after drying. As a result, post-drying treatment like optimum high temperature tempering can be implemented to reduce the incidence of fissuring [9, 10]. Tempering in commercial mills involves holding the grains for a certain time in between drying cycles to decrease the MC gradient and hence reduce the incidence of fissuring and subsequent rice breakage. A modern method of tracking the grains MC gradient and the presence of fissures during the tempering processes can be facilitated by magnetic resonance imaging (MRI) analyses [11]. This technology is one of the more promising techniques that can be employed to track milling quality at every stage of the value chain.

Glass transition temperature (T_g) can be used to explain rice kernel fissuring during the course of seed desiccation where drying and tempering are being accounted [12]. T_g is the temperature range at which polymers shift from a soft "rubbery" state to a hard "glassy" material and vice versa. This can be measured by differential scanning calorimetry (DSC) and dynamic mechanical analysis (DMA), but thermomechanical analysis (TMA) is the most widely-accepted method [13–15]. Taking into account the T_g range during rice drying and tempering is crucial because it affects milling quality [12].

The relationship of kernel MC gradients and T_g to HRY is well established [16, 17]. The major polymer in rice kernel is starch, which can reach up to 90% in milled grains. It is a partially crystalline and partially amorphous polymer of glucose composed of essentially linear amylose and highly branched amylopectin. The amylose and amylopectin branch points form the amorphous layer while the amylopectin outer chains form the crystalline region. The semi-crystalline property of starch dynamically changes according to T and MC [18, 19]. During the drying process, the kernel temperature increases causing each grain to dry unevenly from the surface to the inside [20, 21]. As a result of this moisture gradient, the rice kernel undergoes glass transition at different parts of the grain. During this phase transition, a temperature and moisture gradient can develop along the rice kernel. The temperature gradient has a negligible effect on the integrity of the rice kernel but the MC gradient has a significant impact during and after drying, highlighting the importance of water as plasticizer of rice kernel [15]. Moisture absorption and desorption in the rice grain can significantly affect the resulting HRY [22, 23]. The reversible shifts between the glassy and rubbery states of starch influence the formation of fissures and grain breakage due to differential tensile and compressive stresses resulting from MC gradient within the rice kernel [5, 24]. This glass transition of starch typically occurs during the desiccation process and commercial drying of rice grain [13]. It is possible that the outer layer of a growing grain is already in the glass state while the inner kernel is still in the rubbery state, which can induce fissure and subsequent breakage as a result of differential stresses [9, 13]. Drying the rice grain in the glassy region is preferred because it does not cause significant reduction in HRY [16]. In contrast, drying in the rubbery state and cooling down immediately without tempering can cause increase in MC gradients leading to significant reduction in HRY [16].

The differential tensile and compressive stresses in rice grain during the drying process have viscous and elastic components due to the viscoelastic properties of rice grains. The viscous stress component can be minimized by reducing moisture gradient inside the rice kernel while the elastic component can be reduced by tempering above T_g [25]. To enhance milling recovery, infrared (IR) drying with tempering has been demonstrated to be more effective than conventional drying [26]. However, the associated cost and the feasibility of this technology need to be further explored in large-scale rice milling operations. To circumvent this problem, low-cost optimal drying methods for freshly-harvested rice grains were developed. One recent innovation is the Solar Bubble Dryer (SBD), a low-cost drying method developed by IRRI, Hohenheim University and GrainPro. SBD can process up to one ton of rice compared to mechanical dryer, and has a very low operating cost because it relies on solar energy. Drying occurs in a buffered plastic tunnel which regulates the temperature thereby protecting the grains from overheating. This technology improves upon the traditional sun drying method because the drying process happens in a completely sealed environment. Consequently, postharvest losses are minimized and the grains are protected from contamination and pests.

4 Reducing Grain Chalkiness

Chalk is the opaque white streak in a translucent rice grain. Chalk is influenced by spikelet position [27], it is usually induced by elevated temperature during grain filling [28, 29], and it negatively affects milling quality and HRY [2]. The formation of chalk is related to the disturbed translocation of assimilates [27] leading to the abortion of starch and protein biosynthesis in the developing grain [30]. This produces incompletely filled starch granules with many air spaces in between, which is visible as opaque spots along the translucent grain [31]. These air spaces, which can be visualized by X-ray microtomography [32], are responsible for making chalky grains more brittle. Chalky grains crack more easily compared to translucent grains [33], significantly elevating the proportion of broken grains which are usually discarded [34]. The susceptibility of chalky grains to breakage is due to softer kernel [31, 35] and increased tendency to form fissures [36]. These make chalky grains more prone to mechanical stresses during grain dehulling and polishing. Differences in endosperm compactness and micro-porosity are also thought to explain differences in breakage susceptibility between chalky and non-chalky grains [2]. Even if the chalky grains resist breakage and survive the milling process, their presence significantly lowers the market value and consumer acceptance of milled rice [37] due to alteration in cooking quality [38] and reduction in sensory properties [39]. A more in-depth study on chalk is crucial because it has negative impacts on the value of the rice along the supply chain, from farmer to consumer. This is exhaustively discussed in a recent review article [30].

A more accurate phenotyping method for chalk is needed. The conventional method using Cervitec (Foss) is limited because broken grains are discarded prior to analyses. This introduces bias because only the whole grains are characterized for grain dimension and chalk. Consequently, the percent chalkiness detected by this method is only applicable for the breakage-resistant kernels, as the breakage susceptible proportion of grain is not subjected to imaging. Therefore Cervitec can only detect the chalkiness which affects appearance quality, not the chalkiness that can impact HRY. A newer method based on SeedCount (Next Instrument) is available that uses flatbed scanner for transmission and reflectance image analysis. It offers a better method for chalkiness trait quantification coupled with accurate dimension, color, degree of whiteness, immature grain impurity, and broken grain analyses. Novel quantitative image processing technology that accurately detects location and type of chalk along the grain is available using computer vision which relies on support vector machine (SVM) coupled with principal component analysis (PCA) or convex point matching [40, 41]. However, these methods are not yet commercially available. In addition, they do not simultaneously measure other traits like kernel dimension, percent broken grains, and percent immature grains that are necessary to obtain a holistic understanding of milling quality.

5 Tailoring Grain Maturity and Kernel Dimension

Asynchronous kernel maturation leads to elevated proportion of immature rice grains. This is a problem not just in deterioration of appearance quality but also in milling quality because immature grains tend to be more susceptible to breakage during grain processing. Breakage susceptibility due to immature grains is comparatively less significant for photosensitive rice varieties. This is because photosensitive rice cultivars tend to have more uniform and synchronous anthesis resulting in uniform grain maturation compared to non-photosensitive rice accessions. For non-photosensitive rice cultivars, early-maturing varieties (90–100 days) have more immature grains than medium-maturity cultivars (130–140 days). Hence, tailoring days to flowering (DTF) and synchronizing anthesis are two important targets to minimize HRY and reduce percentage of chalk. There appears to be a link between flowering time, heading date, panicle development, and grain yield in rice [42, 43]. This is an active area of research which is proving to be very challenging because it involves genetic, epigenetic mechanisms, and complex regulatory network [44, 45]. This complexity is better addressed using grain quality genomics and systems genetics techniques as recently demonstrated in rice [46].

With respect to kernel dimension, milled rice can be clustered into four main types based on size: short grain, medium-grain, and long grain (Table 1). Similarly, it can also be grouped into four distinct categories based on shape: slender, medium, bold, and round (Table 2). Kernel size and shape is one of the most stable properties of a rice variety. Kernel dimensions impact HRY, particularly in long slender background. It is hard to isolate the impact of grain size trait from other kernel defects like fissures, chalk, and immature grains [47, 48]. It is therefore crucial to develop a robust method that can simultaneously detect kernel dimension, detects grain breakage, fissure, and chalk percentage of grain that cumulatively affects HRY. Attempts to combine image processing, discriminant analysis, and artificial neural network were successful in morphometric and varietal identification of corn [49] and wheat [50]. In line with this, nondestructive rice quality control, grading, and milling quality scoring is currently being tested in laboratory-scale conditions. For example, machine vision was used to identify grain defects in rice [51]. Breakage and cracks were identified in parboiled rice using a Java program based on an ImageJ improvement which introduces gap-filling segmentation technique [52]. In another study, velocity representation method was used to identity broken kernels using pattern recognition of contour characteristics of rice after image acquisition [53]. Similar studies used flatbed scanning and image analysis to determine size distribution and proportion of broken rice kernels [54], as well as multiple rice quality parameters including grain dimension and aspect ratio, head rice yield, percent chalkiness, and transparency grading [55]. Monitoring of milling quality is also possible by characteristic dimension ratio (CDR) which compares the dimensional features of all head rice kernels to that of broken rice grains in the sample [56]. It is therefore possible to develop a more comprehensive algorithm for integrated detection of rice physical- and milling quality parameters. Grain quality evaluation by computer vision as well as hyperspectral and multispectral imaging techniques as applied in other cereals [57, 58] can now be employed in rice. Toward this goal, hyperspectral imaging was recently used to discriminate rice quality [59]. In this study, the dimension reduction technique was performed on spectral and image information using PCA and back propagation neural

Table 1
Summary of general routine rice grain quality tests [87–89]

Quality parameter	Measure	Principle	Standard method	Alternative method
A. Physical quality				
Moisture content	Moisture content	Evaporation of moisture	Around 2–3 g of rice flour dried in oven at 130 ± 1 °C for an hour. For paddy rice, 10–15 g of samples are dried for 24 h. Samples are cooled in a desiccator for 45–60 min. Moisture content computed by difference between original weight and dry weight.	Moisture meter; Brabender moisture tester; Infrared moisture analyzer
Grain dimension	Length Width Thickness Size Shape	Measurement of grain dimension and appearance	Manual measurement of 50 grains using millimeter ruler and vernier calliper; photographic enlargement or projection of rice grains followed by manual measurement; image scanning followed by image analysis.	Cervitec (Foss); SeedCount (Next Instrument); ImageJ
Grain appearance	Broken grains Chalky grains		Visual observation, counting and scoring; image scanning followed by image analysis.	SeedCount (Next Instrument); Rice translucency meter; ImageJ
Grain color	Grain color	Grain color assessment	Determination of grain color and impurity using Kett whiteness meter.	SeedCount (Next Instrument); color meter or chromameter; milling meter
B. Milling quality				
Milling quality and yield	Head rice yield Broken rice Milled rice yield 1000 grain weight	Milling yield determination by successive grain processing	The paddy is dehulled to obtain brown rice. The hull is separated from the brown rice using a blower. Brown rice is then polished to produce milled rice. Broken rice grains are separated from whole grains by sieving using grain separator. The processed grains are then weighed successively. The percentage weight of brown rice removed as bran is usually used to quantify degree of milling. Grain counter can be used for 1000 grain weight prior to weight for ease of analysis.	None

Grain hardness	Grain hardness	Measurement of hardness of individual grains	The hardness of at least 25 individual grains is measured using Instron or Kiya Hardness Tester. This can also be assessed by ease of grinding using Brabender farinograph with resistograph attachment.	Hand-operated hardness tester; Vicker microhardness distribution
Grain cracks and fissures	Grain cracks and fissures	Visualization of cracks and fissures	Visualization of cracks and fissures in rice grains using transmitted light, hand-held torch, or microbial colony counter. Crack resistance is measured by soaking grains for 1–3 h in 30 °C water bath and subjecting the samples in Kett Pearlest mill.	X-ray visualization
Degree of milling	Degree of milling	Estimation of degree of milling	Five grams of grain is washed three times and stained with 5 mL May-Gruenwald reagent (eosin and methylene blue in methanol solution) for 1–2 min. The grains are destained three times with water. Staining pattern is then measured.	NIRS, Satake milling meter

Table 2
Classification of rice grain based on length

Size	IRRI (1966)	FAO (1972)	USDA	ISO (1995)
Extra long	>7.5	≥7.00	None	None
Long	6.61–7.50	6.00–6.99	Approximately 3× longer than it is wide	≥6.6
Medium	5.51–6.60	5.00–5.99	2.1–2.9 times longer than it is wide	6.2–6.5
Short	≤5.5	<5.00	Less than 2× longer than it is wide	<6.2

network (BPNN). In addition to all of these image analyses techniques, in situ nondestructive methods to characterize rice kernel during development and at the mature paddy level using X-ray, NMR, and MRI technologies can be used to properly screen and identify breakage resistant rice lines prior to dehulling and polishing. Nondestructive detection of cracks in paddy rice is now possible through X-ray imaging without the need to dehull the samples [60]. However, all these techniques are not generally accessible for routine rice grain quality laboratories because they are not yet commercialized and validated by international collaborative testing panels.

Another emerging field in the area of physical and milling quality research related to kernel dimension and maturity involves designing equipment for processing, sorting, and sizing of grains to separate defective from superior quality grains. A new machine called QSorter (Qualisense) is capable of sorting rice grains at 50 kernels per seconds. Sorting is automatically done by the machine to provide faster and more accurate quantification of HRY and milling quality. It is also capable of doing single kernel analyses of grain dimension, chalk and fissures/cracks. In addition, more integrated machine vision grain quality inspection system that sorts out grain according to milling and physical appearance is also now available that can process an average of 1200 kernels per minute [61, 62]. This prototype system that consists of an automatic inspection machine and an image-processing unit is used to classify and sort head rice, chalky and cracked grains. These technological revolutions help overcome the traditional human inspection which is more laborious, inconsistent, and highly subjective. This system can find possible application in export-quality rice to satisfy rice grading system requirements of importing countries.

6 Physiological Mechanisms and Genetic Basis of Grain Breakage Resistance, Chalk, and Grain Dimensions

When rice is subjected to best preharvest and postharvest practices with the aim of minimizing grain loss during grain processing, each variety will still exhibit a cultivar-specific susceptibility to fissuring.

Some cultivars are inherently resistant to fissuring because of endosperm chemical composition, hull and bran diffusivity, and grain size and shape. This indicates that aside from environmental, farming, postharvest, and grain processing factors, inherent genotypic traits can also contribute in avoiding postharvest loss due to fissures. In general, rice grains genetically predisposed to form fissures have critical moisture content (CMC) of 15–16%. Below CMC, the grains will crack. This critical range is lower for fissure-resistant cultivars, which usually crack below 12–14% MC.

The physical and biochemical properties of the hull, bran, and endosperm can affect the movement of moisture through the kernel and are therefore important traits to consider in developing breakage resistant rice grains. The hull is the outer covering of rice composed of interlocking lemma and palea which serves as water barrier protecting the endosperm from rapid changes in moisture content. This depends on hull composition, i.e., amount of silica present, structural integrity of the interlocking lemma and palea. In general, moisture movement is reduced when the hull is thicker and tighter. On the other hand, the bran is a layer composed of aleurone that covers the starchy endosperm. Aside from being nutritious, it acts as a protective moisture barrier between the hull, pericarp, and the endosperm. Thicker bran can reduce the moisture movement and therefore moisture gradient along the endosperm. Lastly, the endosperm is the main component of the rice grain composed of storage starch and seed storage protein. The central core of the grain is mechanically weaker than the periphery and this depends on the shape, orientation, and packing of starch granules.

Fissure resistance is exhibited in different rice cultivars by different structural mechanisms. For example, the fissure resistance of Cypress is due to protective hull barrier while that of Saber is due to high endosperm diffusivity [63]. It is also believed that rice grains with compact starch granules can resist breakage and therefore increase HRY. Hence, the development of integrated methods to measure hull thickness and tightness, bran thickness and diffusivity, and endosperm starch granule packing is crucial in screening for fissure resistance in rice germplasm collection. An induce-fissuring method was developed to identify fissure resistance from early-generation breeding lines [64]. Using this method, studies identified QTLs related to breakage resistance of Cypress, where one is linked to the semidwarf *sd1* locus [63–65]. However, additional follow-up studies are needed to determine the exact genetic basis of fissure resistance based on hull, bran, and endosperm properties. This can be expedited by the creation of novel phenotyping methods that can be used to make detailed analyses for these traits.

Similar to fissure resistance, chalk formation is also influenced by the genotype, aside from environmental, postharvest, and processing factors. Because chalk is easier to phenotype, more studies have been conducted on this subject. The genetic basis of chalk has been investigated both in *japonica* and *indica* germplasm [66–72]. As reviewed by Sreenivasulu et al. [30], more than 100 QTLs and 7 genes are linked with the chalkiness trait [73–79]. However, because of the issue on standardization of method to phenotype chalk discussed above, there is a need to distinguish between chalkiness that only affects appearance quality without resulting in grain breakage from chalkiness that actually reduce milling yield. This can pave the way for determining the actual contribution of chalk in reducing HRY.

Rice grain dimension and shape is also easy to phenotype. As a result, the genetic basis of rice grain size is well characterized from QTL mapping and mutant analyses [80]. Recent studies are revealing a more detailed understanding on the molecular events related to developmental regulation of grain size and shape. Almost all of the identified genes so far (*GW2*, *GS3*, *qGL3*, *GW5*, *GS5*, *GW8*, and *TGW6*) are involved in regulating grain size by modulating the number of cells through enhanced cell division [81]. More interestingly, a newly-identified gene, *GL7/GW7*, is involved in regulating grain size by modulating grain length through cell elongation, and also influences chalk [82, 83]. It appears that genes for cell division have been preferentially selected by rice breeders compared to genes related to cell elongation [84]. It is therefore possible to identify more novel genes for cell elongation in rice diversity panel than in breeding lines. Integrated phenotyping methods that allow for concurrent detection of fissures, chalk, grain dimension, and panicle architecture are crucial. Breeding against chalk and fissures with synchronous flowering, medium-maturity, and optimum grain dimension might be effective in reducing the incidence of breakage susceptibility of rice grain. Moreover, focus should also be on screening for perfect grains, not imperfect grains. In this connection, a method to screen for translucency needs to also be developed and included in the proposed integrated phenotyping method.

7 Additional Notes

The first-ever meeting of the International Network of Quality Rice (INQR) was held at the International Rice Research Institute on December 1–2, 2015. This meeting was attended by an international panel of experts from rice grain quality, breeding, and industry sectors. Based on this 2-day meeting, the panel recommended thrust areas to be addressed in research for enhancing HRY/HRR and to minimize postharvest grain loss. These include:

1. Accessing the 3000 rice genome accession [85] and develop molecular marker database for key traits related to milling quality.

2. Exploring X-ray, NIR, MRI, NMR, polarized light microscopy and other modern technology that can be used to characterize chalk, translucency, cracks, breakage susceptibility, and other grain imperfections that affects head rice and milling yield.
3. Determining the exact location of chalk. Chalk in the dorsal and the ventral side can affect HRY more than chalk in the core. For example, Yamadashi and Arborio rice have core chalkiness but it does not break.
4. Comparing data from commercial, laboratory, and hand mills to determine adjustment factors. Several factors can influence the difference in milling including equipment and brand variability. Various practices that affect storage conditions like moisture content should be factored in.
5. Breeding for higher HRY and translucency instead of breeding alone for kernel dimension and chalk. Focus should be more on the enhancing HRY which can command better price in the market.
6. Using systems biology to link genetics with environmental responses and to identify the underlying biological regulatory networks to improve complex traits such as uniform flowering, synchronous seed filling, finding the ideal panicle architecture, reduce fissures grain breakage and lower chalk. This strategy is already being applied to fruits for regulation of processes and responses, quality prediction, genetic improvement, and development of virtual fruit model [86].

Because HRY and milling quality is becoming a central issue in the hybrid rice industry, the following strategies can be applied to this rapidly expanding rice sector:

7. Germplasm screening for high HRY and translucency in parental lines.
8. Incorporation of high HRY in female and male parents in hybrid breeding programs with fissure-resistant and non-chalky (translucent) grain appearance.
9. Determination of genetic, biochemical, physiological, and physical factors influencing high HRY.
10. Defining the genetics of high HRY across grain types.
11. Understanding the genetic basis of kernel hardness/breaking strength.
12. Development of high-throughput screening methods for quick determination of head rice recovery in small samples (e.g., determination of techniques to identify fissures in paddy grains rather than milled rice, development of nondestructive methods to quantify fissures and chalk).
13. Validated robust markers to hasten the breeding program to reduce grain breakage.

8 Conclusion

Improving the milling recovery of rice has been an active area of research for several decades but focus is mostly on postharvest processing and technology. The genetic improvement of head rice yield is severely unexplored probably because of the lack of cost-effective, reliable, and high-throughput screening method. With the availability of state-of-the-art imaging, visualization and analytical technologies for milling quality, it is envisioned that rice breeding programs will be geared toward improving the head rice recovery of rice. Improving the milling quality of rice is crucial in ensuring food security for half of the world's population in an era of dwindling natural resources and climate unpredictability.

Head rice yield and grain translucency are very important preferred factors universally selected by breeders, traders, millers, and consumers. Therefore, HRY plays an important role in determining the acceptability and price of rice in the market. The aim therefore is to achieve higher HRY. Strategies to improve HRY include breeding against breakage susceptibility, targeting synchronous flowering, reduced chalk, postharvest improvement at the farm level, and grain processing technology improvement. Breeding strategies for improving HRY include optimization of genetic gains for reduced grain breakage with reduced chalk and optimized postharvest handling and storage. For stakeholder strategies, millers should have customized machinery and milling procedures. Parboiling can be considered whenever applicable to enhance HRY in long slender grains, but this strategy is strictly limited to consumer segments that prefer parboiled rice. Stakeholders and breeders should also be working hand-in-hand to find solutions at the genetic level and to tackle better management practices during the post-harvest value chain. Additional strategies include development of methods to correlate laboratory results for HRY with that of commercial mills, and development of procedures to instantly determine HRY of farmers produce and congruence with results from millers.

Acknowledgments

This work has been supported under the CGIAR thematic area Global Rice Agri-Food System CRP, RICE, Stress-Tolerant Rice for Africa and South Asia (STRASA) Phase III, and Australian Centre for International Agricultural Research (Project ID CIM/2016/046) funding.

References

1. Hodges RJ, Buzby JC, Bennett B (2011) Postharvest losses and waste in developed and less developed countries: opportunities to improve resource use. J Agric Sci 149:37–45
2. Buggenhout J, Brijs K, Celus I, Delcour JA (2013) The breakage susceptibility of raw and parboiled rice: a review. J Food Eng 117(3):304–315
3. Bell M, Bakker R, De Padua D, Rickman J (1998) Rice quality management-principles and some lessons. In: ACIAR proceedings, 2000. ACIAR. pp 255–263
4. Otto RK, David LC (2004) Rough-rice drying - moisture adsorption and desorption. In: Champagne ET (ed) Rice: chemistry and technology. Grain science references. American Association of Cereal Chemists, Inc., St. Paul, MN, pp 223–268. https://doi.org/10.1094/1891127349.009
5. Kunze O, Choudhury M (1972) Moisture adsorption related to the tensile strength of rice. Cereal Chem 49:684–696
6. Zhang Q, Yang W, Sun Z (2005) Mechanical properties of sound and fissured rice kernels and their implications for rice breakage. J Food Eng 68(1):65–72
7. Iguaz A, Rodríguez M, Vírseda P (2006) Influence of handling and processing of rough rice on fissures and head rice yields. J Food Eng 77(4):803–809. https://doi.org/10.1016/j.jfoodeng.2005.08.006
8. Sharma AD, Kunze OR (1982) Post-drying fissure developments in rough rice. Trans ASAE 25(2):465–468
9. Cnossen AG, Jimenez MJ, Siebenmorgen TJ (2003) Rice fissuring response to high drying and tempering temperatures. J Food Eng 59(1):61–69
10. Cnossen AG, Siebenmorgan TJ, Yang W, Bautista RC (2001) An application of glass transition temperature to explain rice kernel fissure occurrence during the drying process. Dry Technol 19(8):1661–1682. https://doi.org/10.1081/Drt-100107265
11. Hwang SS, Cheng YC, Chang C, Lur HS, Lin TT (2009) Magnetic resonance imaging and analyses of tempering processes in rice kernels. J Cereal Sci 50(1):36–42. https://doi.org/10.1016/j.jcs.2008.10.012
12. Cnossen AG, Siebenmorgen TJ (2000) The glass transition temperature concept in rice drying and tempering: effect on milling quality. Trans ASAE 43(6):1661–1667
13. Perdon AA, Siebenmorgen TJ, Mauromoustakos A (2000) Glassy state transition and rice drying: development of a brown rice state diagram. Cereal Chem 77(6):708–713
14. Siebenmorgen TJ, Yang W, Sun Z (2004) Glass transition temperature of rice kernels determined by dynamic mechanical thermal analysis. Trans ASAE 47(3):835–839
15. Sun ZH, Yang WD, Siebenmorgen T, Stelwagen A, Cnossen A (2002) Thermomechanical transitions of rice kernels. Cereal Chem 79(3):349–353. https://doi.org/10.1094/Cchem.2002.79.3.349
16. Yang W, Jia CC, Siebenmorgen TJ, Pan Z, Cnossen AG (2003) Relationship of kernel moisture content gradients and glass transition temperatures to head rice yield. Biosyst Eng 85(4):467–476
17. Yang W, Jia CC, Howell TA (2003) Relationship of moisture content gradients and glass transition temperatures to head rice yield during cross-flow drying. Biosyst Eng 86(2):199–206
18. Slade L, Levine H (1991) A food polymer science approach to structure-property relationships in aqueous food systems: non-equilibrium behavior of carbohydrate-water systems. In: Levine H, Slade L (eds) Water relationships in foods, Advances in experimental medicine and biology, vol 302. Springer, New York, pp 29–101. https://doi.org/10.1007/978-1-4899-0664-9_3
19. Slade L, Levine H (1995) Glass transitions and water-food structure interactions. Adv Food Nutr Res 38(2):103–179
20. Sarker NN, Kunze OR, Strouboulis T (1996) Transient moisture gradients in rough rice mapped with finite element model and related to fissures after heated air drying. Trans ASAE 39(2):625–631
21. Yang W, Jia CC, Siebenmorgen TJ, Howell TA, Cnossen AG (2002) Intra-kernel moisture responses of rice to drying and tempering treatments by finite element simulation. J Long Form Workform 45(4):1037–1044
22. Banaszek MM, Siebenmorgen TJ (1993) Individual Rice Kernel Drying Curves. Trans ASAE 36(2):521–528. https://doi.org/10.13031/2013.28368
23. Chen H, Siebenmorgen TJ, Marks BP (1997) Relating drying rate constant to head rice yield reduction of long-grain rice. Trans ASAE 40(4):1133–1139. https://doi.org/10.13031/2013.21331
24. Kunze O (1979) Fissuring of the rice grain after heated air drying. Trans ASEA 22(5):1197–1201. https://doi.org/10.13031/2013.35183

25. Zhang Q, Yang W, Jia C (2003) Preservation of head rice yield under high-temperature tempering as explained by the glass transition of rice kernels. Cereal Chem 80(6):684–688. https://doi.org/10.1094/CCHEM.2003.80.6.684
26. Ding C, Khir R, Zhongli P, Zhang J, El-Mashad H, Tu K (2015) Effect of infrared and conventional drying methods on physicochemical characteristics of stored white rice. Cereal Chem J. https://doi.org/10.1094/CCHEM-11-14-0232-R
27. Fitzgerald MA, Resurreccion AP (2009) Maintaining the yield of edible rice in a warming world. Funct Plant Biol 36(12):1037–1045. https://doi.org/10.1071/fp09055
28. Zakaria S, Matsuda T, Tajima S, Niita Y (2002) Effect of high temperature at ripening stage on the reserve accumulation in seed in some rice cultivars. Plant Prod Sci 5(2):160–168
29. Tabata M, Hirabayashi H, Takeuchi Y, Ando I, Iida Y, Ohsawa R (2007) Mapping of quantitative trait loci for the occurrence of white-back kernels associated with high temperatures during the ripening period of rice(Oryza sativa L.). Breed Sci 57(1):47–52. https://doi.org/10.1270/Jsbbs.57.47
30. Sreenivasulu N, Butardo VM, Misra G, Cuevas RP, Anacleto R, Kavi Kishor PB (2015) Designing climate-resilient rice with ideal grain quality suited for high-temperature stress. J Exp Bot 66(7):1737–1748. https://doi.org/10.1093/jxb/eru544
31. Ashida K, Iida S, Yasui T (2009) Morphological, physical, and chemical properties of grain and flour from chalky rice mutants. Cereal Chem 86(2):225–231. https://doi.org/10.1094/cchem-86-2-0225
32. Zhu LJ, Dogan H, Gajula H, Gu MH, Liu QQ, Shi YC (2012) Study of kernel structure of high-amylose and wild-type rice by X-ray microtomography and SEM. J Cereal Sci 55(1):1–5
33. Lyman NB, Jagadish KSV, Nalley LL, Dixon BL, Siebenmorgen T (2013) Neglecting rice milling yield and quality underestimates economic losses from high-temperature stress. PLoS One 8 (8): ARTN e72157. https://doi.org/10.1371/journal.pone.0072157
34. Lisle AJ, Martin M, Fitzgerald MA (2000) Chalky and translucent rice grains differ in starch composition and structure and cooking properties. Cereal Chem 77(5):627–632
35. Singh N, Sodhi NS, Kaur M, Saxena SK (2003) Physico-chemical, morphological, thermal, cooking and textural properties of chalky and translucent rice kernels. Food Chem 82(3):433–439
36. Swamy YMI, Bhattacharya KR (1982) Breakage of rice during milling. 4. Effect of kernel chalkiness. J Food Sci Technol Mysore 19(3):125–126
37. Fitzgerald MA, McCouch SR, Hall RD (2009) Not just a grain of rice: the quest for quality. Trends Plant Sci 14(3):133–139. https://doi.org/10.1016/j.tplants.2008.12.004
38. Cheng FM, Zhong LJ, Wang F, Zhang GP (2005) Differences in cooking and eating properties between chalky and translucent parts in rice grains. Food Chem 90(1–2):39–46. https://doi.org/10.1016/j.foodchem.2004.03.018
39. Chun A, Song J, Kim K-J, Lee H-J (2009) Quality of head and chalky rice and deterioration of eating quality by chalky rice. J Crop Sci Biotechnol 12(4):239–244. https://doi.org/10.1007/s12892-009-0142-4
40. Yoshioka Y, Iwata H, Tabata M, Ninomiya S, Ohsawa R (2007) Chalkiness in rice: potential for evaluation with image analysis. Crop Sci 47(5):2113–2120
41. Sun CM, Liu T, Ji CX, Jiang M, Tian T, Guo DD, Wang LJ, Chen YY, Liang XM (2014) Evaluation and analysis the chalkiness of connected rice kernels based on image processing technology and support vector machine. J Cereal Sci 60(2):426–432
42. Endo-Higashi N, Izawa T (2011) Flowering time genes heading date 1 and early heading date 1 together control panicle development in rice. Plant Cell Physiol 52(6):1083–1094
43. Huang XH, Zhao Y, Wei XH, Li CY, Wang A, Zhao Q, Li WJ, Guo YL, Deng LW, Zhu CR, Fan DL, Lu YQ, Weng QJ, Liu KY, Zhou TY, Jing YF, Si LZ, Dong GJ, Huang T, Lu TT, Feng Q, Qian Q, Li JY, Han B (2012) Genome-wide association study of flowering time and grain yield traits in a worldwide collection of rice germplasm. Nat Genet 44(1):32–U53
44. Sun CH, Chen D, Fang J, Wang PR, Deng XJ, Chu CC (2014) Understanding the genetic and epigenetic architecture in complex network of rice flowering pathways. Protein Cell 5(12):889–898
45. Tsuji H, Taoka K, Shimamoto K (2011) Regulation of flowering in rice: two florigen genes, a complex gene network, and natural variation. Curr Opin Plant Biol 14(1):45–52
46. Butardo VM Jr, Anacleto R, Parween S, Samson I, de Guzman K, Alhambra CM, Misra G, Sreenivasulu N (2017) Systems genetics identifies a novel regulatory domain of amylose synthesis. Plant Physiol 173(1):887–906. https://doi.org/10.1104/pp.16.01248

47. Sun H, Siebenmorgen TJ (1993) Milling characteristics of various rough rice kernel thickness fractions. Cereal Chem 70(6):727–733
48. Indudhara Swamy YM, Indudhara Swamy YM, Bhattacharya KR (1980) Breakage of rice during milling - effect of kernel defects and grain dimension. J Food Process Eng 3(1):29–42
49. Chen X, Xun Y, Li W, Zhang JX (2010) Combining discriminant analysis and neural networks for corn variety identification. Comput Electron Agric 71:S48–S53
50. Dubey BP, Bhagwat SG, Shouche SP, Sainis JK (2006) Potential of artificial neural networks in varietal identification using morphometry of wheat grains. Biosyst Eng 95(1):61–67
51. Cheng F, Ying YB, Li YB (2006) Detection of defects in rice seeds using machine vision. Trans ASABE 49(6):1929–1934
52. Courtois F, Faessel M, Bonazzi C (2010) Assessing breakage and cracks of parboiled rice kernels by image analysis techniques. Food Control 21(4):567–572
53. Lin P, Chen YM, He Y (2012) Identification of broken rice kernels using image analysis techniques combined with velocity representation method. Food Bioprocess Technol 5(2):796–802
54. van Dalen G (2004) Determination of the size distribution and percentage of broken kernels of rice using flatbed scanning and image analysis. Food Res Int 37(1):51–58
55. Fang CY, Hu XQ, Sun CC, Duan BW, Xie LH, Zhou P (2015) Simultaneous determination of multi rice quality parameters using image analysis method. Food Anal Method 8(1):70–78
56. Yadav BK, Jindal VK (2001) Monitoring milling quality of rice by image analysis. Comput Electron Agric 33(1):19–33
57. Jayas DS, Singh CB (2012) 15 - Grain quality evaluation by computer vision. In: Sun D-W (ed) Computer vision technology in the food and beverage industries. Woodhead Publishing, Cambridge, UK, pp 400–421. https://doi.org/10.1533/9780857095770.3.400
58. Shahin MA, Hatcher DW, Symons SJ (2012) 17 - Development of multispectral imaging systems for quality evaluation of cereal grains and grain products. In: Sun D-W (ed). Computer vision technology in the food and beverage industries. Woodhead Publishing, Cambridge, UK, pp 451–482. https://doi.org/10.1533/9780857095770.3.451
59. Wang L, Liu D, Pu HB, Sun DW, Gao WH, Xiong ZJ (2015) Use of hyperspectral imaging to discriminate the variety and quality of rice. Food Anal Method 8(2):515–523
60. Kumar PA, Bal S (2007) Automatic unhulled rice grain crack detection by X-ray imaging. Trans ASABE 50(5):1907–1911
61. Wan YN, Lin CM, Chiou JF (2002) Rice quality classification using an automatic grain quality inspection system. Trans ASAE 45(2):379–387
62. Wan YN (2002) Kernel handling performance of an automatic grain quality inspection system. Trans ASAE 45(2):369–377
63. Pinson SRM, Jia YL, Gibbons JW (2013) Three quantitative trait loci conferring resistance to kernel fissuring in rice identified by selective genotyping in two tropical Japonica populations. Crop Sci 53(6):2434–2443
64. Pinson SRM, Jia YL, Gibbons J (2012) Response to early generation selection for resistance to rice kernel fissuring. Crop Sci 52(4):1482–1492
65. Nelson JC, McClung AM, Fjellstrom RG, Moldenhauer KA, Boza E, Jodari F, Oard JH, Linscombe S, Scheffler BE, Yeater KM (2011) Mapping QTL main and interaction influences on milling quality in elite US rice germplasm. Theor Appl Genet 122(2):291–309. https://doi.org/10.1007/s00122-010-1445-z
66. Liu X, Hua ZT, Wang Y (2011) Quantitative trait locus (QTL) analysis of percentage grains chalkiness using AFLP in rice (Oryza sativa L.). Afr J Biotechnol 10(13):2399–2405
67. Liu X, Wang Y, Wang SW (2012) QTL analysis of percentage of grains with chalkiness in Japonica rice (Oryza sativa). Genet Mol Res 11(1):717–724. https://doi.org/10.4238/2012.March.22.1
68. Guo T, Liu XL, Wan XY, Weng JF, Liu SJ, Liu X, Chen MJ, Li JJ, Su N, Wu FQ, Cheng ZJ, Guo XP, Lei CL, Wang JL, Jiang L, Wan JM (2011) Identification of a stable quantitative trait locus for percentage grains with white chalkiness in rice (Oryza sativa). J Integr Plant Biol 53(8):598–607. https://doi.org/10.1111/j.1744-7909.2011.01041.x
69. Zhou LJ, Chen LM, Jiang L, Zhang WW, Liu LL, Liu X, Zhao ZG, Liu SJ, Zhang LJ, Wang JK, Wan JM (2009) Fine mapping of the grain chalkiness QTL qPGWC-7 in rice (Oryza sativa L.). Theor Appl Genet 118(3):581–590. https://doi.org/10.1007/s00122-008-0922-0
70. Liu XL, Wan XY, Ma XD, Wan JM (2011) Dissecting the genetic basis for the effect of rice chalkiness, amylose content, protein content, and rapid viscosity analyzer profile characteristics on the eating quality of cooked rice using the chromosome segment substitution line population across eight environments. Genome

54(1):64–80. https://doi.org/10.1139/G10-070

71. Wan XY, Wan JM, Weng JF, Jiang L, Bi JC, Wang CM, Zhai HQ (2005) Stability of QTLs for rice grain dimension and endosperm chalkiness characteristics across eight environments. Theor Appl Genet 110(7):1334–1346
72. Tan YF, Xing YZ, Li JX, Yu SB, Xu CG, Zhang QF (2000) Genetic bases of appearance quality of rice grains in Shanyou 63, an elite rice hybrid. Theor Appl Genet 101(5–6):823–829. https://doi.org/10.1007/s001220051549
73. Kang H-G, Park S, Matsuoka M, An G (2005) White-core endosperm *floury endosperm-4* in rice is generated by knockout mutations in the C_4-type pyruvate orthophosphate dikinase gene (OsPPDKB). Plant J 42(6):901–911. https://doi.org/10.1111/j.1365-313X.2005.02423.x
74. Ryoo N, Yu C, Park C-S, Baik M-Y, Park I-M, Cho M-H, Bhoo SH, An G, Hahn T-R, Jeon J-S (2007) Knockout of a starch synthase gene *OsSSIIIa/Flo5* causes white-core floury endosperm in rice (*Oryza sativa* L.). Plant Cell Rep 26:1083–1095. https://doi.org/10.1007/s00299-007-0309-8
75. Woo M-O, Ham T-H, Ji H-S, Choi M-S, Jiang W, Chu S-H, Piao R, Chin J-H, Kim J-A, Park BS, Seo HS, Jwa N-S, McCouch S, Koh H-J (2008) Inactivation of the UGPase1 gene causes genic male sterility and endosperm chalkiness in rice (Oryza sativa L.). Plant J 54(2):190–204. https://doi.org/10.1111/j.1365-313X.2008.03405.x
76. Wang E, Wang J, Zhu X, Hao W, Wang L, Li Q, Zhang L, He W, Lu B, Lin H, Ma H, Zhang G, He Z (2008) Control of rice grain-filling and yield by a gene with a potential signature of domestication. Nat Genet 40(11):1370–1374. https://doi.org/10.1038/ng.220
77. Li Y, Fan C, Xing Y, Yun P, Luo L, Yan B, Peng B, Xie W, Wang G, Li X, Xiao J, Xu C, He Y (2014) Chalk5 encodes a vacuolar H(+)-translocating pyrophosphatase influencing grain chalkiness in rice. Nat Genet 46(4):398–404. https://doi.org/10.1038/ng.2923
78. She K-C, Kusano H, Koizumi K, Yamakawa H, Hakata M, Imamura T, Fukuda M, Naito N, Tsurumaki Y, Yaeshima M, Tsuge T, Matsumoto K, Kudoh M, Itoh E, Kikuchi S, Kishimoto N, Yazaki J, Ando T, Yano M, Aoyama T, Sasaki T, Satoh H, Shimada H (2010) A novel factor FLOURY ENDOSPERM2 is involved in regulation of rice grain size and starch quality. Plant Cell 22(10):3280–3294. https://doi.org/10.1105/tpc.109.070821
79. Sun WQ, Zhou QL, Yao Y, Qiu XJ, Xie K, Yu SB (2015) Identification of genomic regions and the isoamylase gene for reduced grain chalkiness in rice. PLoS One 10(3):e0122013
80. Zuo JR, Li JY (2014) Molecular genetic dissection of quantitative trait loci regulating rice grain size. Annu Rev Genet 48:99–118
81. Zheng J, Zhang YD, Wang CL (2015) Molecular functions of genes related to grain shape in rice. Breed Sci 65(2):120–126
82. Wang Y, Xiong G, Hu J, Jiang L, Yu H, Xu J, Fang Y, Zeng L, Xu E, Xu J, Ye W, Meng X, Liu R, Chen H, Jing Y, Wang Y, Zhu X, Li J, Qian Q (2015) Copy number variation at the GL7 locus contributes to grain size diversity in rice. Nat Genet. https://doi.org/10.1038/ng.3346 http://www.nature.com/ng/journal/vaop/ncurrent/abs/ng.3346.html - supplementary-information
83. Wang S, Li S, Liu Q, Wu K, Zhang J, Wang S, Wang Y, Chen X, Zhang Y, Gao C, Wang F, Huang H, Fu X (2015) The OsSPL16-GW7 regulatory module determines grain shape and simultaneously improves rice yield and grain quality. Nat Genet. https://doi.org/10.1038/ng.3352 http://www.nature.com/ng/journal/vaop/ncurrent/abs/ng.3352.html - supplementary-information
84. Miura K, Matsuoka M (2015) Rice genetics: control of grain length and quality. Nat Plants 1:15112. https://doi.org/10.1038/nplants.2015.112
85. The 3000 Rice Genomes Project (2014) The 3,000 rice genomes project. GigaScience 3:7. https://doi.org/10.1186/2047-217X-3-7
86. Hertog MLATM, Rudell DR, Pedreschi R, Schaffer RJ, Geeraerd AH, Nicolai BM, Ferguson I (2011) Where systems biology meets postharvest. Postharvest Biol Technol 62(3):223–237
87. Bhattacharya KR (2011) 13 - Analysis of rice quality. In: Bhattacharya KR (ed) Rice quality. Woodhead Publishing, Cambridge, UK, pp 431–530. https://doi.org/10.1533/9780857092793.431
88. Juliano BO (2007) Rice chemistry and quality. Philippine Rice Research Institute, Munoz, Nueva Ecija
89. Bergman CJ, Bhattacharya KR, Ohtsubo K (2004) Rice end-use quality analysis. In: Champagne ET (ed) Rice chemistry and technology, 3rd edn. The American Association of Cereal Chemists, St, Paul. MN

Chapter 2

Improving Rice Grain Quality: State-of-the-Art and Future Prospects

Vito M. Butardo Jr., Nese Sreenivasulu, and Bienvenido O. Juliano

Abstract

Rice grain quality encompasses complex interrelated traits that cover biochemical composition, cooking, eating, nutritional, and sensory properties. Because rice endosperm is composed mainly of starch, rice grain quality is traditionally defined by characterizing starch structure and composition, which is then subsequently correlated with functional properties of the grain. The current proxy tests routinely used to describe rice grain quality preferences are rather limited to the estimation of apparent amylose content, gelatinization temperature, and gel consistency. Additional tests that characterize starch property, viscoelasticity, grain texture, and aroma are also employed in more advanced laboratories. However, these tests are not routinely applied in breeding programs to distinguish cooking quality classes to reflect evolving consumer preference and market demand. As consumer preferences in Asia and all over the world are diverse due to varied demographics and culture, defining uniform attributes to capture regional grain quality preferences becomes more challenging. Hence, novel and innovative proxy tests are needed to characterize rice grain quality to meet the demand for consumer preferences of commercially-released cultivars. In this chapter, the current methods employed in rice grain quality monitoring are succinctly reviewed. Future prospects for improvement are identified, introducing cutting edge technologies that can facilitate high-throughput screening of rice diversity panels and breeding lines. Aside from addressing the requirements for quality improvement in the traditional inbred rice breeding programs, we also tackled the need to enhance grain quality in the hybrid rice sector.

Key words Aroma, Appearance, Cooking, Eating, Grain quality, Milling, Nutrition, Physical, Rice, Sensory

1 Introduction

The current challenge for rice research is not only to increase rice productivity through higher yield (*see* Chapter 1) but also to add value through improved grain quality. Grain quality is a critical component in breeding for high yielding varieties to ensure acceptance by an ever-growing population of rice consumers. In the past, rice is considered a subsistence crop, primarily consumed worldwide as a major caloric source. Hence, the primary focus of rice breeding programs in IRRI during the 1960s up to the 1970s

Nese Sreenivasulu (ed.), *Rice Grain Quality: Methods and Protocols*, Methods in Molecular Biology, vol. 1892,
https://doi.org/10.1007/978-1-4939-8914-0_2, © Springer Science+Business Media, LLC, part of Springer Nature 2019

is on increasing productivity and yield [1]. As developing countries achieve rice self-sufficiency and the ability to export surplus rice, consumers became selective in preferring high quality in the succeeding decades [2]. Consequently, improving rice grain quality has become crucial in ensuring consumer acceptance. It is also important in alleviating the lives of marginalized farmers by providing options for rice cultivars with value-added premium traits. The two factors that can influence sustained farmer adoption of new rice cultivars are high yield and superior grain quality. An emerging challenge therefore is to maintain rice grain quality and yield in an era of dwindling natural resources and uncertain environmental fluctuations.

Rice grain quality is hard to define and quantify, primarily because it depends on regional consumer preference, market demand, and intended functional use. Its definition means different things to different demographics such that what is acceptable in one region may not generally be preferred in another. For instance, South Asians generally prefer long-grain rice with a high amylose content and hard gel consistency, while Southeast Asians favor long grains with intermediate amylose content and soft gel consistency [3]. In some cases, top-quality rice in one region may be considered a low-quality variety in another region. Furthermore, because rice is composed of diverse germplasm, its grains have huge variations in physical, biochemical, cooking, sensory, and processing properties [4], making grain quality classification challenging.

A generally accepted consensus among grain quality experts and cereal chemists is that rice grain quality covers the physicochemical properties of rice kernel which influences their milling, cooking, eating, sensory, and nutritional properties (Fig. 1). All these properties are linked by the biochemical properties of the rice kernel. The general challenge for rice grain quality research is to come up with specific quantitative proxy measures which are simple, accurate, robust and can be universally applicable for all rice varieties for each specific trait. This is critical to distinguish low, medium, and premium classes of rice. Ideally, the method should be portable, high-throughput, and measurable from the farm to the market to facilitate ease of tracking of quality at all stages of the value chain.

The physical and milling quality of rice is reviewed in the first chapter. In this chapter, we review the traditional, advanced, and emerging techniques used to characterize the cooking, eating, sensory, and nutritional traits of rice. We wrap up the chapter by discussing future prospects for improvement in rice grain quality analyses for inbred and hybrid rice.

Fig. 1 The scope of rice grain quality encompasses interrelated physicochemical properties, which influence milling, cooking, eating, sensory, and nutritional traits

2 Conventional Rice Grain Quality Tests

The cooking, eating, textural, and nutritional quality of rice is routinely measured using three main physicochemical tests based primarily on endosperm starch properties. These traditional tests include estimation of apparent amylose content (AAC), gelatinizaton temperature (GT), and gel consistency (GC). Whereas AAC is directly related to the proportion and structure of amylose, GT and GC are amylopectin properties. Low GT starch has shorter chain amylopectin branches, whereas intermediate/high GT have more long-chain branches. They will be discussed one by one in the following subsections.

2.1 Apparent Amylose Content

The first and most important physicochemical test is apparent amylose content (AAC) estimation which relies on iodine binding to gelatinized rice flour to produce a blue iodine-amylose complex which is routinely quantified using a visible-light spectrophotometer [5]. AAC of milled rice grains can be ranked as waxy (0–2%), very low (3–9%), low (10–19%), intermediate (20–25%), and high (>25%) using the AAC classification [6]. Milled rice can also be classified as non-glutinous (non-waxy) or glutinous (waxy), depending on the presence or absence of amylose chains, respectively. Improvements were proposed in the 1970s to reduce sources of variations identified in the original assay [7, 8]. The reproducibility of the simplified method was thoroughly validated by several international testing laboratories and additional enhancements were proposed, including the introduction of an automated colorimetric method using AutoAnalyzer (Technicon and Bran Luebbe)

[9]. This simplified method gained widespread adoption in several routine rice breeding programs to screen for AAC of thousands of samples per year due to its generally strong correlation with cooking, textural, and digestibility attributes of cooked milled rice [10]. For instance, the method played prominent role in characterizing the grain quality of world rice accessions in the germplasm collection of IRRI in the 1990s [11].

The IRRI Grain Quality and Nutrition Center (GQNC) currently uses a Continuous Flow Analyzer (CFA) to estimate amylose content based on the simplified method using San++ Automated Wet Chemistry Analyzer Model 3000 (Skalar Analytical, The Netherlands). This new high-throughput system consists of an auto-sampler, an amylose chemistry unit (manifold, proportioning pump, and colorimeter with 620 nm filter), an interface, and a computer for data processing. It offers flexibility for increasing daily sample throughput by optionally installing additional amylose chemistry units. The current system in GQNC can estimate the amylose content of more than 300 samples per run using two chemistry units (*see* Chapter 8).

There are two approved methods for the estimation of amylose content in milled rice. One was established by the Association for American Cereal Chemists (AACC) International (AACCI Method 61-03.01) and the other by the International Organization for Standardization (ISO Method 6647-1:2015) [12, 13]. Despite these standard methods, one major drawback of iodine binding for amylose determination is the increased iodine affinity to longer amylopectin chains that overestimates actual amylose values [14–16]. This is the reason why the result obtained from this method is technically known as "apparent" amylose content, to signify that it just provides a blue value (BV) estimate of the true amylose content of the flour sample. An international cooperative testing in 2009 identified additional sources of inter-laboratory variations in the conduct of the amylose assay [17]. Additional improvements were proposed including the use of calibrated rice standards, the use of longer wavelengths (e.g., 720 nm) to minimize the interference of iodine-amylopectin complex [17], and the replacement of acetate with ammonium buffer to reduce amylopectin-iodine absorbance [18]. Using ammonium buffer (pH 9) for amylose estimation provides more accurate result compared to the traditional acetate buffer because it reduces amylopectin-iodine absorption and gives AAC values close to the DSC method (see below). Proposal to improve the ease, portability, and affordability of the assay are also published, including the use of standard color charts [19], portable colorimeter [20], and direct iodine staining of milled rice grains [21]. It must be pointed out, however, that no international cooperative testing has been published on these new proposed modifications to date. Additional modifications might be harder to implement because these radical changes in the procedure can alter the amylose classification of previously-released commercial rice varieties. For example, the use of ammonium buffer pH 9 in place of acetate buffer pH 4.5 lowered apparent amylose content of milled rice by about 3%, but closer to DSC values [18, 22]. Implementation of this new method can potentially have historical ramifications on rice breeding programs worldwide.

2.2 Gelatinization Temperature

The second physicochemical test is gelatinization temperature (GT) to estimate the time and temperature of cooking of milled rice grains. GT is the temperature at which the semicrystalline structure of starch begins to melt upon cooking of rice flour in a specific ratio of water. It is traditionally measured by determining the loss of birefringence of starch granules as viewed under a polarized light microscope. The GT of milled rice grains are ranked as low (55–69.5 °C), intermediate (70–74 °C), and high (74.5–80 °C) [23]. There are several techniques to measure GT in rice based on the changes in optical, hydration, swelling, crystallinity, and viscosity parameters of flour slurry during cooking [24]. The two methods that gained popularity are based on grain digestion by alkali and differential scanning calorimetry (DSC). The method based on alkali digestion is an indirect proxy measure which relies on the tendency of starch granules of milled rice to disintegrate when soaked in dilute alkali such as potassium hydroxide [25]. This is numerically represented as alkali spreading value (ASV) [25] or alkali score (AS) [26]. In comparison, the DSC method is more accurate and direct because it relies on measuring the first peak of the endotherm in real time as the starch granules gelatinize [27, 28]. This peak endotherm is associated with variations in amylopectin fine structure [29, 30] which was validated using more than 4000 rice samples [31]. However, the DSC method cannot be routinely used for screening in rice breeding programs because the aluminum pans and lids used in the assay are expensive. Consequently, the ASV is the routine method of choice to screen thousands of lines in the breeding pipeline. Validation of GT using DSC is usually conducted only on advanced breeding lines before their release as new rice varieties. Waxy and low-AC rices with ASV score of 4–5 typically have high instead of intermediate GT [32].

The gelatinization temperature of rice can also be estimated by measuring the viscosity of starch pastes. A method based on Brabender amylograph viscosity can estimate final GT of rice flour paste as validated by an international cooperative test [33]. However, this method is low-throughput and requires large quantities of samples (50 g) which is not suitable for routine breeding programs. This method is replaced by Rapid Visco Analyzer (RVA) which offers speed, convenience, and lower sample size requirement (4 g). More recently, the AACC International approved an RVA method (AACCI Method 61-04.01) to precisely estimate the GT of milled rice flour based on a recent collaborative inter-laboratory study. This method provides more precise and accurate results compared to those obtained from parallel testing using the amylograph and the DSC methods [34].

2.3 Gel Consistency

The third physicochemical method is the measurement of gel consistency (GC) to reflect the range of texture observed in high amylose rice grains. The eating quality of rice can vary between amylose classes. The original test was developed to exclusively distinguish between textural attributes of high amylose rice grains with 24–30% AAC [35]. The test relies on the horizontal migration of cold rice paste after cooking, cooling, and incubation using 13 × 100 mm glass test tubes. Gel consistency is ranked as hard (27–40 mm), medium (41–60 mm), and soft (61–100 mm) depending on the migration of rice starch paste. Hard GC is due to high content of long-chain amylopectin while soft GC is because of a higher proportion of short-chain amylopectin. The GQNC uses high (27–35 mm), intermediate (36–49 mm), and low (≥50 mm) scoring system. Using this test, rice grains with softer gel consistency produce tender cooked grains which remain soft even after cooling [36]. Because consumers tend to prefer rice grains with softer grain consistency in Southeast Asia, this trait has been the target of most rice breeding programs in the region [37]. Hard gel consistency correlates with a high content of long-chain amylopectin with high iodine affinity, found in high-AC rices [14, 38]. The "hot-water insoluble amylose" of Bhattacharya [39] is probably long-chain amylopectin that remains in the gelatinized starch granule.

3 Current Advanced Rice Grain Quality Tests

The three proxy tests mentioned above are insufficient to classify the diversity of grain quality observed in rice varieties. This is because cooking, eating, and textural quality of rice is a complex multifactorial attribute which is best characterized by a panel of rice consumers through sensory evaluation. This will be discussed in a subsequent section. However, sensory evaluation is a time-consuming, expensive, and generally subjective method. Hence, a combination of sensory and proxy measures is routinely employed to infer the cooking, eating, and textural properties of rice. For example, the eating quality of rice can be extrapolated based on Instron hardness, gel and amylograph consistency, and final gelatinization of rice [40]. This highlights the need to look at grain quality using multiple parameters at systems level [16]. It also points out to the need to develop additional proxy measures to exhaustively characterize the complexity of grain quality of cooked rice kernels.

In order to gain a deeper understanding of cooking quality, starch-based parameters need to be defined from single- to multi-dimensional parameters. There is a need to move beyond amylose content estimation to more advanced rice grain quality tests involving starch structure, viscoelasticity, rheometry, texture, and aroma (Table 1). All of these assays are tedious, low- to medium-throughput, expensive and time-consuming. These traits need to be quantified through semi-throughput techniques to derive new rice cultivars screened by advanced proxy measures and validated through consumer panels. Ideally, molecular markers associated with grain quality traits should also be developed to facilitate multiple screening using molecular techniques (see below). In this section, advanced state-of-the-art techniques that can measure multiple grain quality parameters are described, including the use of molecular markers associated with rice grain quality.

3.1 Apparent Amylose Content

Several analytical methods have been proposed to more accurately determine the amylose content of rice grains which do not rely on iodine binding. This includes the use of differential scanning calorimetry (DSC), near-infrared spectroscopy (NIRS), and debranched starch size exclusion chromatography (SEC). The DSC method is based on measuring the melting endotherm of the amylose-lipid complex in non-glutinous rice samples [27]. Subsequent studies revealed a solid linear relationship between amylose content and the enthalpies of melting of the amylose-lipid complex as observed in rice flour and other major cereals and root crops [43]. This calorimetric method is not affected by long-chain amylopectin interference or the presence of endogenous lipids, offering high accuracy, reproducibility, and convenience [44]. On the other hand, the near-infrared spectroscopy (NIRS) method relies on near-infrared reflectance (NIR) or transmittance (NIT) spectrum of ground milled rice [45, 46]. NIRS gained popularity not just for estimating amylose content but also in determining other rice quality parameters such as protein content, whiteness, transparency, and degree of milling [47]. However, this method requires significant calibration from training population before it can reach a significant level of accuracy. Lastly, amylose estimation based on debranched starch structure offered a more direct measurement of molecular weight distribution (MWD) of amylose chains [48, 49]. Mechanistic information from MWD [50] was successfully used to characterize rice kernel mutants with defects in the starch biosynthetic pathway [15, 51, 52]. This technique was recently employed to quantify the amylose content of a rice diversity panel with superior accuracy compared to the traditional iodine method [16]. However, this technique relies on debranching of rice starch, hence co-elution of amylose and long-chain amylopectin is possible. This issue can be indirectly addressed by isobutanol sub-fractionation of amylose and amylopectin fractions prior to debranching and SEC analyses [53]. Long-chain amylopectins are

Table 1
Summary of general routine rice grain quality tests [4, 22, 23]

Quality parameter	Measure	Principle	Standard method	Alternative method
A. Cooking and eating quality				
Apparent amylose content	Apparent amylose content	Determination of apparent amylose content by iodine binding	A 100 mg milled rice sample is dispersed in 9 mL of 1 M NaOH. The solution is then neutralized to pH 4.5–4.8 with acetic acid. Apparent amylose content is measured by iodine colorimetry at 620 nm. A variation exists where the hot-water-insoluble amylose content is determined by using hot water as solvent instead of dilute alkali. Hot-water-insoluble amylose content is highly correlated with cooked-rice texture.	HPLC-SEC, DSC, NIRS, NIR-FTR
Cooking time	Cooking time	Estimation on the amount of time it takes for a grain to cook	Rice grains are cooked in preheated water in a test tube immersed under boiling water bath. A few grains are removed at regular time intervals. The grains are pressed between two glass slides. Cooking time is determined from the exact time at which the opaque central core disappears. Amount of water needed to cook the rice can also be estimated using similar technique but using a small automatic rice cooker. Rice to water ratio (w/v) is estimated.	Cooking time is indirectly measured by gelatinization temperature using DSC
Grain elongation	Grain elongation	Determination of the ratio between cooked and uncooked grain	This test is specifically for Basmati-type rices. Raw rice grains are pre-soaked in water for 30 min. The rice grains are then cooked for 10 min in boiling water. After cooling, grain elongation ratio between the average lengths of the cooked grain to that of the uncooked grain is then estimated.	None
Grain volume expansion	Grain volume expansion	Determination of volume expansion of rice grain after cooking	Exactly 20 g of rice is cooked with 50 mL water in graduated cylinder. The setup is placed in boiling water bath and cooked. The initial and final volume of the cooked rice is determined.	None

Gelatinization temperature	Alkali spreading value or alkali digestion score	Determination of extent of disintegration of rice grain under alkali solution	Six grains of rice are placed in rectangular plastic box containing 10 mL 1.7% KOH. The extent of grain digestion is noted after 20 h. Scores range from 1 (no effect) to 7 (all grains are completely disintegrated). Commonly done in triplicates.	Measurement of starch gelatinization by differential scanning calorimeter, photometric method, alkali photometry, or RVA pasting curve
Pasting properties	Starch paste viscosity profiling	Determination of swelling, pasting, viscosity of starch paste when cooked and how it congealed when cooled	Rice flour sample is mixed with water. The slurry is subjected to a controlled cycle of heating and cooling under continuous mixing. The starch viscosity curve is determined where the peak, hot-paste and cold-paste viscosity is determined. Gelatinization temperature, breakdown, setback and breakdown are then derived from the viscogram.	Brabender viscograph or visco-amylograph, micro Viscoanalyser
B. Textural and sensory quality				
Gel consistency	Gel mobility test	Estimation of cooked rice grain texture by mobility of starch gel after cooking in dilute alkali	A 100 mg fine rice flour powder (100-mesh) is dispersed in a test tube with small volume of alcohol. The flour is then dispersed in 2 mL dilute KOH and cooked in boiling water bath. The gel dispersion is immediately cooled in ice and then laid horizontally on a table. The distance of the gel mobility after an hour is noted. Normally run in duplicates.	None
Texture profiling	Texture profiling	Characterization of cooked rice grain texture using texture analyser	Rice grains are cooked by boiling and three individual rice grains parallel with each other are placed in an aluminium plate. Some methods only require single grain. Texture profile analysis is done using 35 mm aluminium cylinder probe and a 5 kg load cell. Force-deformation curve is obtained using 2-cycle compression test and post-test speed of 0.5 mm/s. Usually done in duplicates. Selected alternative methods itemized in the next column have been systematically compared.	Instron hardness testing. Parallel plate plastometer, consistometer, texturometer, hardness tester, viscoelastograph, tensipresser, surface tensiometer, Kramer shear or texture press, extrusion and back extrusion, puncture test

(continued)

Table 1
(continued)

Quality parameter	Measure	Principle	Standard method	Alternative method
Sensory evaluation	Sensory evaluation of cooked rice grain	Characterization of appearance, aroma, texture, and taste of cooked rice	Untrained consumer panel or trained expert panel are employed to score the appearance, aroma, texture, and taste of cooked rice. Raw rice can also be subjected to sensory evaluation. A standardized attribute and definition has been devised for descriptive sensory analysis of cooked rice [41, 42].	None
Aroma profiling	Aroma	Determination of aroma of cooked or raw rice grains	The rice is cooked and scored for aroma. Raw rice grains can be cracked in the mouth and scored for aroma. Rice grains can also be placed in dilute alkali and the evolved aroma smelled directly.	Detection and quantification of 2-acetyle-1-pyrroline by GC-MS Detection of total volatile metabolome by GC-MS
Rancidity test	Rancidity	Test of hydrolytic rancidity by increased in acidity of milled rice	Ten mL of dilute methyl red and bromothymol blue are placed in 5 g of milled rice in a test tube. The stained rice grain changes with increasing acidity from yellow-green to yellow to orange. Freshly milled rice are greenish while aged rice are orange to reddish. A solution of bromothymol blue and phenol red can also be used. Freshly milled rice will stain deep violet while old rice will be yellow for this test.	Detection of free fatty acids by titration or colorimetry
C. Nutritional quality				
Protein content	Protein content determination	Determination of protein content by Kjeldahl method	Amount of protein in the flour is determined by semimicro or micro Kjeldahl digestion, distillation and titration of ammonia nitrogen. Colorimetric method is also available.	NIRS
Lipid content	*Crude fat and fatty acid determination*	*Determination of total lipids and fatty acid composition*	TLC, GC, GLC *determination of lipid and fatty acid* composition.	*Metabolomics approach using LC-MS or GC-MS*

Resistant starch content	Quantification of total resistant starch	Determination of total resistant starch using enzyme digestion	Flour samples are digested overnight for 16 h using alpha-amylase and amyloglucosidase. The undigested starch fraction is hydrolysed into glucose and quantified as total resistant starch.	None
Nonstarch polysaccharide content and dietary fiber content	Quantification of total dietary fiber	Determination of total dietary fiber after enzyme digestion	Starch, proteins and lipids are removed using enzymatic hydrolysis. Total, soluble and insoluble dietary fiber can then be determined. Dietary fiber composition can also be done for major non-starch polysaccharides.	CE, HPLC coupled with mass spec detector
Micronutrients	Elemental analysis	Atomic absorption spectroscopy, X-ray fluorescence imaging	Micronutrient identification and quantification using AAS, ICP-MS, LIBS, ICP-OES.	None
Digestibility	Estimation of digestibility kinetics	In vitro determination of estimated glycemic scores (EGS) and digestibility rate coefficient (*k-value*)	Rice grain, flour, or starch samples are digested with enzymes that digests starch, proteins and lipids. The amount of sugar released or starch hydrolysed are quantified. EGS or *k*-values are then computed.	Time-resolved NMR

present in all high-AC cereal starches [54]. They can be separated from amylose by butanol crystallization of amylose as a complex, leaving amylopectin in solution [14]. The decrease in the first peak by SEC after amylose removal is true amylose content [54]. However, adopting this protocol for routine amylose screening will make the method more laborious and time-consuming. Alternatively, a more accurate method that relies on direct analyses of native (i.e., undebranched) starches may be necessary, as will be discussed in the subsequent section.

3.2 Amylopectin Fine Structure

Quantifying the proportion of amylose and amylopectin in rice starch does not give a complete picture. The amylopectin fine structure has to also be considered to explain functional properties of rice as it is related to thermal, viscometric, textural, and nutritional traits. Besides being the major starch fraction, the proportion and fine structure of amylopectin contributes to many properties of cooked rice. One method based on fluorophore-assisted carbohydrate electrophoresis (FACE) using capillary electrophoresis (CE) is currently used for this purpose [55, 56]. This method replaced an older technique that involves gel filtration chromatography (GFC) of debranched starch samples [57]. The chain length distribution (CLD) of amylopectin can also be quantified using high-performance size-exclusion chromatography attached to multiangle laser-light scattering (MALLS) and refractive index (RI) detectors (HPSEC-MALLS-RI) or high-performance anion-exchange chromatography coupled with a pulsed amperometric detector (HPAEC-PAD). Characterizing the CLD of amylopectin of rice grains revealed polymodal distribution indicating several groups of B-chains (B1, B2, B3 and B4) consistent with the cluster model of amylopectin structure [58, 59]. Amylopectin chain ratio (ACR) or molar ratio of Σ degree of polymerization (DP) 6–10/Σ DP 6–24 $<$ 0.20 of CLD of debranched amylopectin denotes intermediate/high GT and ACR of $>$0.23 is low GT [29].

3.3 Starch Paste Viscosity

As has been mentioned above, starch paste viscosity measurement using Brabender Viscoamylograph (AACC International Method 61-01.01) was replaced by the Rapid Visco Analyzer (RVA, Newport Scientific, Australia). The RVA method (AACC International Method 61-02.01) uses a stirring and heating viscometer to mimic the cooking process of rice grain by exposing a flour-water suspension to a heat-hold-cool-hold temperature cycle while continuously subjecting the sample to mechanical shear stress [60, 61]. The onset of viscosity begins when amylopectin crystals begin to melt and hydrate in excess water at elevated temperature. As the suspension is heated further, viscosity increases as a result of starch granule swelling, while amylose and small amylopectin molecules leach out [62]. The pasting profiles of rice flour are important predictors of rice eating, cooking, and processing quality characteristics. Viscosity curves are the most useful tool available for assessing cooking quality of rice, with the outcome of data associated with sensory or processing attributes [63]. Aside from gel consistency, pasting property of rice is one of the secondary differences that arises within varieties with the same amylose content [64, 65]. In addition, most RVA parameters were significantly correlated with amylose content [66, 67]. For instance, a low setback value is associated with softness after cooking and a high viscosity breakdown value is related to good palatability [63, 67, 68]. In addition, cohesiveness of mass, the maximum degree to which a sample of cooked rice holds together during chewing, is correlated with lift off and weakly correlated with break down [69].

3.4 Starch Gel Rheology

The rheological response of the system can be resolved into elastic and a viscous component, the complete characterization of which provides a better idea of the gelation behavior of rice flour [70]. The gelatinization and pasting properties of rice flour described above are indicators of the ease or difficulty of cooking flour, while rheological profiling gives a clear picture of its viscoelastic tendencies [71]. Rheological parameters are increasingly correlated with starch granule rigidity which can cause differences in the texture of cooked rice [70]. The viscoelastic property of rice flour is commonly measured using an oscillating rheometer [72]. A pioneering study using this instrument reveals that the storage modulus (G′) of non-glutinous rice starch gels is exponentially correlated with AAC and that it is greater than the loss modulus (G″) [73]. The G′ measures elasticity while the G″ measures viscosity. Additionally, the same study revealed that loss tangent (tangent δ), which measures the liquid-like properties of a gel, is inversely correlated with AAC. Mechanistically, a two-phase gelation (staling) process is observed using this technique. The first rapid phase is due to amylose, while the second slower phase is due to the recrystallization of amylopectin [74].

3.5 Textural Profiling of Cooked Rice Grains

The textural properties of rice have been increasingly recognized as an important characteristic that can influence their acceptability and consumption. Measurement of the texture of cooked rice grain is traditionally conducted using Instron food tester, Texturometer, and viscoelastograph [75–78]. Milled rice is cooked with either the amount of water that produce the optimum texture (<75% water), or in excess boiling water until the core is translucent (ca. 75% water) [79]. The main concern is how to cook rice grains differing in starch properties to the same degree that will make them intercomparable. Interestingly, an international cooperative testing to measure cooked rice grain texture revealed that hardness and stick-

Table 2
Contribution of true amylose, amylopectin chains, and protein to rice quality [82]

		Amylopectin		
Starch property	**True amylose**	**Short-chain**	**Long-chain**	**Protein**
Crystallinity		X		
Gelatinization temperature		X		
Apparent amylose content	X		X	
Staling during cooling	X		X	
Slow staling on cooling (reversible)		X		
Gel consistency			X	X
RVA/Amylograph consistency/setback	X		X	X

iness values measured by textural profiling instruments are more accurate than sensory panel scores [78]. The use of AAC was also validated as a highly correlated proxy measure for texture. This led to the use of AAC estimation as the primary screening tool to classify rice based on texture, with the instrumental methods used only for verification of selected advanced breeding lines [80]. The current method for textural characterization involves Texture Profile Analysis (TPA) TA.XTplus Texture Analyzer (Stable Microsystems, UK) to measure multiple textural parameters of cooked milled rice grains in just one experiment (*see* Chapter 9). It is a double-compression test used to describe the hardness and stickiness of cooked milled rice grains. During TPA analysis, rice grains are compressed twice by a texture analyzer to imitate the action of the jaw while measuring several attributes during simulated chewing.

Limited studies have been currently conducted to validate the relationships of rheological parameters with instrumental texture profiles and other quality indicators. Attempts have been made to correlate instrumental texture profiles with rice quality predictors such as RVA measurements and amylose content [63, 80, 81], as will be discussed below. For example, biochemical composition such as apparent amylose content and protein content was demonstrated to influence pasting properties to explain the differences in cooked rice texture [80]. It is becoming clear that true amylose content, amylopectin chain lengths, and protein content influence the texture of cooked rice [82] (Table 2).

3.6 Aroma Profiling

Rice fragrance is an eating and a sensory trait that significantly increases the market value of rice. There are more than 100 volatile compounds responsible for rice grain aroma but 2-acetyl-1-pyrroline (2-AP) is the most potent and is responsible for the *pandan* scent in rice kernels [83–85]. It is present in jasmine and basmati rice, two types of rice from Thailand and India respectively, which are highly valued worldwide for superior aroma and grain quality. Because 2-AP has a high odor threshold, it is traditionally detected by smelling after reaction with 0.1 M KOH. However, this method is harmful to the nasal cavity of the analyst upon continuous and prolonged exposure. To circumvent this problem, methods using gas chromatography coupled with flame ionization detector (GC-FID) or mass spectrometry (GC-MS) were developed [86, 87]. However, not all rice breeding facilities have GC-MS instrument due to its expensive capital investment, running and maintenance costs. Furthermore, it cannot be used to routinely screen thousands of parental and breeding lines because it is low-throughput. Due to these restrictions, molecular markers related to 2-AP are currently used in rice breeding programs for aroma (see below).

Aroma in rice grain is not just due to 2AP. Some volatile compounds in rice grain are known to enhance rice grain aroma, while others can decrease sensory quality due to off-flavors and rancidity. Hence, comprehensive detection and comparison of aroma compounds are necessary to identify other volatile compounds to determine how they affect sensory properties. One effective metabolomics method is a solvent-free technique using solid-phase microextraction (SPME) which allows for simultaneous extraction of volatile compounds during cooking of samples [88]. This technique was able to differentiate the volatile compounds present in aromatic and non-aromatic rice grains [89]. However, jasmine- and basmati-type aroma was not resolved using this technique [89]. The differences in volatile profile were resolved by combining GC-MS, GC-olfactometry, and dynamic headspace system. Volatile compounds from rice grains can be identified using odor threshold and odor activity values to distinguish varietal differences in rice grain aroma [90]. The complexity of rice grain aroma and flavor can be dissected using integrated multidisciplinary approach involving aroma profiling, flavor metabolomics, and sensory analyses. These integrated approaches can be employed to further understand the sensory properties of rice grains [91].

3.7 Sensory Evaluation

Sensory evaluation involves the assessment, characterization, measurement, analysis, and interpretation of the sensory properties of food products. Sensory evaluation for rice grain quality usually involves comparative, descriptive, or quantitative profiling of human perception to rice grain consumption using trained sensory panellists or untrained market consumers. Linking sensory and textural properties to quality attributes lies in the deeper understanding of the physicochemical properties and the biochemical composition of rice grain that influence the consumer perception of cooking, eating, and sensory quality of a particular rice cultivar. As pointed out in the previous section, routine biochemical tests are currently being used primarily in the evaluation of breeding lines as proxy indicators of grain quality. Understanding rice grain quality has reached an upper limit using these traditional techniques. New approaches have to be used to unravel multiple facets of rice cooking and eating quality that were not previously captured by these routine physicochemical tests. A good starting point is the use of sensory and consumer panels to describe and quantify various types of rice aroma and flavor. This can be done in parallel with the exploration of other novel analytical tools to understand and quantify traits that influence consumer acceptance and preference.

Descriptive sensory evaluation and instrumental analyses revealed that the aroma and flavor of rice grains are very diverse [92]. Consumers define various classes of rice quality at the local, regional, and international context [93–95]. Because quality preference for rice is at least as diverse as the culture and palate of the consumers, market segmentation has to be properly identified based on regional influence and profile of the rice varieties. It must be pointed out, however, that there is still a huge difference on how sensory and quality preferences are measured in the laboratory compared to how actual consumer preference is estimated in target markets. In this connection, international standards for grain quality and sensory properties should be established so that each rice variety can be benchmarked against a set of defined criterias. An international collaboration among rice grain quality and sensory specialists need to be forged to accomplish this task. Activities can include exhaustive textural and aroma profiling using traditional sensory and modern instrumental techniques to improve our understanding of cooking quality. An internationally-accepted lexicon for taste, flavor, and aroma descriptors should also be standardized [96, 97]. Because traditional sensory methods cannot be used as routine selection tool to breed superior grain quality varieties due to its inherent intricacy and subjectivity, instrumental proxy methods need to be used to bridge the gap. The results obtained from these instrumental sensory platforms should be highly correlated with the major traits of interest. Sensory evaluation can be used to calibrate instrumental tests which can provide quantitative values to avoid the inherent subjectivity of the sensory evaluation process.

3.8 Molecular Markers for Rice Grain Quality

Advanced rice chemistry laboratories also employ molecular markers to classify grain quality of rice. These molecular markers are used to improve the eating, cooking, and nutritional quality of rice [98, 99]. One limitation of this approach is that only a few molecular markers have been associated with functionally validated rice grain quality traits by far (Table 3). These molecular markers are primarily restricted to explaining variations in rice amylose composition [104, 114], amylopectin fine structure [66, 113, 115], and fragrance [100, 101, 116].

Allelic variations for starch synthesis-related genes (SSRGs) coding for starch synthases, branching enzymes, and debranching enzymes have been associated with physicochemical properties of cooked rice grains [67, 117–121]. Of all the SSRGs highly expressed in the rice endosperm, the *Granule-Bound Starch Synthase I* (*GBSSI*) gene that regulates amylose synthesis is the most comprehensively studied. Allelic variations in the structural gene for *GBSSI* are associated not just with amylose content but also with other functional properties of cooked rice grain. For instance, alleles of this gene are also associated with gel consistency [122], starch pasting, and digestibility in rice [123]. Another major starch synthase gene that influences starch and functional properties is *Starch Synthase IIa* (*SSIIa*) which determines amylopectin fine structure and gelatinization property of rice [71, 123, 124]. Allelic variations in *SSIIa* were also found to influence gel consistency in rice [124].

The use of molecular markers to track for grain fragrance (*fgr*) in rice is made possible with the discovery of 8-bp deletion in exon 7 of a gene that codes for betaine aldehyde dehydrogenase (*Badh2*) [100, 125]. This method is more cost-effective and higher-throughput compared to the GC-MS-based volatile profiling method described above. This paved the way for high-throughput screening of aromatic rice accessions by PCR amplification of the functional mutation in *Badh2* associated with presence or absence of 2AP. However, not all aromatic rice accessions were found to have this deletion based on screening of a rice diversity panel, hinting that other alleles for this trait must be present [87]. This was verified by subsequent studies which revealed the existence of novel alleles in *Badh2* responsible for 2AP accumulation [101, 102]. Currently, several molecular markers are needed to detect all variant alleles present in *Badh2* as listed in Table 3. Additionally, other nonfunctional haplotypic variants of *Badh2* are detected in diverse rice germplasm collections [122, 126]. It is possible that other yet uncharacterized mutations exist in *Badh2* that were favorably selected during the course of domestication of cultivated rice.

Since the establishment of GQNC in IRRI in 2004, we routinely detect allelic variants of genes related to apparent amylose content [106], gelatinization temperature [30], and aroma by 2-AP [87, 122, 126, 127] using polymerase chain reaction (PCR) and gel electrophoresis. These molecular markers are also employed in routine marker-assisted breeding (MAB) programs for rice all over the world.

3.9 Nutritional Quality

The nutritional quality of rice covers a fairly complex set of traits related to health-promoting macronutrient, micronutrient, and phytochemicals present in the whole or milled grains [128]. Macronutrients consist of biopolymers such as carbohydrates, proteins, and lipids which make up the bulk of foods primarily consumed worldwide for energy homeostasis [129]. On the other

Table 3
Molecular markers associated with rice grain quality

Trait	Gene/alleles	PCR primers	PCR conditions	Reference
Fragrance from 2-acetyl-1-pyrroline	*Betaine aldehyde dehydrogenase* (*Badh2*)			
	badh2 8-bp deletion on exon 7	TTGTTTGGAGCTTGCTGATG CATAGGAGCAGCTGAAATATATACC CTGGTAAAAAGATTATGGCTTCA AGTGCTTTACAAAGTCCCGC	94 °C, 5 min 35× of 94 °C, 45 s 55 °C, 45 s 72 °C, 45 s 72 °C, 10 min	Bradbury et al. [100]
	badh2 7-bp deletion in exon 2	CCTCTGCTTCTGCCTCTGAT GATTGCGCGGAGGTACTTG PCR = 200/207 bp or CTTCTGCCTCTGATTAGCCT GCCGTGAGCCATATACACTT PCR = 643/650 bp	94 °C, 2 min 30× of 94 °C, 30 s 50–60 °C, 30 s 72 °C, 1 min 72 °C, 10 min	Shi et al. [101]
	*badh2*803 bp deletion between exon 4 and 5	TGCTGGATGCTTTGAGTA GTTTAGCACACCTGAAGGAAGACCA PCR = 321/1123 bp	94 °C, 2 min 30× of 94 °C, 45 s 55 °C, 45 s 72 °C, 1 min 72 °C, 8 min	Shao et al. [102]

Amylose content	*Granule-bound starch synthase I (GBSSI)*			
	$(CT)_n$ Repeat	TTGCAGATGTTCTTCCTGATG CTTTGTCTATCTCAAGACAC	95 °C, 5 min 35× of 94 °C, 45 s 52 °C, 45 s 72 °C, 45 s 72 °C, 10 min	Bligh et al. [103] Ayres et al. [104] Bergman et al. [105] Roferos et al. [106] Tuaño et al. [32]
	Exon 2 23-bp duplication	TGCAGAGATCTTCCACAGCA GCTGGTCGTCACGCTGAG	95 °C, 5 min 35× of 94 °C, 30 s 60 °C, 30 s 72 °C, 1 min 72 °C, 5 min	
	Intron 1 G/T SNP	GCTTCACTTCTCTGCTTGTG ATGATTTAACGAGAGTTGAA	94 °C, 5 min 35× of 94 °C, 40 s 55 °C, 40 s 72 °C, 1 min 72 °C, 10 min	Ayres et al. [104] Tuaño et al. [32]
	Exon 4			Larkin and Park [107] Mikami et al. [108]
	Exon 6 A/C SNP	GGTTGGAAGCATCACGAGTT TCTTCAGGTAGCTCGCCAGT CCCATACTTCAAAGGAACATA	94 °C, 5 min 32× of 94 °C, 30 s 60 °C, 30 s 72 °C, 1 s 72 °C, 5 min	Cai et al. [109] Larkin and Park [110] Chen et al. [111] Tuaño et al. [32]
	Exon 10 C/T SNP	GCATCACCGGCATCGTC GCTCCGGCCATGATGAGATG	94 °C, 5 min 35× of 94 °C, 40 s 55 °C, 40 s 72 °C, 1 min 72 °C, 10 min	Larkin and Park [110] Dobo et al. [112] Tuaño et al. [32]

(continued)

Table 3 (continued)

Trait	Gene/alleles	PCR primers	PCR conditions	Reference
Gelatinization temperature	*Starch synthase IIa* (*SSIIa*)			
	SSIIa SNP2 (A/G)	AGAACGACTGGAAGATGAACG GATGTCCACACCTTTCTGCC CTTGCACCGCGGCTTGCC	94 °C, 5 min 35× of 94 °C, 30 s 62 or 68 °C, 30 s 72 °C, 1 min 72 °C, 7 min	Waters et al. [113] Cuevas et al. [31] Tuaño et al. [32]
	SSIIa SNP3 (A/G)	GGGTGGGTGGGGTTCTCG CACCATTGGTACTTGGCCTTGAC GCGGGCTGAGGGACAGCA		
	SSIIa SNP 4 (GC/TT)	GGGTGGGTGGGGTTCTCG CACCATTGGTACTTGGCCTTGAC GCCGCGCACCTGGAAA		

hand, micronutrients are small organic (e.g., vitamins) or inorganic (e.g., minerals) molecules present in food required in trace amounts but are essential for normal growth and metabolism [130, 131]. In contrast, phytochemicals are naturally-occurring bioactive secondary metabolites responsible for organoleptic properties in plants that also have potential health benefits in humans, but may not necessarily be considered as essential for normal growth and metabolism [132]. Food consumption involving balanced intake of macronutrients, micronutrients, and phytochemicals will ensure general health and well-being. Current efforts are underway to tailor the nutritional properties of rice grains due to its potential as a nutrient delivery system to bring optimal health to half of the global population [128].

Just like the other grain quality traits discussed so far, the nutritional quality of rice grain is also primarily influenced by starch since it is the major macronutrient in the milled rice grain, composed of around 80–90% of grain weight. Interestingly, starch is not digested at equal rate in the human gastrointestinal tract. The gold standard in estimating the starch digestibility of starch-containing food is by measuring its glycemic impact. It is assayed under clinical settings by determining the blood glucose response of human volunteers upon consumption of a food containing 50 g available carbohydrates compared to a standard solution containing 50 g glucose [133, 134]. The glycemic response (GR) is reported as glycemic index (GI) based on a pioneering study by Jenkins et al. in 1981 [135] which was validated by several laboratories in 2003 [136]. The concept of glycemic load (GL) was subsequently introduced in 1997 to account for the observation that the proportion of available carbohydrates in the food can have a profound impact on overall glycemia [137]. Unfortunately, determining the glycemic impact of food cannot be routinely employed in screening for low GI in rice breeding programs because it is low-throughput and requires expensive clinical assays. Consequently, the nutritional fractions of starch can be estimated in vitro by quantifying total sugar, total starch, rapidly digestible starch (RDS), slowly digestible starch (SDS), and resistant starch (RS) [138, 139]. GI can also be estimated in vitro using starch hydrolysis [140], estimated glycemic score (EGS) [141], and hydrolysis index (HI) values [15, 123]. In addition, the rate and extent of starch digestibility can also be estimated by amylolysis using log of slope (LOS) plot for first-order kinetics [142]. The in vitro methods routinely employed in GQNC to estimate cooked grain digestibility are presented in Chapter 13.

Aside from starch, the other macronutrients in rice are storage proteins, storage lipids, and nonstarch polysaccharides (NSPs). Although they are present as minor components in the rice grain, these biopolymers play a major role in influencing nutritional quality, textural and sensory traits, and functional properties [128]. For instance, the average storage protein composition of milled rice grain is just around 7% but it is a major source of protein in developing countries. Rice storage protein is gaining popularity worldwide because it is hypoallergenic and has superior amino acid composition [143]. The Kjeldahl Method [144] is the internationally-accepted protocol widely used for estimating total protein content (AACCI Method 46-16.01), with modifications to accommodate smaller sample sizes (AACCI Method 46-13.01) [12]. Individual amino acids can also be quantified after acid hydrolysis using pre-column derivatization using a fluorescent derivatizing reagent followed by HPLC separation [145, 146].

Similar to storage proteins, the storage lipids in rice are just less than 1% in milled rice grain but they significantly influence nutritional and functional properties [120]. The compositional analyses of storage lipids in rice are interesting, especially from a nutritional and grain quality perspective [147]. For instance, minor lipids in rice grains are found to influence pasting properties and to confer protection against chronic ailments including cardiovascular diseases and cancer [148]. Crude fat in rice grains is routinely analyzed using a standard method (AACCI Method 30-10.01). Classical method employs thin-layer and gas chromatography to determine lipid composition in rice grains [149]. A standard method is routinely available to determine total, saturated, unsaturated, and monounsaturated fats after acid hydrolysis (AACCI Method 58-19.01). The proportion of nonstarch (free or bound) and starch lipids (i.e., amylose-complexed) can also be determined in milled rice. The fatty acid composition of the bran layer can also be analyzed using gas-liquid chromatography (GLC) [150]. More recently, a liquid chromatography method coupled to a mass spec detector (LC-MS) was developed to characterize the phospholipid content of rice [151] and to determine variation in composition among cultivars [152].

NSPs constitute the dietary fiber in rice grains which are concentrated in the bran layer. Although they are just present in trace amounts in milled grains, NSPs are interesting from a nutritional point of view because they are compositionally unique in rice compared to other cereals [135]. The total dietary fiber (TDF) in the grain can be assayed after enzymatic removal of starch, proteins, and lipids (AACCI Method 32-05.01). A rapid gravimetric method is also available using parallel determinations of water-soluble and neutral detergent-insoluble fractions (AACCI Method 32-06.01). The total, soluble and insoluble dietary fiber can also be estimated using a single assay (AACCI Method 32-07.01). These compositional assays should be supplemented with structural determinations to identify the diversity, size, and shape of NSPs in rice grain.

Micronutrients and phytochemicals in rice grains are also primarily concentrated in the rice bran [153]. Polishing to produce milled rice grains removes most of the micronutrients in rice which could otherwise be used to mitigate the impact of micronutrient deficiency, a form of hidden hunger that continues to plague underdeveloped and developing countries worldwide. Micronutrient deficiency can lead to growth stunting characterized by growth retardation, developmental delay, impairment in the immune system, reduction in cognitive function, and increased risk of metabolic disorders such as obesity and hypertension [154]. Efforts are underway to increase the micronutrient density and composition in rice endosperm to circumvent the problem of micronutrient deficiency in rice-consuming countries [155, 156]. Hence, innovative methods are required to simultaneously identify and quantify micronutrients in milled rice grains. This is usually accomplished using atomic absorption spectrophotometry (AAS), inductively coupled plasma–mass spectrometry (ICP-MS), laser-induced breakdown spectroscopy (LIBS), and inductively coupled plasma–optical emission spectrometry (ICP-OES) [157, 158]. In GQNC, we use ICP-OES to simultaneously quantify 12 elements: Na, K, Ca, Mg, Fe, Mn, Zn, Cu, Al, P, S, and Mo (Chapter 14). In contrast, there is no routine method for the quantification and characterization of bioactive phytochemicals in rice grain. Readers are referred to original research articles for analytical procedures [159–166].

4 Emerging Rice Grain Quality Tests

Most of the emerging state-of-the-art rice grain quality tests involve the use of nondestructive, multi-parameter assays that are highly correlated with traditional methods. Most of these tests have not been substantially validated using large sample population of rice grains but they offer significant improvement in sample throughput and accuracy. However, most of these methods would require significant capital investments.

4.1 Grain Structural Traits

The accurate and repeatable separation of highly branched polysaccharides such as rice starch in SEC remains challenging even when coupled with multiple detectors [167]. A range of instrumental techniques are currently employed to analyze several levels of starch structural organization. The main challenge in this area is to characterize starch structure in its native conformation. Toward this goal, combining flow field-flow fractionation (FFF) with multi-angle laser light scattering (MALLS) and differential refractometer index (DRI) were successful in fractionating starch in aqueous conditions [168]. More recently, two separation methods using asymmetrical flow field flow fractionation (A4F) and hydrodynamic and size-exclusion chromatography (HDC-SEC) were used to determine the molecular size and mass distributions of native starches including rice [169]. Similar results were also accomplished with A4F in tandem with MALLS and refractive index (RI) detectors [170]. In addition, free solution capillary electrophoresis (CE) can provide an alternative to the conventional debranched SEC technique because of its ability to determine native starch structure, composion, and degree of branching [171]. In terms of higher granular organization, a distinct nonlinear spectral shift in hydrated starches was detected as starch molecular order increases using a combination of DSC, ^{13}C nuclear magnetic resonance (NMR), X-ray diffraction (XRD), and Fourier transform infrared spectroscopy (FTIR) techniques [172].

4.2 Grain Cooking Traits

Spectroscopic techniques such as NIRS, Fourier transform infrared spectroscopy (FTIR), and Raman spectroscopy show promising potential in characterizing the cooking quality of rice grains. Several spectroscopic techniques, mostly using NIRS, were employed to characterize apparent amylose [45, 46, 173], moisture [174], protein [175], and fat contents [176]; gelatinization [174], pasting [63, 177], and textural properties [178, 179]; as well as solid loss, volumetric expansion, and water uptake of rice during cooking [180].

4.3 Grain Textural Traits

Chemometric method employing various parameters from starch structure and rice grain composition highlighted the importance of amylose-to-amylopectin ratio (AAR) and crude protein (CP) in influencing the texture of rice grains [181]. A combination of TPA and dynamic rheological measurement revealed that consistency coefficient (K^*) and loss tangent ($\tan \delta$) are appropriate proxy measures for rice grain hardness and stickiness, respectively [182]. Predictive model associating textural properties of rice with the physicochemical parameters of rice grains have also been done in the past. One such model revealed that AAC, blue value (BV), and moisture content (MC) significantly influence the texture of rice [183].

4.4 Grain Sensory Traits

As mentioned above, some sensory traits such as hardness and stickiness are better measured instrumentally using TPA. However, not all sensory traits have instrumental correlations. Hence, one of the main challenges for sensory profiling of rice grain is to identify instrumental methods that are well correlated with sensory traits. Toward this goal, a single-compression TPA test coupled with novel multivariate data analysis successfully predicted rice sensory texture for cooked grains [184]. In the same study, seven sensory attributes (cohesion, adhesion, hardness, cohesiveness of mass, roughness of mass, toothpull, and toothpack) were predicted using the novel step-wise data analysis method. Sensory preference and taste scores were predicted by combining taste analyzer with routine grain quality data and utilizing fuzzy logic algorithm [185]. Metabolome profiling coupled with descriptive sensory evaluation identified metabolite biomarkers related to excellent rice flavor quality that can differentiate jasmine-type rice grains [186]. Visible and shortwave near-infrared spectral data from NIRS were used to predict sensory quality of rice [187]. Modeling for sensory property of *indica* rice was successful using support vector machine (SVM) [188]. Lastly, the use of electronic nose (e-nose) and electronic tongue (e-tongue) to respectively detect volatile and non-volatile compounds in food and beverages are gaining popularity due to advancements in sensor and separation technologies [189]. These advanced sensor and separation technology can potentially be used in rice grain sensory evaluation in the future. In addition, the oral processing of food is also receiving considerable attention to describe the pleasure of taste, texture, and flavor during consumption [190–192]. Toward this end, the rheology and tribology of food texture sensation are increasingly becoming important tools in food consumption and sensory evaluation [193, 194]. Techniques from sensory science perspective can be combined with concepts in material science to capture the dynamic aspects of oral processing [195]. However, no systematic study to date has been conducted to characterize the tribology of rice grain consumption. This is an important area for future study considering that chewing behavior varies more among consumers than rice types [196]. Systematic research on oral processing of cooked rice grain is also necessary in altering grain composition to make them healthier without compromising sensory quality including texture and mouthfeel. Lastly, limited studies have also been done on short-term and long-term satiety indices of cooked milled and brown rice (such as satiety quotient for hunger, fullness, desire to eat and prospective consumption; overall satiety index; and 2-h postmeal rice intake) [41, 197].

4.5 Grain Digestibility Traits

In vitro carbohydrate digestion methods are available but substantial improvements and validation are needed [42, 198]. The kinetics of in vitro digestibility of starch has been tracked by time-resolved NMR [199]. Free solution CE also offers the potential to track release of sugars and short oligosaccharides in real time by simple UV detection during in vitro amylolysis [200].

5 Future Prospects

Rice cereal chemists have continued to develop innovative proxy methods to estimate grain quality traits to reflect constantly evolving and rapidly expanding consumer preference. In this last section, future prospects for rice grain quality assessment is discussed by considering what methods are routinely used in the past to the present and by taking into account available emerging technologies.

5.1 Accurate Assessment of Consumer Preference

Actual sensory evaluation will only be a critical component in rice breeding in characterizing parental lines and prior to releasing the newly-developed rice cultivar. It is therefore necessary to integrate instrumental proxy measures for sensory evaluation during the early stages of the breeding pipeline. The extent to which correlations can be made between instrumental and panel-based sensory profiling can be validated using different samples capturing regional preferences of various market quality segments. Toward this goal, it would be interesting to merge methods in rheology, rheometry, and textural analyses with taste and flavor metabolomics as well as systems genetics [201, 202].

Instrumental and panel-based characterization of sensory attributes in the private rice sector is becoming more crucial because it is market and demand-driven. If there is no repeat demand for the newly-released rice cultivar, the variety will not succeed in penetrating the target market. In this connection, quality and sensory profiling of the rice mega varieties using instrumental and sensory tests are important to understand the reasons why these lines continue to be widely adopted by farmers in major rice growing regions and preferred by majority of rice consumers despite the availability of newer rice varieties. Market research and market intelligence will also be crucial for preference mapping and clustering of consumer demands for various quality and sensory attributes.

5.2 Including Nutritional Quality in the Rice Improvement Agenda

Rice is a diverse species with normal apparent amylose range from 0 to 30% [16]. We therefore expect diversity not only in starch structure but also in rice grain digestibility. Rice has an important role to play to mitigate the impact of non-communicable diseases. Available germplasm can be screened for resistant starch, amylose content, digestibility, and other health-promoting properties. Since clinical evaluation of rice digestibility is difficult, an accurate method of in vitro estimation of digestibility should be developed and validated. In vitro hydrolysis index [15, 123] can be used for screening but its suitability should be further validated in vivo. Cooking process that can alter digestibility should also be investigated. This should also be coupled with industrial processes to modify rice digestibility such as parboiling and processing of high amylose-based products. In connection with this, regional specifications and demands should also be considered. More importantly, some of the cooking quality traits can also affect nutritional traits. These tradeoffs need to be tackled to identify the key determinants of rice grain cooking quality while ensuring health benefits.

5.3 Associating Additional Molecular Markers to Rice Grain Quality

A deeper understanding of the genetic bases of rice grain quality attributes can ultimately help breeding programs predict the quality of their newly developed breeding lines at a faster rate through marker-assisted breeding (MAB) and genomic selection (GS). Rice grain quality traits are controlled by major and minor quantitative trait loci (QTLs), implying that the environmental and genetic mechanisms underlying quality traits are complex [195]. It is also influenced by biochemical composition of grain, storage, and processing parameters. In this era where next-generation sequencing (NGS) is commonly employed to generate thousands of rice genome sequences [203], phenotyping these accessions and performing genome-wide association studies to identify key regulatory regions of grain quality and nutritional traits will be of added value. In this context, the development of robust and accurate grain quality and nutritional phenotyping methods elaborated above should keep pace with the rapid advancements in genomics and genome-wide association tools freely available to characterize rice diversity population.

Associating molecular markers to grain quality traits can facilitate predictive classification of rice at the genotype level and it can assist in the molecular breeding of rice cultivars targeting for superior grain quality in higher yielding background. These markers can aide in rice breeding by marker-assisted selection (MAS). However, employing robust genomic selection for rice grain quality is still in its infancy. This is because a molecular understanding of how various physical and biochemical properties are related to processing, cooking, and sensory quality is generally lacking. Development of additional functional diagnostic markers related to rice grain quality is therefore imperative to breed for rice cultivars with novel combinations of traits while ensuring higher acceptability by consumers. Modern phenotyping methods such as understanding comprehensive starch structure influence on cooking quality and extracting the value of visco-elastic properties and texture in selecting premium cooking quality classes during prebreeding are crucial to accurately detect the genotypic basis of the specific trait of interest. These novel traits can then be easily tracked during the course of introgression and subsequently pyramided into locally-adapted cultivars using a suite of functional markers. To achieve this, a collaborative team of breeders, food scientists, cereal chemists, and molecular geneticists need to work on a holistic approach to reflect on multi-dimension traits to distinguish low-, medium-, and premium-quality segments. The use of genotyping-by-sequencing (GBS), targeted resequencing, and next-generation sequencing (NGS) coupled with genome-wide association studies (GWAS) and other systems genetics tools are expected to significantly increase the identification and utilization of single-nucleotide polymorphisms (SNPs) associated with rice grain quality and nutrition.

5.4 Integrating Grain Quality and Nutrition in Hybrid Rice Technology

The success of hybrid rice breeding pipelines is normally gauged based on yield potential, as well as tolerances and resistances to various biotic and abiotic stresses. As rice consumers become increasingly particular about the quality of the rice they are eating and how rice contributes to human health, rice hybrid breeders need to look at how to improve the quality in a more holistic way. As a first step, initial quality indicators of cooking quality should be linked with texture and visco-elastic properties to capture distinct cooking grain classes. Subsequently, the genetic basis of the variation of cooking and eating quality traits should be established through genome-wide association studies (GWAS) by studying both diverse set of rice core collection and pre-breeding materials. Overall emphasis has to be given to holistic understanding of vital molecular-physiological mechanisms of grain quality traits through gene discovery via QTL cloning and structural-functional genomics strategies and utilizing this information in the hybrid rice breeding sector to develop superior quality rice lines. There is an urgent need to improve grain quality preferences in the hybrid rice sector because many programs currently use parental materials with divergent grain quality traits. This leads to the production of hybrid rice varieties that segregate for grain quality traits.

In summary, more holistic strategies are needed to incorporate value-added rice grain quality traits by multi-QTL pyramiding approaches both in the inbred and hybrid rice breeding pipelines. Toward this goal, a recent study has demonstrated that high yielding elite rice lines can be pyramided with multiple complex traits to ensure superior grain quality [204]. The systematic design of high-yielding and superior-quality elite rice lines in the future will rely not only on the rapid advancements in genetics and genomics technologies but also on the current and emerging phenotyping methods for rice grain quality.

Acknowledgments

This work has been supported under the CGIAR thematic area Global Rice Agri-Food System CRP, RICE, Stress-Tolerant Rice for Africa and South Asia (STRASA) Phase III, and Australian Centre for International Agricultural Research (Project ID CIM/2016/046) funding.

References

1. International Rice Research Institute (1985) Rice grain quality and marketing. Papers presented at the 1985 International Rice Research Conference, Los Banos, Laguna. International Rice Research Institute, Manila, Philippines, p 74
2. International Rice Research Institute (1979) Proceedings of the workshop on chemical aspects of Rice Grain Quality, IRRI, 1978. International Rice Research Institute, Los Banos, Laguna, Philippines, p 390
3. Khush GS, Juliano BO (1985) Breeding for high-yielding rices of excellent cooking and eating qualities. In: Rice grain quality and marketing. Papers presented at the International Rice 1985 Research Conference. International Rice Research Institute, Manila, Philippines, pp 61–69
4. Bergman CJ, Bhattacharya KR, Ohtsubo K (2004) Rice end-use quality analysis. In: Champagne ET (ed) Rice chemistry and technology, 3rd edn. AACC International, St. Paul, MN, pp 415–472

5. Williams VR, Wu W-T, Tsai HY, Bates HG (1958) Rice starch, varietal differences in amylose content of rice starch. J Agric Food Chem 6(1):47–48
6. Kumar I, Khush GS (1986) Genetics of amylose content in rice (*Oryza-sativa*-L). J Genet 65(1–2):1–11. https://doi.org/10.1007/Bf02923530
7. Juliano BO (1971) A simplified assay for milled-rice amylose. Cereal Sci Today 16(10):334–338, 340, 360
8. Perez CM, Juliano BO (1978) Modification of the simplified amylose test for milled rice. Starch 30(12):424–426
9. Juliano BO, Perez CM, Blakeney AB, Castillo T, Kongseree N, Laignelet B, Lapis ET, Murty VVS, Paule CM, Webb BD (1981) International cooperative testing on the amylose content of milled rice. Starch 33(5):157–162
10. Juliano BO, Oñate LU, del Mundo AM (1972) Amylose and protein contents of milled rice as eating quality factors. Phil Agric 56(1–2):44–47
11. Juliano BO, Villareal C (1993) Grain quality evaluation of world rices. International Rice Research Institute, Los Baños, Philippines
12. AACC International (2009) Approved methods of analysis. AACC Inter'l, St. Paul, MN
13. International Organization for Standardization (2007) ISO 6647-2:2007–Rice—Determination of amylose content—Part 2: Routine methods. International Organization for Standardization, Geneva
14. Takeda Y, Hizukuri S, Juliano BO (1987) Structures of rice amylopectins with low and high affinities for iodine. Carbohydr Res 168:79–88
15. Butardo VM Jr, Fitzgerald MA, Bird AR, Gidley MJ, Flanagan BM, Larroque O, Resurreccion AP, Laidlaw HK, Jobling SA, Morell MK, Rahman S (2011) Impact of down-regulation of starch branching enzyme IIb in rice by artificial microRNA- and hairpin RNA-mediated RNA silencing. J Exp Bot 62(14):4927–4941
16. Butardo VM Jr, Anacleto R, Parween S, Samson I, de Guzman K, Alhambra CM, Misra G, Sreenivasulu N (2017) Systems genetics identifies a novel regulatory domain of amylose synthesis. Plant Physiol 173(1):887–906
17. Fitzgerald MA, Bergman CJ, Resurreccion AP, Moller J, Jimenez R, Reinke RF, Martin M, Blanco P, Molina F, Chen M-H, Kuri V, Romero MV, Habibi F, Umemoto T, Jongdee S, Graterol E, Reddy KR, Bassinello PZ, Sivakami R, Rani NS, Das S, Wang Y-J, Indrasari SD, Ramli A, Ahmad R, Dipti SS, Xie L, Lang NT, Singh P, Toro DC, Tavasoli F, Mestres C (2009) Addressing the dilemmas of measuring amylose in rice. Cereal Chem 86(5):492–498
18. Juliano BO, Tuaño APP, Monteroso DN, Aoki N, Mestres C, Duldulao JBA, Bergonio KB (2012) Replacement of acetate with ammonium buffer to determine apparent amylose content of milled rice. Cereal Foods World 57(1):14–19
19. Avaro MRA, Tong L, Yoshida T (2009) A simple and low-cost method to classify amylose content of rice using a standard color chart. Plant Prod Sci 12(1):97–99
20. Khoomtong A, Noomhorm A (2015) Development of a simple portable amylose content meter for rapid determination of amylose content in milled rice. Food Bioprocess Technol 8(9):1938–1946
21. Ambardekar AA, Siebenmorgen TJ, Pereira T (2011) Colorimetric method for rapidly predicting rice amylose content. Cereal Chem 88(6):560–563
22. Tuaño APP, Perez LM, Padolina TF, Juliano BO (2015) Survey of grain quality of Philippine farmers' specialty rices. Phil Agric Scientist 98:446–456
23. Juliano BO (2007) Rice chemistry and quality. Philippine Rice Research Institute, Munoz, Nueva Ecija
24. Bhattacharya KR (2011) Analysis of rice quality. In: Bhattacharya KR (ed) Rice quality. Woodhead Publishing Ltd, Cambridge, UK, pp 431–530
25. Little RR, Hilder GB, Dawson EH (1958) Differential effect of dilute alkali on 25 varieties of milled white rice. Cereal Chem 35:111–126
26. Bhattacharya KR, Sowbhagya CM (1972) An improved alkali reaction test for rice quality. J Food Technol 7:323–331
27. Biliaderis CG, Page CM, Maurice TJ, Juliano BO (1986) Thermal characterization of rice starches: a polymeric approach to phase transition of granular starch. J Agric Food Chem 34:6–24
28. Normand FL, Marshall WE (1988) Thermal-properties of whole grain milled rice and milled rice flour studied by differential scanning calorimetry. Cereal Foods World 33(8):663–663
29. Nakamura Y, Sakurai A, Inaba Y, Kimura K, Iwasawa N, Nagamine T (2002) The fine structure of amylopectin in endosperm from Asian cultivated rice can be largely classified into two classes. Starch 54(3–4):117–131

30. Umemoto T, Aoki N, Lin H, Nakamura Y, Inouchi N, Sato Y, Yano M, Hirabayashi H, Maruyama S (2004) Natural variation in rice *starch synthase IIa* affects enzyme and starch properties. Funct Plant Biol 31:671–684
31. Cuevas RP, Daygon VD, Corpuz HM, Reinke RF, Waters DLE, Fitzgerald MA (2010) Melting the secrets of gelatinisation temperature in rice. Funct Plant Biol 37:439–447
32. Tuaño APP, Aoki N, Fujita N, Oitome NF, Merca FE, Juliano BO (2014) Grain and starch properties of waxy and low-apparent amylose Philippine rices and of NSIC Rc222. Phil Agric Sci 97:329–339
33. Juliano BO, Perez CM, Alyoshin EP, Romanov VB, Bean MM, Nishita AB, Blakeney AB, Welsh LA, Delgado L, El Baya AW, Fossati G, Kongseree N, Mendes FP, Brilhante S, Suzuki H, Tada M, Webb BD (1985) Cooperative test on amylography of milled-rice flour for pasting viscosity and starch gelatinization temperature. Starch 37(2):40–50
34. Dang JMC, Bason ML (2014) AACCI Approved Methods Technical Committee report: Collaborative study on a method for determining the gelatinization temperature of milled rice flour using the rapid visco analyser. Cereal Foods World 59(1):31–34
35. Cagampang GB, Perez CM, Juliano BO (1973) A gel consistency test for eating quality in rice. J Sci Food Agric 24:1589–1594
36. Juliano BO (1979) The chemical basis of rice grain quality. In: Workshop on chemical aspects of rice grain quality 1978. International Rice Research Institute, Los Baños, Laguna, pp 69–90
37. Khush GS, Paule CM, de la Cruz N (1979) Rice grain quality evaluation and improvement at IRRI. In: Workshop on chemical aspects of rice grain quality, Los Baños, 1978. International Rice Research Institute, Los Baños, Laguna, pp 22–31
38. Horibata T, Nakamoto M, Fuwa H, Inouchi N (2004) Structural and physicochemical characteristics of endosperm starches from rice cultivars recently bred in Japan. J Appl Glycosci 51:303–313
39. Bhattacharya KR (2009) Physicochemical basis of eating quality of rice. Cereal Foods World 54:18–28
40. Juliano BO, Perez CM (1976) Gel consistency tests for eating quality of milled rices. Cereal Foods World 21(8):422–422
41. Felix AD, Tuaño APP, Mallillin AC, Trinidad TP, Juliano BO (2016) Short-term satiety of brown rice and milled rice. Phil J Crop Sci 41:52–59
42. Woolnough JW, Monro JA, Brennan CS, Bird AR (2008) Simulating human carbohydrate digestion in vitro: a review of methods and the need for standardisation. Int J Food Sci Technol 43(12):2245–2256
43. Sievert D, Holm J (1993) Determination of amylose by differential scanning calorimetry. Starch 45(4):136–139
44. Mestres C, Matencio F, Pons B, Yajid M, Fliedel G (1996) A rapid method for the determination of amylose content by using differential-scanning calorimetry. Starch 48(1):2–6
45. Delwiche SR, Bean MM, Miller RE, Webb BD, Williams PC (1995) Apparent amylose content of milled rice by near-infrared reflectance spectroscopy. Cereal Chem 72(2):182–187
46. Villareal CP, De La Cruz NM, Juliano BO (1994) Rice amylose analysis by near-infrared transmittance spectroscopy. Cereal Chem 71(3):292–296
47. Delwiche SR, McKenzie KS, Webb BD (1996) Quality characteristics in rice by near-infrared reflectance analysis of whole-grain milled samples. Cereal Chem 73(2):257–263
48. Castro JV, Ward RM, Gilbert RG, Fitzgerald MA (2005) Measurement of the molecular weight distribution of debranched starch. Biomacromolecules 6(4):2260–2270
49. Ward RM, Gao Q, de Bruyn H, Gilbert RG, Fitzgerald MA (2006) Improved methods for the structural analysis of the amylose-rich fraction from rice flour. Biomacromolecules 7(3):866–876
50. Castro JV, Dumas C, Chiou H, Fitzgerald MA, Gilbert RG (2005) Mechanistic information from analysis of molecular weight distributions of starch. Biomacromolecules 6(4):2248–2259
51. Butardo V Jr, Daygon VD, Colgrave ML, Campbell PM, Resurreccion AP, Cuevas RP, Jobling SA, Tetlow IJ, Rahman S, Morell MK, Fitzgerald MA (2012) Biomolecular analyses of starch and starch granule proteins in the high-amylose rice mutant Goami 2. J Agric Food Chem 60(46):11576–11585
52. Dhital S, Butardo VM Jr, Jobling SA, Gidley MJ (2015) Rice starch granule amylolysis—differentiating effects of particle size, morphology, thermal properties and crystalline polymorph. Carbohydr Polym 115:305–316
53. Jane J-L, Chen J-F (1992) Effect of amylose molecular size and amylopectin branch chain length on paste properties of starch. Cereal Chem 69(1):60–65
54. Hanashiro I, Ohta K, Takeda C, Mizukami H, Takeda Y (2005) Structural characteriza-

tion of long unit-chains amylopectin. J Appl Glycosci 52:233–237

55. O'Shea MG, Samuel MS, Konik CM, Morell MK (1998) Fluorophore-assisted carbohydrate electrophoresis (FACE) of oligosaccharides: efficiency of labelling and high-resolution separation. Carbohydr Res 307(1–2):1–12
56. Morell MK, Samuel MS, O'Shea MG (1998) Analysis of starch structure using fluorophore-assisted carbohydrate electrophoresis. Electrophoresis 19(15):2603–2611
57. Biliaderis CG, Grant DR, Vose JR (1981) Structural characterization of legume starches. I. Studies on amylose, amylopectin, and β-limit dextrins. Cereal Chem 58(6):496–502
58. Hizukuri S (1986) Polymodal distribution of the chain lengths of amylopectins, and its significance. Carbohydr Res 147:342–347
59. Ball SG, Guan H-P, James MG, Myers AM, Keeling PL, Mouille G, Buleon A, Colonna P, Preiss J (1996) From glycogen to amylopectin: a model for the biogenesis of the plant starch granule. Cell 86(3):349–352
60. Juliano BO (1996) Rice quality screening with the Rapid Visco Analyser. In: Walker CE, Hazelton JL (eds) Applications of the Rapid Visco Analyser, Proc AACC RVA symposium, Nashville, 1994. Newport Scientific, Warriewood, NSW, pp 19–24
61. Blakeney A (1996) Rice. In: Henry RJ, Kettlewell PS (eds) Cereal grain quality, 1st edn. Chapman Hall, London, UK, pp 55–76
62. Batey IL, Curtin BM (2000) Effects on pasting viscosity of starch and flour from different operating conditions for the Rapid Visco Analyser. Cereal Chem 77(6):754–760
63. Champagne ET, Bett KL, Vinyard BT, McClung AM, Barton FE, Moldenhauer KA, Linscombe S, McKenzie KS (1999) Correlation between cooked rice texture and Rapid Visco Analyser measurements. Cereal Chem 76(5):764–771
64. Allahgholipour M, Ali AJ, Alinia F, Nagamine T, Kojiima Y (2006) Relationship between rice grain amylose and pasting properties for breeding better quality rice varieties. Plant Breed 125:357–362
65. Yan C, Li X, Zhang R, Sui JM, Liang GH, Shen XP, Gu SL, Gu MH (2005) Performance and inheritance of rice starch RVA profile characteristics. Rice Sci 12:39–47
66. Bao J-S, Corke H, Sun M (2006) Nucleotide diversity in *starch synthase IIa* and validation of single nucleotide polymorphisms in relation to starch gelatinization temperature and other physicochemical properties in rice (*Oryza sativa* L.). Theor Appl Genet 113(7):1171–1183
67. Bao J-S, Shen S, Sun M, Corke H (2006) Analysis of genotypic diversity in the starch physicochemical properties of nonwaxy rice: apparent amylose content, pasting viscosity and gel texture. Starch 58:259–267
68. Suwannaporn P, Pitiphunpong S, Champangern S (2007) Classification of rice amylose content by discriminant analysis of physicochemical properties. Starch 59(3–4):171–177
69. Kurasawa H, Takei K, Ogawa S, Okabe T, Kanauti Y, Igaue I, Hayakawa T (1972) Some physicochemical properties of starch from non-glutinous paddy rice cultured in niigata prefecture. 2. Correlation analysis between eating quality, rheological property and amylose content of starch. Agric Biol Chem 36(10):1809
70. Reddy KR, Subramanian R, Ali SZ, Bhattacharya KR (1994) Viscoelastic properties of rice-flour pastes and their relationship to amylose content and rice quality. Cereal Chem 71(6):548–552
71. Lii C-Y, Tsai M-L, Tseng K-H (1996) Effect of amylose content on the rheological property of rice starch. Cereal Chem 73(4):415–420
72. Biliaderis CG, Zawistowski J (1990) Viscoelastic behavior of aging starch gels: effects of concentration, temperature, and starch hydrolysates on network properties. Cereal Chem 67:240–246
73. Biliaderis CG, Juliano BO (1993) Thermal and mechanical properties of concentrated rice starch gels of varying composition. Food Chem 48(3):243–250
74. Biliaderis CG, Tonogai JR (1991) Influence of lipids on the thermal and mechanical-properties of concentrated starch gels. J Agric Food Chem 39(5):833–840
75. Blakeney A (1979) Instron measurement of cooked rice texture. In: Proceedings of the workshop on chemical aspects of rice grain quality 1978. International Rice Research Institute, Los Banos, Laguna, Philippines, pp 343–353
76. Suzuki H, Ikehashi H, Kushibuchi K (1979) Rice grain quality evaluation in Japan. In: Proceedings of the workshop on chemical aspects of rice grain quality. International Rice Research Institute, Los Banos, Laguna, Philippines, pp 123–127
77. Laignelet B, Feillet P (1979) Use of the visco-elastograph for measuring the texture of

cooked rice. In: Chemical aspects of rice grain quality 1978. International Rice Research Institute, Los Banos, Laguna, pp 355–361

78. Juliano BO, Perez CM, Barber S, Blakeney AB, Iwasaki TA, Shibuya N, Keneaster KK, Chung S, Laignelet B, Launay B, del Mundo AM, Suzuki H, Shiki JI, Tsuji S, Yama JT, Tatsumi K, Webb BD (1981) International cooperative comparison of instrument methods for cooked rice texture. J Texture Stud 12(1):17–38
79. Juliano BO, Perez CM, Alyoshin EP, Romanov VB, Blakeney AB, Welsh LA, Choudhury NH, Delgado LL, Iwasaki T, Shibuya N, Mossman AP, Siwi B, Damardjati DS, Suzuki H, Kimura S (1984) International cooperative test on texture of cooked rice. J Texture Stud 15:357–376
80. Champagne ET, Bett-Garber KL, McClung AM, Bergman CJ (2004) Sensory characteristics of diverse rice cultivars as influenced by genetic and environmental factors. Cereal Chem 81(2):237–243
81. Perez CM, Pascual CG, Juliano BO (1979) Eating quality indicators for waxy rices. Food Chem 4(3):179–184
82. Juliano BO (2010) Grain quality of Philippine rice. Philippine Rice Research Institute, Muñoz, Nueva Ecija
83. Buttery RG, Ling LC, Juliano BO, Turnbaugh JG (1983) Cooked rice aroma and 2-acetyl-1-pyrroline. J AgriC Food Chem 31(4):823–826
84. Buttery RG, Ling LC, Juliano BO (1982) 2-Acetyl-1-pyrroline: an important aroma component of cooked rice. Chem Ind 23:958–959
85. Buttery RG, Ling LC, Mon TR (1986) Quantitative analysis of 2-acetyl-1-pyrroline in rice. J AgriC Food Chem 34(1):112–114
86. Chen S, Wu J, Yang Y, Shi W, Xu M (2006) The *fgr* gene responsible for rice fragrance was restricted within 69 kb. Plant Sci 171(4):505–514
87. Fitzgerald MA, Hamilton NRS, Calingacion MN, Verhoeven HA, Butardo VM (2008) Is there a second fragrance gene in rice? Plant Biotechnol J 6(4):416–423
88. Zeng Z, Zhang H, Chen JY, Zhang T, Matsunaga R (2007) Direct extraction of volatiles of rice during cooking using solid-phase microextraction. Cereal Chem J 84(5):423–427
89. Bryant RJ, McClung AM (2011) Volatile profiles of aromatic and non-aromatic rice cultivars using SPME/GC-MS. Food Chem 124(2):501–513
90. Yang DS, Shewfelt RL, Lee KS, Kays SJ (2008) Comparison of odor-active compounds from six distinctly different rice flavor types. J Agric Food Chem 56(8):2780–2787
91. Calingacion M, Fang L, Quiatchon-Baeza L, Mumm R, Riedel A, Hall RD, Fitzgerald M (2015) Delving deeper into technological innovations to understand differences in rice quality. Rice 8(1):43
92. Champagne ET (2008) Rice aroma and flavor: a literature review. Cereal Chem 85(4):445–454
93. Calingacion M, Laborte A, Nelson A, Resurreccion A, Concepcion JC, Daygon VD, Mumm R, Reinke R, Dipti S, Bassinello PZ, Manful J, Sophany S, Lara KC, Bao J, Xie L, Loaiza K, El-hissewy A, Gayin J, Sharma N, Rajeswari S, Manonmani S, Rani NS, Kota S, Indrasari SD, Habibi F, Hosseini M, Tavasoli F, Suzuki K, Umemoto T, Boualaphanh C, Lee HH, Hung YP, Ramli A, Aung PP, Ahmad R, Wattoo JI, Bandonill E, Romero M, Brites CM, Hafeel R, Lur H-S, Cheaupun K, Jongdee S, Blanco P, Bryant R, Thi Lang N, Hall RD, Fitzgerald M (2014) Diversity of global rice markets and the science required for consumer-targeted rice breeding. PLoS One 9(1):e85106
94. Unnevehr LJ, Juliano BO, Perez CM, Marciano E (1985) Consumer demand for grain quality in Thailand, Indonesia and the Philippines. IRRI Res Paper Ser 116:20
95. Unnevehr LJ, Duff B, Juliano BO (1992) Consumer demand for rice grain quality: introduction and major findings. In: Unnevehr LJ, Duff B, Juliano BO (eds) Consumer demand for rice grain quality. International Rice Research Institute, Manila, Philippines, pp 5–20
96. Limpawattana M, Yang DS, Kays SJ, Shewfelt RL (2008) Relating sensory descriptors to volatile components in flavor of specialty rice types. J Food Sci 73(9):456–461
97. Limpawattana M, Shewfelt RL (2010) Flavor lexicon for sensory descriptive profiling of different rice types. J Food Sci 75(4):S199–S205
98. Lau WCP, Rafii MY, Ismail MR, Puteh A, Latif MA, Ramli A (2015) Review of functional markers for improving cooking, eating, and the nutritional qualities of rice. Front Plant Sci 6:832
99. Bao J (2014) Genes and QTLs for rice grain quality improvement. In: Yan W, Bao J (eds) Rice - germplasm, genetics and improvement. IntechOpen, Croatia
100. Bradbury LMT, Henry RJ, Jin Q, Reinke RF, Waters DLE (2005) A perfect marker for fra-

grance genotyping in rice. Mol Breed 16(4):279–283

101. Shi W, Yang Y, Chen S, Xu M (2008) Discovery of a new fragrance allele and the development of functional markers for the breeding of fragrant rice varieties. Mol Breed 22(2):185–192
102. Shao GN, Tang A, Tang SQ, Luo J, Jiao GA, Wu JL, Hu PS (2011) A new deletion mutation of fragrant gene and the development of three molecular markers for fragrance in rice. Plant Breed 130(2):172–176
103. Bligh HFJ, Till RI, Jones CA (1995) A microsatellite sequence closely linked to the *Waxy* gene of *Oryza sativa*. Euphytica 86(2):83–85
104. Ayres NM, McClung AM, Larkin PD, Bligh HFJ, Jones CA, Park WD (1997) Microsatellites and a single-nucleotide polymorphism differentiate apparent amylose classes in an extended pedigree of US rice germplasm. Theor Appl Genet 94(6–7):773–781
105. Bergman CJ, Delgado JT, McClung AM, Fjellstrom RG (2001) An improved method for using a microsatellite in the rice *waxy* gene to determine amylose class. Cereal Chem 78(3):257–260
106. Roferos LT, Butardo V Jr, Fitzgerald MA, Juliano BO (2008) Association between alleles of the *Waxy* gene and traits of grain quality in Philippine Seed Board Rice Varieties. Phil Agric Scientist 91(3):334–337
107. Larkin PD, Park WD (1999) Transcript accumulation and utilization of alternate and non-consensus splice sites in rice granule-bound starch synthase are temperature-sensitive and controlled by a single-nucleotide polymorphism. Plant Mol Biol 40(4):719–727
108. Mikami I, Aikawa M, Hirano H-Y, Sano Y (1999) Altered tissue-specific expression at the *Wx* gene of the opaque mutants in rice. Euphytica 105(2):91–97
109. Cai XL, Wang ZY, Xing YY, Zhang J-L, Hong M-M (1998) Aberrant splicing of intron 1 leads to the heterogeneous 5′ UTR and decreased expression of waxy gene in rice cultivars of intermediate amylose content. Plant J 14(4):459–465
110. Larkin PD, Park WD (2003) Association of *waxy* gene single nucleotide polymorphisms with starch characteristics in rice (*Oryza sativa* L.). Mol Breed 12(4):335–339
111. Chen M-H, Bergman CJ, Pinson SRM, Fjellstrom RG (2008) *Waxy* gene haplotypes: associations with apparent amylose content and the effect by the environment in an international rice germplasm collection. J Cereal Sci 47:536–545
112. Dobo M, Ayres N, Walker G, Park WD (2010) Polymorphism in the *GBSS* gene affects amylose content in US and European rice germplasm. J Cereal Sci 52:450–456
113. Waters DLE, Henry RJ, Reinke RF, Fitzgerald MA (2006) Gelatinization temperature of rice explained by polymorphisms in *starch synthase*. Plant Biotechnol J 4(1):115–122
114. Mikami I, Uwatoko N, Ikeda Y, Yamaguchi J, Hirano H-Y, Suzuki Y, Sano Y (2008) Allelic diversification at the *wx* locus in landraces of Asian rice. Theor Appl Genet 116(7):979–989
115. Umemoto T, Aoki N (2005) Single-nucleotide polymorphisms in rice *starch synthase IIa* that alter starch gelatinisation and starch association of the enzyme. Func Plant Biol 32(9):763–768
116. Chen SH, Yang Y, Shi WW, Ji Q, He F, Zhang ZD, Cheng ZK, Liu XN, Xu ML (2008) Badh2, encoding betaine aldehyde dehydrogenase, inhibits the biosynthesis of 2-acetyl-1-pyrroline, a major component in rice fragrance. Plant Cell 20(7):1850–1861
117. Hsu Y-C, Tseng M-C, Wu Y-P, Lin M-Y, Wei F-J, Hwu K-K, Hsing Y-I, Lin Y-R (2014) Genetic factors responsible for eating and cooking qualities of rice grains in a recombinant inbred population of an inter-subspecific cross. Mol Breed 34(2):655–673
118. Asante MD, Offei SK, Gracen V, Adu-Dapaah H, Danquah EY, Bryant RJ, McClung AM (2013) Starch physicochemical properties of rice accessions and their association with molecular markers. Starch 65:1–7
119. Kharabian-Masouleh A, Waters DLE, Reinke RF, Ward RM, Henry RJ (2012) SNP in starch biosynthesis genes associated with nutritional and functional properties of rice. Sci Rep 2:557
120. Yu GQ, Olsen KM, Schaal BA (2011) Molecular evolution of the endosperm starch synthesis pathway genes in rice (Oryza sativa L.) and its wild ancestor, O-rufipogon L. Mol Biol Evol 28(1):659–671
121. Tran NA, Daygon VD, Resurreccion AP, Cuevas RP, Corpuz HM, Fitzgerald MA (2011) A single nucleotide polymorphism in the Waxy gene explains a significant component of gel consistency. Theor Appl Genet 123(4):519–525
122. Kovach MJ, Calingacion MN, Fitzgerald MA, McCouch SR (2009) The origin and evolution of fragrance in rice (*Oryza sativa* L.). Proc Natl Acad Sci U S A 106(34):14444–14449

123. Fitzgerald MA, Rahman S, Resurreccion AP, Concepcion JC, Daygon VD, Dipti SS, Kabir KA, Klingner B, Morell MK, Bird AR (2011) Identification of a major genetic determinant of glycaemic index in rice. Rice 4:66–74
124. Gao Z, Zeng D, Cheng F, Tian Z, Guo L, Su Y, Yan M, Jiang H, Dong G, Huang Y, Han B, Li J, Qian Q (2011) *ALK*, the key gene for gelatinization temperature, is a modifier gene for gel consistency in rice. J Integr Plant Biol 53(9):756–765
125. Bradbury LMT, Fitzgerald TL, Henry RJ, Jin Q, Waters DLE (2005) The gene for fragrance in rice. Plant Biotechnol J 3(3):363–370
126. Shao GN, Tang SQ, Chen ML, Wei XJ, He JW, Luo J, Jiao GA, Hu YC, Xie LH, Hu PS (2013) Haplotype variation at Badh2, the gene determining fragrance in rice. Genomics 101(2):157–162
127. Bounphanousay C, Jaisil P, Sanitchin J, Fitzgerald MA, Sackville-Hamilton NR (2008) Chemical and molecular characterization of fragrance in black glutinous rice from Lao PDR. Asian J Plant Sci 7(1):1–7
128. Butardo VM Jr, Sreenivasulu N, Jeon K (2016) Tailoring grain storage reserves for a healthier rice diet and its comparative status with other cereals. Int Rev Cell Mol Biol 323:31–70
129. Carreiro AL, Dhillon J, Gordon S, Higgins KA, Jacobs AG, McArthur BM, Redan BW, Rivera RL, Schmidt LR, Mattes RD (2016) The macronutrients, appetite, and energy intake. Annu Rev Nutr 36(1):73–103
130. Shenkin A (2006) Micronutrients in health and disease. Postgraduate Med J 82(971):559–567
131. DellaPenna D (1999) Nutritional genomics: manipulating plant micronutrients to improve human health. Science 285(5426):375–379
132. Johnson I, Williamson G (2003) Phytochemical functional foods. CRC, Cambridge, England
133. Wolever TMS (2006) Glycaemic index : a physiological classification of dietary carbohydrate. CAB International Book Archive, titles from 2005, 2006, 2007. CABI Publishing, Wallingford, UK
134. Trinidad TP, Mallillin AC, Encabo RR, Sagum RS, Felix AD, Juliano BO (2013) The effect of apparent amylose content and dietary fiber on the glycemic response of different varieties of cooked milled rice and brown rice. Int J Food Sci Nutr 64:89–93
135. Jenkins DJ, Wolever TM, Taylor RH, Barker H, Fielden H, Baldwin JM, Bowling AC, Newman HC, Jenkins AL, Goff DV (1981) Glycemic index of foods: a physiological basis for carbohydrate exchange. Am J Clin Nutr 34(3):362–366
136. Wolever TMS, Vorster HH, Bjorck I, Brand-Miller JC, Brighenti F, Mann JI, Ramdath DD, Granfeldt Y, Holt S, Perry TL, Venter C, Wu XM (2003) Determination of the glycaemic index of foods: interlaboratory study. Eur J Clin Nutr 57(3):475–482
137. Salmeron J, Ascherio A, Rimm EB, Colditz GA, Spiegelman D, Jenkins DJ, Stampfer MJ, Wing AL, Willett WC (1997) Dietary fiber, glycemic load, and risk of NIDDM in men. Diabetes Care 20(4):545–550
138. Englyst HN, Kingman SM, Cummings JH (1992) Classification and measurement of nutritionally important starch fractions. Eur J Clin Nutr 46:S33–S50
139. Englyst KN, Englyst HN, Hudson GJ, Cole TJ, Cummings JH (1999) Rapidly available glucose in foods: an in vitro measurement that reflects the glycemic response. Am J Clin Nutr 69(3):448–454
140. Goñi I, Garcia-Alonso A, Saura-Calixto F (1997) A starch hydrolysis procedure to estimate glycemic index. Nutr Res 17(3):427–437
141. Frei M, Siddhuraju P, Becker K (2003) Studies on the in vitro starch digestibility and the glycemic index of six different indigenous rice cultivars from the Philippines. Food Chem 83(3):395–402
142. Butterworth PJ, Warren FJ, Grassby T, Patel H, Ellis PR (2012) Analysis of starch amylolysis using plots for first-order kinetics. Carbohydr Polym 87(3):2189–2197
143. Shih FF (2004) Rice proteins. In: Champagne ET (ed) Rice: chemistry and technology, 3rd edn. American Association of Cereal Chemistry, Inc, St. Paul, MN, pp 143–162
144. Kjeldahl J (1883) Neue methode zur bestimmung des stickstoffs in organischen körpern. Fresenius' J Analyt Chem 22(1):366–382
145. Cohen SA, Michaud DP (1993) Synthesis of a fluorescent derivatizing reagent, 6-aminoquinolyl-N-hydroxysuccinimidyl carbamate, and its application for the analysis of hydrolysate amino acids via high-performance liquid chromatography. Anal Biochem 211(2):279–287
146. Cohen SA (2000) Amino acid analysis using precolumn derivatization with 6-aminoquinolyl-N-hydroxysuccinimidyl carbamate. In: Cooper C, Packer N, Williams K (eds) Amino acid analysis protocols. Humana Press, Totowa, NJ, pp 39–47

147. Liu L, Waters DLE, Rose TJ, Bao J, King GJ (2013) Phospholipids in rice: significance in grain quality and health benefits: a review. Food Chem 139(1–4):1133–1145
148. Godber JS, Juliano BO (2004) Rice lipids. In: Champagne ET (ed) Rice: chemistry and technology, 3rd edn. American Association of Cereal Chemistry, St. Paul, MN, pp 163–190
149. Maniñgat CC, Juliano BO (1980) Starch lipids and their effect on rice starch properties. Starch 32(3):76–82
150. Lugay JC, Juliano BO (1964) Fatty acid composition of rice lipids by gas-liquid chromatography. J Am Oil Chem Soc 41(4):273–275
151. Liu L, Tong C, Bao JS, Waters DLE, Rose TJ, King GJ (2014) Determination of starch lysophospholipids in rice using liquid chromatography-mass spectrometry (LC-MS). J Agric Food Chem 62(28):6600–6607
152. Tong C, Liu L, Waters DLE, Rose TJ, Bao JS, King GJ (2014) Genotypic variation in lysophospholipids of milled rice. J Agric Food Chem 62(38):9353–9361
153. Kamal-Eldin A (2007) Micronutrients in cereal products: their bioactivities and effects on health. In: Hamaker BR (ed) Technology of functional cereal products. Elsevier, Amsterdam. https://doi.org/10.1533/9781845693886.1.86
154. Branca F, Ferrari M (2002) Impact of micronutrient deficiencies on growth: the stunting syndrome. Ann Nutr Metab 46:8–17
155. Bashir K, Takahashi R, Nakanishi H, Nishizawa NK (2013) The road to micronutrient biofortification of rice: progress and prospects. Front Plant Sci. https://doi.org/10.3389/fpls.2013.00015
156. Welch RM, Graham RD (2004) Breeding for micronutrients in staple food crops from a human nutrition perspective. J Exp Bot 55(396):353–364
157. Hansen TH, Laursen KH, Persson DP, Pedas P, Husted S, Schjoerring JK (2009) Micro-scaled high-throughput digestion of plant tissue samples for multi-elemental analysis. Plant Methods 5:12
158. Wheal MS, Fowles TO, Palmer LT (2011) A cost-effective acid digestion method using closed polypropylene tubes for inductively coupled plasma optical emission spectrometry (ICP-OES) analysis of plant essential elements. Anal Methods 3(12):2854–2863
159. Lai P, Li KY, Lu S, Chen HH (2009) Phytochemicals and antioxidant properties of solvent extracts from Japonica rice bran. Food Chem 117(3):538–544
160. Heuberger AL, Lewis MR, Chen MH, Brick MA, Leach JE, Ryan EP (2010) Metabolomic and functional genomic analyses reveal varietal differences in bioactive compounds of cooked rice. PLoS One 5(9):e12915
161. Moongngarm A, Daomukda N, Khumpika S (2012) Chemical compositions, phytochemicals, and antioxidant capacity of rice bran, rice bran layer, and rice germ. APCBEE Proc 2:73–79
162. Niu YG, Gao BY, Slavin M, Zhang XW, Yang F, Bao JS, Shi HM, Xie ZH, Yu LL (2013) Phytochemical compositions, and antioxidant and anti-inflammatory properties of twenty-two red rice samples grown in Zhejiang. Lwt-Food Sci Technol 54(2):521–527
163. Pereira-Caro G, Cros G, Yokota T, Crozier A (2013) Phytochemical profiles of black, red, brown, and white rice from the Camargue region of France. J Agric Food Chem 61(33):7976–7986
164. Pereira-Caro G, Watanabe S, Crozier A, Fujimura T, Yokota T, Ashihara H (2013) Phytochemical profile of a Japanese black-purple rice. Food Chem 141(3):2821–2827
165. Pintha K, Yodkeeree S, Pitchakarn P, Limtrakul P (2014) Anti-invasive activity against Cancer cells of phytochemicals in red Jasmine rice (Oryza sativa L.). Asian Pac J Cancer Prev 15(11):4601–4607
166. Reddy CK, Kimi L, Haripriya S (2016) Variety difference in molecular structure, functional properties, phytochemical content and antioxidant capacity of pigmented rice. J Food Meas Charact 10(3):605–613
167. Gaborieau M, Castignolles P (2011) Size-exclusion chromatography (SEC) of branched polymers and polysaccharides. Anal Bioanal Chem 399(4):1413–1423
168. Roger P, Baud B, Colonna P (2001) Characterization of starch polysaccharides by flow field-flow fractionation–multi-angle laser light scattering–differential refractometer index. J Chromatogr A 917(1):179–185
169. Rolland-Sabaté A, Guilois S, Jaillais B, Colonna P (2011) Molecular size and mass distributions of native starches using complementary separation methods: asymmetrical flow field flow fractionation (A4F) and hydrodynamic and size exclusion chromatography (HDC-SEC). Anal Bioanal Chem 399(4):1493–1505
170. Chiaramonte E, Rhazi L, Aussenac T, White DR (2012) Amylose and amylopectin in starch by asymmetric flow field-flow fraction-

ation with multi-angle light scattering and refractive index detection (AF4-MALS-RI). J Cereal Sci 56(2):457–463

171. Mantovani V, Galeotti F, Maccari F, Volpi N (2018) Recent advances in capillary electrophoresis separation of monosaccharides, oligosaccharides, and polysaccharides. Electrophoresis 39(1):179–189
172. Warren FJ, Gidley MJ, Flanagan BM (2016) Infrared spectroscopy as a tool to characterise starch ordered structure—a joint FTIR–ATR, NMR, XRD and DSC study. Carbohydr Polym 139:35–42
173. Bao J-S, Cai YZ, Corke H (2001) Prediction of rice starch quality parameters by near-infrared reflectance spectroscopy. J Food Sci 66(7):936–939
174. Barton FE, Himmelsbach DS, McClung AM, Champagne E (2002) Two-dimensional vibration spectroscopy of rice quality and cooking. Cereal Chem 79(1):143–147
175. Himmelsbach DS, Barton FE, McClung AM, Champagne ET (2001) Protein and apparent amylose contents of milled rice by NIR-FT/Raman spectroscopy. Cereal Chem 78(4):488–492
176. Wang HL, Wan XY, Bi JC, Wang JK, Jiang L, Chen LM, Zhai HQ, Wan JM (2006) Quantitative analysis of fat content in rice by near-infrared spectroscopy technique. Cereal Chem 83(4):402–406
177. Meadows F, Barton FE (2002) Determination of Rapid Visco Analyser parameters in rice by near-infrared spectroscopy. Cereal Chem 79(4):563–566
178. Bao JS, Wang YF, Shen Y (2007) Determination of apparent amylose content, pasting properties and gel texture of rice starch by near-infrared spectroscopy. J Sci Food Agric 87(11):2040–2048
179. Champagne ET, Bett-Garber KL, Grimm CC, McClung AM, Moldenhauer KA, Linscombe S, McKenzie KS, Barton FE (2001) Near-infrared reflectance analysis for prediction of cooked rice texture. Cereal Chem 78(3):358–362
180. Cho MJ, Morawicki R, Guerra C (2009) Ability of near-infrared spectroscopy to predict solid loss, volumetric expansion, and water uptake of cooked rice. Cereal Chem 86(1):113–116
181. Patindol J, Gu XF, Wang YJ (2010) Chemometric analysis of cooked rice texture in relation to starch fine structure and leaching characteristics. Starch 62(3–4):188–197
182. Li H, Prakash S, Nicholson TM, Fitzgerald MA, Gilbert RG (2016) Instrumental measurement of cooked rice texture by dynamic rheological testing and its relation to the fine structure of rice starch. Carbohydr Polym 146(1879–1344):253–263
183. Zhu B, Li B, Zheng XD, Xu BL, Liang S, Kuang X, Ma MH (2010) Study on predictive models relating physicochemical properties to texture of cooked rice and the application in rice blends. J Texture Stud 41(2):101–124
184. Sesmat R, Meullenet JF (2001) Prediction of rice sensory texture attributes from a single compression test multivariate regression, and a stepwise model optimization method. J Food Sci 66(1):124–131
185. Champagne ET, Richard OA, Bett KL, Grimm CC, Vinyard BT, Webb BD, McClung AM, Barton FE, Lyon BG, Moldenhauer KA, Linscombe S, Mohindra R, Kohlwey DE (1996) Quality evaluation of US medium-grain rice using a Japanese taste analyzer. Cereal Chem 73(2):290–294
186. Daygon VD, Prakash S, Calingacion M, Riedel A, Ovenden B, Snell P, Mitchell J, Fitzgerald M (2016) Understanding the Jasmine phenotype of rice through metabolite profiling and sensory evaluation. Metabolomics 12(4):63. https://doi.org/10.1007/s11306-016-0989-6
187. Lapchareonsuk R, Sirisomboon P (2015) Sensory quality evaluation of rice using visible and shortwave near-infrared spectroscopy. Int J Food Prop 18(5):1128–1138
188. Lu L, Zhu ZW (2014) Prediction model for eating property of indica rice. J Food Qual 37(4):274–280
189. Baldwin EA, Bai JH, Plotto A, Dea S (2011) Electronic noses and tongues: applications for the food and pharmaceutical industries. Sensors 11(5):4744–4766
190. Chen JS (2009) Food oral processing - a review. Food Hydrocoll 23(1):1–25
191. Pascua Y, Koc H, Foegeding EA (2013) Food structure: roles of mechanical properties and oral processing in determining sensory texture of soft materials. Curr Opin Colloid Interface Sci 18(4):324–333
192. Pereira LJ, Van der Bilt A (2016) The influence of oral processing, food perception and social aspects on food consumption: a review. J Oral Rehabil 43(8):630–648
193. Chen JS, Stokes JR (2012) Rheology and tribology: two distinctive regimes of food texture sensation. Trends Food Sci Technol 25(1):4–12
194. Prakash S, Tan DDY, Chen JS (2013) Applications of tribology in studying food oral processing and texture perception. Food Res Int 54(2):1627–1635

195. Stokes JR, Boehm MW, Baier SK (2013) Oral processing, texture and mouthfeel: from rheology to tribology and beyond. Curr Opin Colloid Interface Sci 18(4):349–359
196. Moongngarm A, Bronlund JE, Grigg N, Sriwai N (2012) Chewing behavior and bolus properties as affected by different rice types. Int J Med Biol Sci 6:51–56
197. Felix AD, Trinidad TP, Tuaño APP, Juliano BO (2013) Short-term satiety of cooked Philippine rices of varying apparent amylose content and glycemic index. Phil Agric Sci 96:179–186
198. Hasjim J, Lavau GC, Gidley MJ, Gilbert RG (2010) In vivo and in vitro starch digestion: are current in vitro techniques adequate? Biomacromolecules 11(12):3600–3608
199. Dona AC, Pages G, Gilbert RG, Gaborieau M, Kuchel PW (2009) Kinetics of *in vitro* digestion of starches monitored by time-resolved ^{1}H nuclear magnetic resonance. Biomacromolecules 10(3):638–644
200. Toutounji MR, Van Leeuwen MP, Oliver JD, Shrestha AK, Castignolles P, Gaborieau M (2015) Quantification of sugars in breakfast cereals using capillary electrophoresis. Carbohydr Res 408:134–141
201. Concepcion JCT, Ouk M, Zhao D, Fitzgerald MA (2015) The need for new tools and investment to improve the accuracy of selecting for grain quality in rice. Field Crops Res 182(0):60–67
202. Sreenivasulu N, Butardo VM, Misra G, Cuevas RP, Anacleto R, Kavi Kishor PB (2015) Designing climate-resilient rice with ideal grain quality suited for high-temperature stress. J Exp Bot 66(7):1737–1748
203. The 3000 Rice Genomes Project (2014) The 3,000 rice genomes project. GigaScience 3:7
204. Zeng D, Tian Z, Rao Y, Dong G, Yang Y, Huang L, Leng Y, Xu J, Sun C, Zhang G, Hu J, Zhu L, Gao Z, Hu X, Guo L, Xiong G, Wang Y, Li J, Qian Q (2017) Rational design of high-yield and superior-quality rice. Nat Plants 3:17031

Chapter 3

Morphology of Rice Seed Development and Its Influence on Grain Quality

Paul A. Counce and Karen A. K. Moldenhauer

Abstract

Various quality attributes of rice seed are affected by the wide array of biochemical products accumulated during the course of reproductive development and the environmental conditions which impact the grain composition. The staging of rice plant reproductive development is needed in experiments to define phase transitions of seed biology. The application of the nomenclature and criteria of the rice growth staging system can facilitate recording the reproductive development by distinct stages. The meaningful progression from one stage to another in time can then be evaluated in a tabular or graphic manner. In order to determine the developmental stages of rice in experiments, it is desirable to select a representative group of plants and to record the development of those plants. We provide procedures for efficiently (1) observing and recording development of rice plants and (2) collecting, storing and seperating seed by developmental stages. It is necessary to divide seeds into differing groups to track development from fertilization until maturity. The earliest seeds to be fertilized on a panicle are superior grains and the latter seeds to develop are inferior grains. In some cases, it is necessary to divide individual seeds into the aforementioned groups and the different stages of development for various analyses. A procedure for dividing seeds into differing stages of development is presented to more appropriately select seeds for further analysis. The developmental record can then be statistically and graphically analyzed to better understand responses to treatments and interactions among treatments, years, and locations.

Key words Rice, *Oryza sativa* L., Reproductive development, Seed, Grain, Seed development, Growth staging system, Sampling, Timing, Inferior grains, Superior grains

1 Introduction

The rough rice kernel includes the husks or hulls and pedicel, as well as the caryopsis (Fig. 1). "The weight distribution of the various parts of the rice caryopsis [at maturity] is as follows: pericarp (1–2%), seed coat and aleurone (5%), starchy endosperm (89–91%), and embryo (2–3%)" [1]. The rice caryopsis (the rice seed or brown rice kernel) accumulates mass rapidly during the R6 stage of development, which lasts approximately from 5 to 15 days after fertilization under optimum conditions for development (Fig. 2). Starch is accumulated in the greatest concentration in the starchy

Nese Sreenivasulu (ed.), *Rice Grain Quality: Methods and Protocols*, Methods in Molecular Biology, vol. 1892,
https://doi.org/10.1007/978-1-4939-8914-0_3,

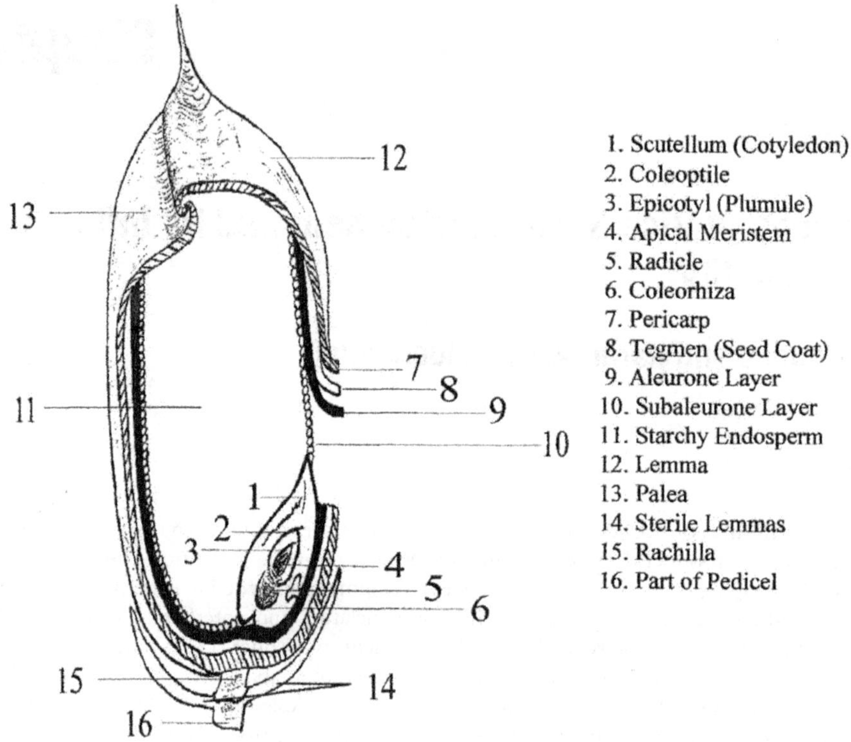

Fig. 1 Parts of rough rice grain

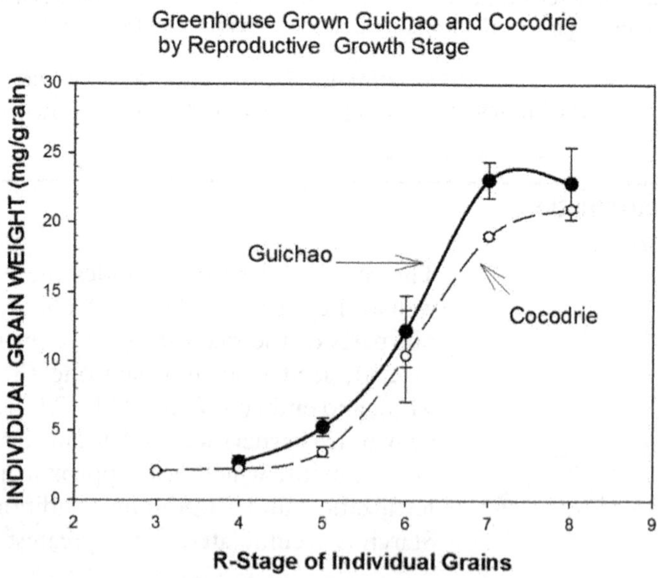

Fig. 2 Rice mass accumulation by growth stage for Cocodrie and Guichao2 rice

endosperm. Small amounts of starch are found in the subaleurone layer and very small amounts are present in the embryo and aleurone layer [2]. Starch synthesis is carried out in the plastids of these tissues and is aggregated into highly structured starch granules [3]. From approximately 4 to 18 days after anthesis, starch accumulates rapidly to a maximum in the endosperm tissue [4]. At maturity, protein concentration declines gradually from the outer layer of the caryopsis to the center of the grain in the starchy endosperm. Functional proteins are found in the embryo, aleurone and subaleurone layers, starchy endosperm and maternal tissues during development; storage proteins accumulate in these tissues as well [5]. The quantity of storage proteins are greatest in the starchy endosperm but concentration of protein is greater in the aleurone layer, less in the subaleurone layer, and least in the starchy endosperm [1]. Storage proteins begin to accumulate in the starchy endosperm slightly later than starch. At maturity, lipid concentrations are greatest in the embryo, followed by the aleurone layer, the subaleurone layer, and the starchy endosperm [1]. Lipids, in the form of lipid bodies, are reported to begin to accumulate around 5 days after anthesis and increase in content in conjunction with starch and storage protein accumulations but may continue to accumulate over a somewhat longer period [6]. During development, the biological activity of the pericarp and seed coat is important for barley (and likely other cereals including rice) but the synthetic activity of the seed covering maternal tissue begins to decline prior to endosperm and embryo maturity [7].

The panicles composing the rice crop and the individual grains within each panicle develop together beginning at different times and proceeding at different rates. The rice staging system was developed to improve production, research, extension, and education communication regarding the plants and crop [8], but it can also be adapted to differentiate the individual seeds at different stages of development. Reproductive development consists of ten growth stages based on discrete morphological criteria (Figs. 3, 4, and 5, Table 1): panicle initiation (RO), panicle differentiation (R1), flag leaf collar formation (R2), panicle exertion (R3). Stages R4–R8 refer to the first grain on a panicle to reach the following criteria: anthesis (R4), grain length and width expansion (R5), grain depth expansion (R6), grain dry down (R7), single-grain maturity (R8). When all grains that reach R6 have reached R8 then complete panicle maturity (R9) has been reached. The grains change visibly during their development both externally and internally Fig. 6). Consequently, the first grain for a panicle to reach the criteria for R4 denotes the R4 stage. Consequently, when the panicle is at the R6 stage, the individual grains on the panicle will frequently be at the R4, R5, and R6 stages of development and at the R8 stage, individual grains

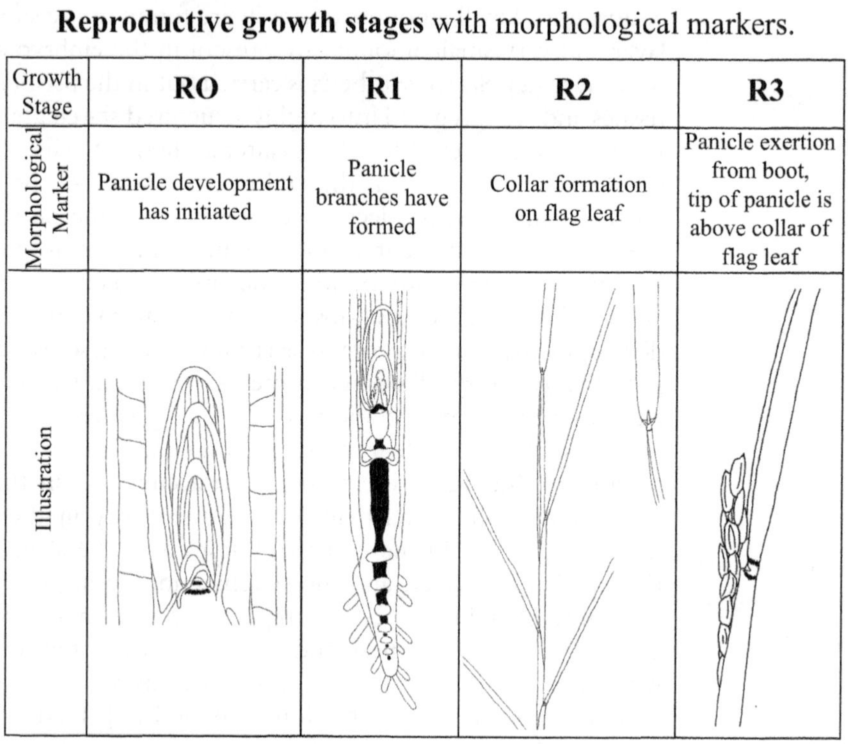

Reproductive growth stages with morphological markers.

Growth Stage	RO	R1	R2	R3
Morphological Marker	Panicle development has initiated	Panicle branches have formed	Collar formation on flag leaf	Panicle exertion from boot, tip of panicle is above collar of flag leaf
Illustration				

Fig. 3 Rice growth stages R0–R3

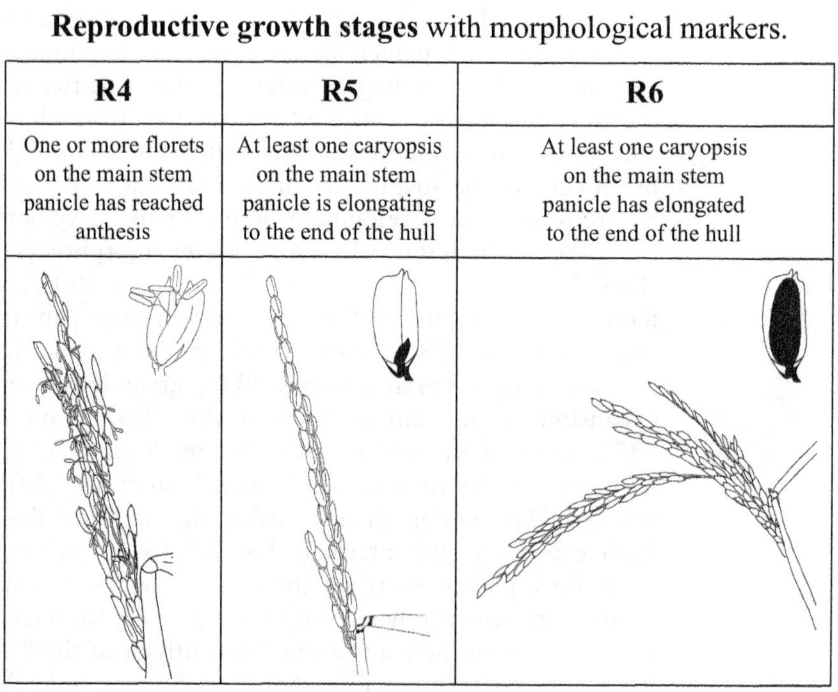

Reproductive growth stages with morphological markers.

R4	R5	R6
One or more florets on the main stem panicle has reached anthesis	At least one caryopsis on the main stem panicle is elongating to the end of the hull	At least one caryopsis on the main stem panicle has elongated to the end of the hull

Fig. 4 Rice growth stages R4–R6

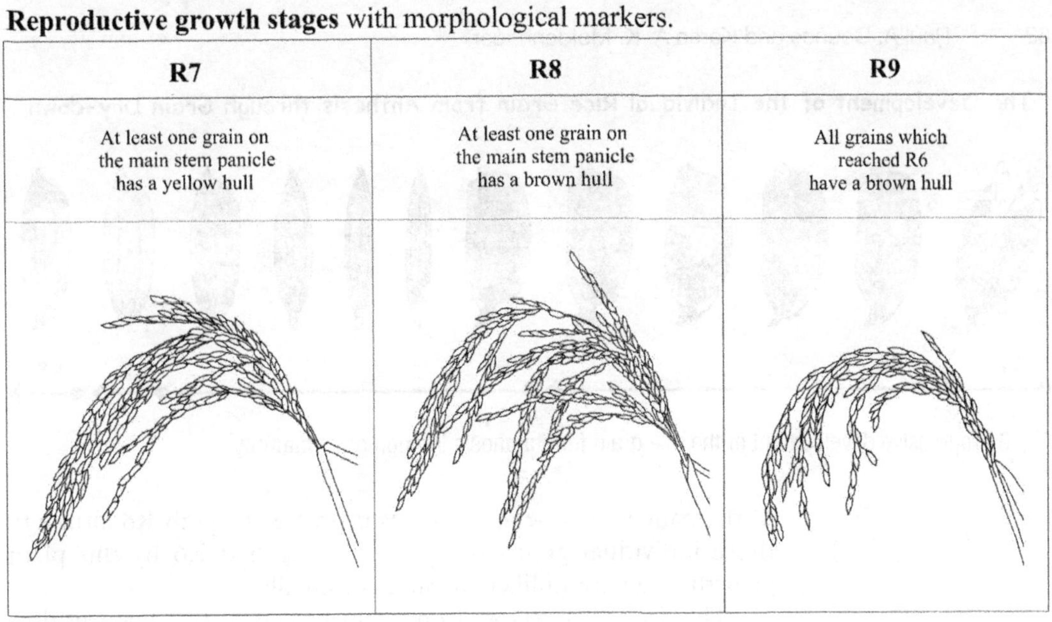

Fig. 5 Rice growth stages R7–R9

Table 1
Rice reproductive stages of development, external visibility, description, and criteria

Rice reproductive growth stage	Externally visible	Brief description	Criteria
R0	Not visible	Panicle differentiation	Initial formation of panicle
R1	Not visible	Panicle initiation	Panicle Branches differentiate
R2	Visible	Flag leaf collar formation	Collar has formed on Flag Leaf
R3	Visible	Panicle exertion	Panicle has emerged from flag leaf sheath
R4	Visible	Anthesis	Anthers and filaments exert from one floret on panicle
R5	Visible	Grain length and width expansion	Elongating caryopsis visible but not extending to the length of hull
R6	Visible	Grain Depth expansion	Caryopsis of one grain on panicle has elongated to the length of the hull
R7	Visible	Grain dry down	One hull with a filled caryposis (length, width and depth expanded) on the panicle is not visibly green but is yellow over the entire hull compared to floret color during R6
R8	Visible	Single grain maturity	One hull has turned visibly brown and the luster of the hull is more flat or dull compared to that of R7
R9	Visible	Complete panicle maturity	Last grain to reach R6 on panicle has turned brown with same criteria for that kernel as for R8

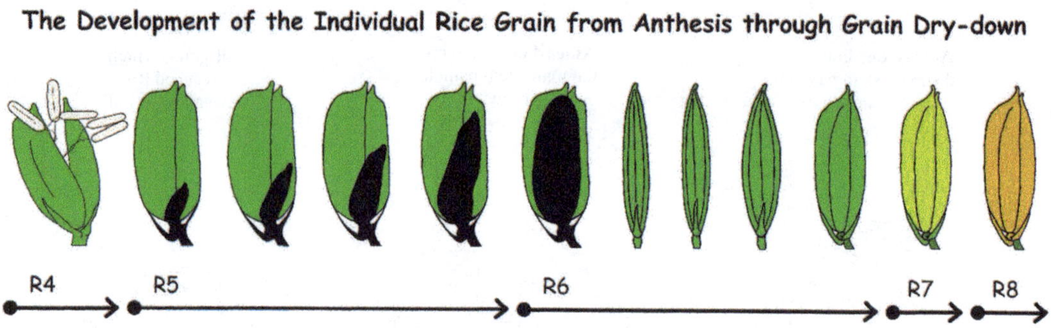

Fig. 6 Progressive development of the rice grain from anthesis through grain maturity

of the panicle will be at all stages from R3 through R8 although those individual grains that have not reached R6 by this plant growth stage are unlikely to successfully fill.

The rice seed develops from the panicle progressively over time with the upper grain developing first and the lower grain developing later. The seed per se develops after pollination and fertilization of the egg and the polar nuclei by the sperm. Subsequently the fertilized egg cell develops into the embryo, while the fertilized polar nuclei form the endosperm. The surrounding maternal tissue including the seed covering, the lemma, and palea also develop concurrently. The florets are exposed to light and air as the panicle exerts (R3), the floret opens forcefully and the anther and filament project out of the floret (R4) facilitating pollination and fertilization. Cell division in both the embryo and the endosperm proceeds during R4. Subsequently, the caryopsis expands visibly (R5); and the cells in the embryo continue to develop as the endosperm cells fill (primarily with starch) and increase in mass (R6); after the endosperm cells have filled, the chlorophyll in the various maternal tissues covering the seed is degraded. The lemma and palea lose chlorophyll and these tissues turn from green to yellow (R7). As the process continues, maternal tissue chlorophyll completes degradation and the lemma and palea, seed coat, endosperm, and embryo dry out and the seed enters a quiescent state (R8). Usually, the crop is harvested during the late part of the R8 stage for the crop but R9 is reached when all grains that reach R6 have reached R8.

The panicle on rice has branches off the central axis called primary branches while branches from the primary branches are termed secondary branches. Branches arising from secondary branches are the tertiary branches. The seeds on the panicle progressively develop from florets to fully developed seeds and a hierarchy among the seeds of the panicle becomes progressively more visible: the florets that develop first (usually the upper branches, outer periphery of the branches and on the primary branches) are followed progressively by the later florets (usually those on the lower branches, the inner parts of the branches and at the secondary and tertiary branches). The first florets to develop into seeds are called *superior grains* and the later developing seeds are called *inferior grains*. The superior grains on the panicle fill rapidly, are larger and denser while the inferior grains fill slower, are smaller and less dense [9]. Many scientists have divided the seeds into more detailed delineations: Reports include varying levels of distinction among various delineations of superior and inferior grains [10–12]. While most scientists simply report the aggregate of harvested grains without distinguishing position on the panicle, subdividing the grains on panicle samples for supplementary analyses can supply valuable information to aid in interpreting treatment effects. Subdividing grains, in particular, can eliminate sources of variation which can be critical to enzyme and starch analyses [10, 11].

1.1 Individual Grain Development Within the Panicles

For R4–R8, the staging system is primarily keyed to when the first grains on the panicle begin to open and develop. Within the panicle, individual grains are at different stages of development. The multiple stages of development of the individual grains within the panicles during crop development (R6–R8) have distinct implications for rice grain quality [13–16], while earlier reproductive stages have large effects on pollen viability [17, 18] and consequently on rice crop grain yield [19].

During the process of grain filling storage products accumulate in distinct phases with starch followed by storage proteins and lipid contents. Grain filling, in terms of mass, is done in the R6 stage of the individual grains. The R7 is a transitional stage from filling grain to seed drying. The R8 stage is also critical to seed quality because during this stage, the seed dries and can absorb water after drying. When this occurs fissures can form in the rice grains. During the R8 stage fissures often form in the superior grains while the inferior grains are still filling. Fissure formation leads to reduced head rice yield (unbroken kernels or kernels ¾ or more of their original length which is one of the most critical measures of grain quality).

2 Materials

2.1 Determining and Recording Crop, Panicle, and Individual Seed Development

1. R2 stage rice plants.
2. White plastic tags (Kaywood Distributing Company, 4399 Westgrove Drive, Addison, TX 75001; kaywood.com).
3. Waterproof marker.
4. Field book ("Rite in the Rain" All Weather Field Book 310, 614 Pacific Highway, Tacoma, WA 98424) (or alternative electronic device) for recording data.
5. Pen with stable ink after drying.

2.2 Collecting, Storing and Separating Individual Seed by Stages of Development

1. Rice reproductive stage plants.
2. Containers for seed of different stages of development.
3. Table lamp or flash light.
4. Liquid nitrogen in Dewar or thermos.
5. Mortars for holding liquid nitrogen for seeds at different stages of development.
6. Tweezers.
7. 10 or 50 mL Falconer tubes.
8. −80 ° C freezer.
9. Freezer boxes for organizing stored seeds.

3 Methods

Experimental treatments are imposed to determine the effects of the treatments on various seeds on the entire panicle and on grain at various stages of development. In this chapter we address methods for recording grain development, subdividing grain from the panicle for purposes of enzyme assay and quantification, gene expression, starch, protein, and lipid analysis.

3.1 Determining and Recording Crop, Panicle, and Individual Seed Development

For some studies it is essential to know the timing of seed development. The methods we present are the result of trial and error over several years of recording rice reproductive development. These methods can be modified depending on (1) the objectives of the study, (2) the uniformity of plants in the study, and (3) the use to be made of the staging information. The staging system utilized in these methods is presented by Counce et al. [8].

Culms at the R2 stage are easily identified by the emergence of the final leaf or flag leaf and the formation of the collar (Fig. 7). The culm is marked with a plastic numbering tag and waterproof ink (Fig. 8). The numbered tag is placed below the penultimate leaf (second leaf from the top) to prevent dislodging or tangling when the panicle exerts. Alternatively, the leaf itself can be numbered with an ultrafine point permanent marker but finding the culm is often more difficult without the tag because the tag can be seen more easily. For this reason, we use white tags so that the culms can be located and notes more efficiently made. Therefore, flags or visible markers in the field plots are useful for locating the culms on the plants to be sampled. Subsequently, the culms are monitored daily for the dates of R3, R4, R5, R6, R8, and R9 and are noted for each panicle (Table 2). Determining R3 is relatively easy and the exertion of the panicle is highly visible but the panicle may exert through the flag leaf collar or below the flag leaf collar (Fig. 9). Growth stage R4 is also very visible with the anthers now present on the outside of the floret and the anthers being turgid

Fig. 7 R2 rice culms viewed from different angles when the leaf is definitely the final or flag leaf and after tag attachment below the second leaf on left photo

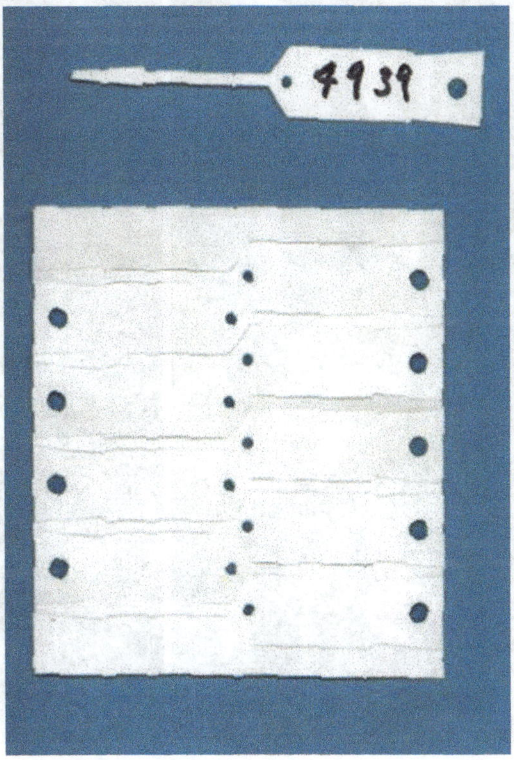

Fig. 8 Tags for growth staging along with sheet of tags

Table 2
Excerpt from the field book for a rice field test conducted in 2015 at Stuttgart, Arkansas, U.S.A

Test	C15S1	.	C15S1	.	C15S1	.	C15S1	.	C15S1	.	C15S1
Plot	403	.	307	.	301	.	203	.	107	.	101
Cultivar	Presidio	.	Cypress	.	LaGrue	.	RoyJ	.	Presidio	.	LaGrue
Rep	4	.	3	.	3	.	2	.	1	.	1
Panicle	4721	.	4735	.	4748	.	4762	.	4777	.	4790
Stage	*Day of the year in 2015*										
R3	*205*	.	*211*	.	*208*	.	*215*	.	*208*	.	*209*
R4	*208*	.	*215*	.	*209*	.	*216*	.	*209*	.	*212*
R5	*210*	.	*216*	.	*212*	.	*219*	.	*212*	.	*216*
R6	*211*	.	*217*	.	*215*	.	*223*	.	*215*	.	*217*
R7	*222*	.	*229*	.	*223*	.	*229*	.	*219*	.	*226*
R8	*224*	.	*230*	.	*226*	.	*230*	.	*226*	.	*229*
R9	*238*	.	*258*	.	*253*	.	*257*	.	*239*	.	*254*

The dates are days of the year numbering from Day 1 on January 1 and ending with Day 365 on non-leap years. The table layout is in bold and the dates recorded for stages are in italics. The columns have the identification and plot information for each panicle and the rows are for each respective growth stage. The columns with periods indicate breaks in the data from the field book and the panicle numbers are continuous in the actual field book

Fig. 9 Freshly exerted panicles (rice growth stage R3). Panicle on left exerted through the collar of the flag leaf sheath while the panicle on the right exerted through the flag leaf sheath below the collar

and standing out just after anthesis (Fig. 10). Determining R5 requires close scrutiny of the upper florets on the panicle (ordinarily) with the floret interposed between the sunlight and the observer to allow the light to shine through the hull to view caryopsis elongation. Likewise R6 requires close scrutiny (normally of the upper florets) to determine when the caryopsis has elongated to the end of the hull (Fig. 11). When the grain begins to visibly have both depth and width, R6 is also visible but this is usually sometime after the start of R6. Photographs of the R7 stages are misleading but easily visible by comparing the green color in the R6 grains and the gradual yellowing leading to R7 (Fig. 12). R8 is usually a phase change and different cultivars differ in the luster and shade of brown color indicating R8 but in this case also the comparative difference among grains (green vs. yellow) makes this much less subjective than the verbal description suggests. Usually, the luster of the R8 grains goes from bright to dull as well as brown (Fig. 13). Comparing the grains within the panicle on a daily basis is essential to accurately note the transitions from R6 through R8. Relying on photographs or written descriptions must be augmented with observation, especially during R8 determination because the lemma and palea of rice lines differ with some being almost white to light tan or tan at R8 while other cultivars may turn dark brown at R8: progressive observations over days will allow these differences to be noted. Determination of the R9 stage requires close scrutiny of the final grains to mature on the panicle: many unfilled grains at the bottom of the panicle will remain green after the filled grains have matured (Fig. 14). The best way, in many cases, is simply to press the last green florets between the fingers to determine if the hull is actually flat. Extreme care is needed in handling the panicles at this stage as the pedicles of the earliest superior grains will often easily break leading to loss of those kernels but this is unavoidable in our experience for some panicles. If you require all of the grains to be collected for the experiment a bag can be placed over the panicle to avoid loss. Loss of the earliest grains during growth staging is usually only a problem if the panicles are to be collected for further analysis at R9.

The interval between making notes can, of course, be done at varying frequencies but many stages, especially intervals R3–R4, R4–R5, and R5–R6, are frequently less than 4 days. It is important to clearly determine whether staging data is needed, the purpose intended for the data, and the precision desired. While longer intervals for making notes will lessen the precision of the information, the precision needed for a particular study may be sufficient with less frequent note taking, i.e., taking notes twice a week rather than daily.

The number of culms needed for sampling is dependent on the plant uniformity and the potential use for the information. For uniform field plots, we usually sample 7–10 culms per plot. For pots in a greenhouse or in controlled climate conditions we sample one culm per plant if one pot is the experimental unit.

Fig. 10 Anthers and filaments on panicle at the R4 stage of development shortly after dehiscence

In cases in which the vegetative development is monitored, main stems are progressively monitored through reproductive development as well. In pots with single plants per pot or with a few plants physically separated within the pot, main steins can usually be identified after the fact even in cases in which the main stem is not identified and monitored beginning at early vegetative development. Identifying main stems in late vegetative and early reproductive development in separated plants can normally be done by observing the culms and surmising which is the main stem. This is time consuming, requires some judgment and experience, and is not certain. In dense field plots with either drill-seeded rows or transplanted rice with more one plant per hill, determining main stems can be nearly impossible without uprooting plants (which would defeat the purpose intended for sequentially observing the same undisturbed plants in the crop). Consequently in field plots, selecting the earliest culms allows assessment of panicles most likely to produce most of the yield. In field plots, it is normally best to mark the earliest culms (in bordered areas within the plots) to reach R2 for monitoring panicles. The earliest panicles generally

Fig. 11 Rice panicle (Growth Stage R6) with typically different stages of development at the different positions within the panicle. Note that florets at the top of the panicle are at R6 and in some grains the caryopsis completely fills the enclosure of the hull while many of the lower grains are still at R3 (anthers visible within the hull)

Fig. 12 Panicle just reaching growth stage R7 with several grains reaching the critical stage of the same day. Note that grains on nearby panicles are also reaching the criterion for R7 (loss of visible green color) and their easily visible difference from the green R6 grains on the same panicles

Fig. 13 Rice panicle at Growth Stage R8

contribute a disproportionally greater part of crop yield than later culms and plants (assuming uniform genetic material such as a crop stand derived from pure seed). The tags should be placed below the penultimate leaf on the culm. If the tags are placed only below the flag leaf, the exerting panicle can become entangled in the tag.

At the completion of the study, an analysis of variance can be applied to the staging data and, if appropriate given the analysis, the data can be averaged within treatments to produce a tabular (Table 3) or graphic timeline of development [20]. The intervals can be by either calendar days, thermal time units, or both. Combining the timeline for development (dates for R3–R9) with environmental conditions during that development can be extremely useful in understanding the many responses which differ under different environments or treatments. Specifically treatment by year and treatment by location interactions for responses can often become clearer by thoughtful analysis of varying environmental conditions at particular stages of development in different years and locations. Moreover, even if only one stage of reproductive development is determined at one point in time for the various treatments, projections of subsequent reproductive stages can be predicted using calendar days or thermal time and relevant data sets such as Counce et al. [20] used can be obtained.

Fig. 14 Panicle observed as R9 although a number of unfilled but green florets are present on the panicle

Table 3
Thermal-unit accumulation (DD10 values) and duration in days for "Wells" rice from field experiments conducted at Stuttgart, Arkansas, U.S.A. from 2007 to 2010 [20]

	2007		2008		2009		2010	
	DD10	Days	DD10	Days	DD10	Days	DD10	Days
R3	10.4	1.2	15.0	1.9	25.9	3.0	19.6	2.6
R4	35.5	3.8	24.4	3.5	20.6	2.4	24.7	3.3
R5	19.9	2.1	13.9	2.0	18.3	2.2	19.8	2.6
R6	57.7	6.1	56.5	8.1	58.9	7.3	46.8	6.0
R7	34.1	3.6	29.6	3.8	17.4	2.6	28.2	3.8
R8	149.6	17.2	204.4	30.7	178.4	26.3	172.8	23.5

3.2 Collecting, Storing and Separating Individual Seed by Stages of Development

A familiarity with the progress of rice plant development in field plots, greenhouse, or controlled climate environments is essential to avoid wasted effort or even misleading data. Critical timing of data collection is frequently determined too late to be of value. Thus samples are frequently missed because the stage for sampling is already passed. This is particularly true of sampling seeds which are actively filling. The superior grains within the crop reach R6 within 7–12 days of the R3 for the plants. The inferior grains normally reach R6 over a much longer period. It is critical to be able to identify which grains are selected and to sample those grains at the rapid period of grain filling. Various enzymes involved in grain filling increase rapidly in R6 and have declined by several folds in its activity by R7. Moreover, the enzyme activity of critical enzymes for grain filling is different for the same stage of development for superior and inferior grains. Consequently, to meaningfully compare treatment effects, seeds should be sampled in such a manner as to eliminate these potential sources of variation.

For many enzyme activity determinations, fresh seed samples can be removed from the plants, immediately placed on ice, rapidly separated into the stage needed, quickly extracted and promptly assayed. If the seeds are kept cold from removal from the plant through assay, such a procedure often yields the maximum enzyme activity. If seeds cannot be assayed within 2–4 h, the best procedure is to quickly freeze them in liquid nitrogen and store them frozen. For most enzymes, activity is rapidly reduced at temperatures above freezing. The fresh panicles should be removed from plants, placed in falconer tubers (50 mL volume often works well), lids should be screwed on tightly, and the tubes dropped in liquid nitrogen. On delivery to the lab, tubes can be placed in containers and stored in a −80 °C freezer.

To separate grains into growth stages, the seeds are removed from the panicles and inspected for individual grain growth stages using the criteria for determining the plant growth stage (Table 1 and Fig. 6). The determination of R3 can be made for individual grains by inspection against a suitable light source to see if the anthers and filaments are still within the hull. R4 can be determined if the anthers or filaments are visible outside of the hull or are absent while the caryopsis is still not visible; R5 can be determined if the caryopsis is elongating within the hull, R6 is determined if the caryopsis has elongated to the length of the hull; R7 is determined when the hull is yellow, lacking any visible green areas such as veins in the hull; R8 is determined when the hull has proceeded from yellow to the mature grain color. If the panicles are observed daily during development: the distinctions between R7 and R8 are quite obvious by the hue as well as the color of the hull. The divisions of individual grains by growth stages can yield valuable information by itself (Table 4) as well as providing the grains for further analysis.

Table 4
Individual grain growth stage percentages at the beginning of three reproductive plant growth stages from field experiments conducted at Stuttgart, Arkansas, U.S.A. in 2011

	Individual grain growth stage				
	R3 and R4	R5	R6	R7	R8
Plant growth stage	Percentage of grains per panicle				
R6	74.1	11.7	14.1	0	0
R7	29.3	23.9	39.7	7.1	0
R8	15.5	24.9	33.1	15	11.6

If required, the process described above can be done with frozen seeds and tweezers, working rapidly from removing the frozen seed from panicles individually, determining growth stage and placing seeds for each growth stage present into separate, labeled mortars containing liquid nitrogen (*see* **Note 1**). At the end of collection, seeds can be removed from the mortars and placed into labeled falconer tubes, attaching tops to tubes quickly and freezing those tubes (*see* **Note 2**). This is critical for both enzyme assay and expression analysis of seed.

4 Notes

1. To avoid various forms of contamination, the seeds collected for enzyme analysis and enzyme expression analysis should either be collected in such a manner as to avoid skin contact with seed or be handled with sterile gloves or tweezers. In many field situations, with limited assistance, sterile gloves are impractical and the best alternative is simply to remove the panicle below the peduncle and drop the panicle into the falconer tube. Usually even then a suitable instrument is required to force fit the panicle branches into the tube. Ziploc bags can be used for transport and storage. Fitting the panicles into such bags is easier than for falconer tubes. On freezing, however, the plastic bags become brittle with further handling and break. This can lead to spills, lost samples, or contamination.
2. Care should be taken to avoid introducing liquid nitrogen into the falconer tubes for storage. Since any liquid nitrogen easily transitions to gas, a risk of exploding the falconer tube exists.

References

1. Lu S, Luh BS (1991) Properties of the rice caryopsis. In: Luh BS (ed) Rice. Springer, New York, pp 389–419
2. Champagne E, Wood DF, Juliano BO, Bechtel DB (2004) The rice grain and its gross composition. In: Champagne ET (ed) Rice chemistry and technology. AACC International, St. Paul, MN, pp 77–100
3. Smith AM, Denyer K, Martin C (1997) The synthesis of the starch granule. Annu Rev Plant Biol 48(1):67–87
4. Singh R, Juliano BO (1977) Free sugars in relation to starch accumulation in developing rice grain. Plant Physiol 59(3):417–421
5. Yamagata H, Sugimoto T, Tanaka K, Kasai Z (1982) Biosynthesis of storage proteins in developing rice seeds. Plant Physiol 70(4):1094–1100
6. Ichihara K, Kobayashi N, Saito K (2003) Lipid synthesis and acyl-CoA synthetase in developing rice seeds. Lipids 38(8):881–884
7. Sreenivasulu N, Radchuk V, Strickert M, Miersch O, Weschke W, Wobus U (2006) Gene expression patterns reveal tissue-specific signaling networks controlling programmed cell death and ABA-regulated maturation in developing barley seeds. Plant J 47(2):310–327
8. Counce PA, Keisling TC, Mitchell AJ (2000) A uniform, objective, and adaptive system for expressing rice development. Crop Sci 40(2):436–443
9. Mohapatra P, Patel R, Sahu S (1993) Time of flowering affects grain quality and spikelet partitioning within the rice panicle. Func Plant Biol 20(2):231–241
10. Cheng F, Zhang Q, Zhu H, Zhao N, Wang F, Chen K, Zhang G (2007) The difference in amylose content within a panicle as affected by the panicle morphology of rice cultivars. J Cereal Sci 46(1):49–57
11. Umemoto T, Nakamura Y, Ishikura N (1994) Effect of grain location on the panicle on activities involved in starch synthesis in rice endosperm. Phytochemistry 36(4):843–847
12. Wang F, Cheng F, Zhang G (2008) Impact of cultivar variation in grain density of rice panicle on grain weight and quality. J Sci Food Agric 88(5):897–903
13. Counce PA, Bryant RJ, Bergman CJ, Bautista RC, Wang YJ, Siebenmorgen TJ, Moldenhauer KAK, Meullenet JFC (2005) Rice milling quality, grain dimensions, and starch branching as affected by high night temperatures. Cereal Chem 82(6):645–648
14. Cooper NTW, Siebenmorgen TJ, Counce PA, Meullenet JF (2006) Explaining rice milling quality variation using historical weather data analysis. Cereal Chem 83(4):447–450
15. Ambardekar AA, Siebenmorgen TJ, Counce PA, Lanning SB, Mauromoustakos A (2011) Impact of field-scale nighttime air temperatures during kernel development on rice milling quality. Field Crops Res 122(3):179–185
16. Lanning SB, Siebenmorgen TJ, Counce PA, Ambardekar AA, Mauromoustakos A (2011) Extreme nighttime air temperatures in 2010 impact rice chalkiness and milling quality. Field Crops Res 124(1):132–136
17. Coast O, Ellis RH, Murdoch AJ, Quiñones C, Jagadish KS (2015) High night temperature induces contrasting responses for spikelet fertility, spikelet tissue temperature, flowering characteristics and grain quality in rice. Func Plant Biol 42(2):149–161
18. Krishnan P, Ramakrishnan B, Reddy KR, Reddy V (2011) Chapter three-High-temperature effects on rice growth, yield, and grain quality. In: Sparks DL (ed) Advances in agronomy, vol 111. Academic, Burlington, pp 87–206
19. Peng S, Huang J, Sheehy JE, Laza RC, Visperas RM, Zhong X, Centeno GS, Khush GS, Cassman KG (2004) Rice yields decline with higher night temperature from global warming. Proc Natl Acad Sci U S A 101(27):9971
20. Counce P, Siebenmorgen T, Ambardekar A (2015) Rice reproductive development stage thermal time and calendar day intervals for six US rice cultivars in the Grand Prairie, Arkansas, over 4 years. Ann Appl Biol 167(2):262–276. https://doi.org/10.1111/aab.12226

Chapter 4

Novel Imaging Techniques to Analyze Panicle Architecture

Erstelle Pasion, Roinand Aguila, Nese Sreenivasulu, and Roslen Anacleto

Abstract

Panicle architecture is known to directly influence grain yield in rice, and thus is an important trait for rice varietal improvement. However, spike branching consequences trigger variation in number of superior and inferior grains and thus affect grain quality. The genetics behind the length of both primary and secondary branches were studied resulting in the identification of cloned genes. Extending this knowledge to include other physiological parameters of panicle architecture is not yet well studied, and it requires high-throughput imaging techniques that are accurate. In this chapter we put the spotlight on Panicle Trait Phenotyping Tool (P-TRAP), a freely available platform independent software to analyze the panicle architecture of rice, as one of such methods that can be used to generate a comprehensive and reproducible panicle architecture data and identify superior breeding lines. P-TRAP measures 15 panicle structure and nine spikelet traits. These quantitative traits can be used in genome-wide association studies to understand their genetic basis.

Key words Panicle architecture, Grain yield, Rice breeding, Grain quality

1 Introduction

One of the major challenges in post-green revolution rice breeding is to maximize the yield of stress-resilient semi-dwarf rice plants to ensure global food security. If the rice-consuming world was to grow by another billion people in 2030, then breeding has to produce high-yielding superior varieties with good grain quality features to catch up with the estimated additional requirement of 65 million tons of milled rice by then [1]. New plant type was conceptualized in the mid-1980s as a strategy for genetic improvement to increase the yield potential in rice [2]. Varietal improvement addressed wide adaptation, optimal growth duration, resistance to biotic and abiotic stresses, and necessary modifications to the components of plant architecture, particularly panicle structure, to increase the number and density of grains that the panicle can produce [3, 4].

Genetic improvement of panicle architecture is correlated with yield and grain quality traits [5]. It directly affects the number of

Nese Sreenivasulu (ed.), *Rice Grain Quality: Methods and Protocols*, Methods in Molecular Biology, vol. 1892, https://doi.org/10.1007/978-1-4939-8914-0_4, © Springer Science+Business Media, LLC, part of Springer Nature 2019

grains per panicle and therefore the final rice yield [6]. The number of grains per panicle is further determined by the number of spikelets per panicle that is mainly determined by the primary and secondary branching as well as the seed setting rate of the spikelets [7]. New plant type studies conducted in the 1990s however have shown that large sink size results in poor grain filling and thus negatively impact grain quality. From lax panicles of wild-type varieties, new cultivars were developed with compact panicles bearing more spikelets [8]. The increased number of spikelets per panicle for most cultivars is mainly attributed to extra grains located at the secondary rachis branches where spikelets are inefficiently filled [9]. Greater variation in terms of spikelet development and grain filling was also observed among rachis branches within a compact panicle type compared to loose panicle type [9]. Moreover, higher percentage of chalkiness was found in grains at the upper rachis branches and primary rachis branches compared to those grains located at the bottom rachis branches and secondary rachis branches [10]. Initial reports also showed that inferior grains have lower palatability, higher protein content, lower amylose content, and less weight compared to superior grains [11, 12]. Because of the large variation among grains within a panicle, the overall grain yield and quality is differentially impacted [13, 14]. Negative correlations are also usually observed between the length of panicle and number of panicles per plant [5], as well as between the size of panicles and overall size of spikelets per plant [15]. These tradeoffs between assimilates and sinks compromise both the quantity and quality of grains within a panicle [5], factors contributing to poor grain filling have been identified [16] and addressed throughout IRRI's recent breeding history.

A reliable method to measure panicle architecture and a few derived traits are presented in this chapter. Analyzing panicle architecture traits is performed using 2D image-based panicle phenotyping software. A prototype developed by Duan et al. [17] automatically evaluates grain yield-related traits such as total number of spikelets, number of filled spikelets, grain length, grain width, and 1000-grain weight. This method was only intended for one rice plant at a time. Smart-PL system based on dual-camera imaging was developed [18] to measure only the rice panicle length including the length from the first rachis node to the panicle neck. Additional 2D-image-based panicle phenotyping software includes PASTAR, P-TRAP, and PANorama. Panicle TRAits Phenotyping Tool (P-TRAP) is a free software written in Java with a graphical user interface [6] to capture images of panicles taken using a camera at a fixed distance, lighting, shutter speed, ISO, and minimum camera resolution of 1024×768 pixels, where images can be batch loaded into the software as a "project." From then on, users can preprocess the images within P-TRAP by cropping out non-panicle areas without altering the resolution of the images. P-TRAP can be calibrated to the precise cm/pixel ratio (the supported scale of P-TRAP) to accurately measure the panicle and grain parameters. The novelty of P-TRAP that is not in other software of similar function is that it can simultaneously measure panicle structure parameters and quantify grain related traits such as the spikelet number per panicle and several spikelet dimensional traits such as length, width, circularity, ellipticity, perimeter, area, compactness, and aspect ratio. Unlike P-TRAP that can measure numerous traits of spikelets whether or not the spikelets are attached to rachis branches, SmartGrain can only determine the shape and size of spikelets threshed out from the panicles [19]. PASTAR (Panicle Structure Analyzer for Rice) [20] is similar to P-TRAP in a way that it can measure the number of spikelets and length and number of rachis branches, but it is not available for public use and it does not measure spikelet dimensional traits. PANorama [5] is faster and more accurate in panicle traits analysis than P-TRAP version 20130213220. However, PANorama measures only attributes of rachis and rachis branches but not spikelet-related traits, so in that sense P-TRAP provides more value.

An integration of an image analysis method and a 5-point calibration model was done [21] in order to rapidly estimate the number of spikelets per panicle. Although this approach eliminates the need for high-resolution images, it provides measurement for only one panicle trait. Another method focusing on only one trait is number of panicles per plant during heading is based on multi-angle imaging developed [22]. A simple and noninvasive technique was developed [23] based on X-ray computed tomography for rice panicle post-anthesis development analysis. This method ultimately aims to provide precise prediction of the optimal crop harvest time and to facilitate selection in breeding for higher grain yield. One more recent method for phenotyping rice panicles is the Panicle-SEG [24]. This software however focuses on panicle segmentation and generates data about heading period and panicle development based on maturity test and changes in area and color of panicles in the field. No definite measurements about rachis branches and spikelets can be measured by Panicle-SEG.

Although new methods are available, P-TRAP was still used in the panicle architecture analysis performed in order to consider both panicle structural variation and spikelet traits. Recent genome-wide association studies also used P-TRAP in generating rice panicle phenotype data [25].

2 Materials Used in Panicle Architecture Measurement

1. Nikon D90 DSLR camera—chosen mainly because of its compatibility with DCamCapture—the software for tethered

shooting. Any camera make and model may be used provided that it works with the software chosen for tethered shooting.

2. USB Cable connector—Type A to Type mini-B cable (Nikon D90 uses a mini-B USB connector).
3. MS Windows XP (or higher) for compatibility to both the image acquisition and panicle architecture analysis software.
4. DCamCapture—free software for tethered shooting downloadable from http://www.bernd-peretzke.de/index.php/dcamcapture-en). Other software for tethered shooting that is compatible with the DSLR camera may also be used.
5. GIMP—open source software for faster image editing that can be freely downloaded from http://www.gimp.org/downloads.
6. Panicle trait phenotyping (P-TRAP) tool—downloadable from http://bioinfo.mpl.ird.fr/index.php/resources?id=102.
7. Caliper (optional)—to be used in generating a reference image for P-TRAP.
8. White 11" × 16" mounting sheet or board.
9. Putty-like adhesive (e.g., Blu-Tack)—to be used to attach the panicles to the mounting sheet.
10. Rice panicles ~24 days after fertilization stored individually in labeled containers (*see* **Note 1**).

3 Procedure

1. Trim the panicle 1 cm below the first primary branch node.
2. Securely mount the trimmed panicle onto the white mounting sheet (e.g., A3 heavy bond paper) (*see* **Note 2**) using enough inconspicuously placed glue tacks. Make sure that the glue tacks are small enough to be undetectable during imaging. Mount the panicles such that the branches (primary, secondary, etc.) are as straight as possible without any grains from different branches (or parts of different branches) touching each other, and that no panicle part extends beyond the border of the mounting sheet (Fig. 1a). Consecutive primary rachis branches should also be mounted alternately on opposite sides to attain maximum distance from neighboring rachis branches. Higher order branches should be well distinguished from the rachis branches from which they originated in order to attain an accurate count of the number of spikelets per rachis branch.
3. Mount the Nikon D90 camera on a tabletop tripod and tether it to the computer using the specified USB cable. Choose a

Fig. 1 Image acquisition. (**a**) The panicle mounted on a white board with sufficient margins. (**b**) The image acquisition setup where the camera is positioned at an optimal and consistent distance to the magnetic board. Constant diffused office lighting was used when capturing the images. This image also shows the Microsoft Windows workstation tethered to the camera via USB cable and the image acquisition software (DCamCapture) already open. (**c**) How the viewfinder should appear when capturing images. Different camera models use different viewfinder focus points

fixed ISO, shutter speed, and image resolution (*see* **Note 3**) making sure that this will be the same for all images taken.

4. Open the DCamCapture software and make sure that the computer detects the camera. Take a test shot, if necessary. Decide on a naming convention of the image files in the "Filename" setting (*see* **Note 4**).
5. Position a clipboard some distance opposite the camera enough to frame the whole mounting sheet. This clipboard shall be used to securely hold the mounted panicle when taking images (Fig. 1b).
6. Clip a mounted panicle onto the clipboard and then manipulate the positions and orientations of the camera and the clipboard so that the edges of the mounting sheet are framed as

tightly as possible in the camera's view finder (Fig. 1c). Maintain this framing for all images.

7. By now, the imaging setup is ready—camera system and the position and distance of the clipboard. Take images of all the panicles to be analyzed. Ensure that shadowing is absent during image capture. Complete the imaging first before doing any analysis to minimize batch effect (*see* **Note 5**).
8. Using the same imaging setup, take a reference image of an object of known size (e.g., the caliper mentioned in the list of materials) (*see* **Note 6**).
9. Open the P-TRAP software.
10. Create a new project (File>New Project) for this analysis and then set it as the main project (Run>Set Main Project).
11. Highlight the project's name at the project navigation window (by clicking on it) and then import the images of the panicles to this project (by clicking the toolbar labeled "+"). Verify that the names of the images appear when you expand the images node in the project's tree view (*see* **Note** 7).
12. Using the reference image (whose name should appear in the list of images), calibrate the scaling factor of P-TRAP. The goal is to set the correct "Real Units/Pixels" calibration ratio. Refer to the P-TRAP manual for calibration details. Use the reference image obtained in a previous step (*see* **Note 8**).
13. Open one of the images (by double-clicking its name in the list of images) and then set the "grain color" and "background color" using the context menu (right-click on a spikelet and on the background, respectively). Please see the P-TRAP manual on how to do color selection. In cases where color is not uniform within the panicle, all the possible colors of spikelets may be selected as "grain color" in order to ensure all spikelets are detected by P-TRAP (*see* **Notes 9, 10**).
14. Detect the structure of the panicle and edit manually to ensure correctness. Refer to the P-TRAP manual for more details on structure detection (*see* **Note 11**).
15. Detect the grains and edit the wrongly indicated number of grains in the grain clusters. Also, mark the grains that were not detected. Refer to the P-TRAP manual on how to edit the number of spikelets initially indicated (*see* **Note 11**).
16. Collect the data for both the panicle structure and the grains. P-TRAP generates two files for the panicle analysis results: MainTraits.csv and GrainsTraits.csv, respectively. MainTraits.csv has the following columns: file_name, PA_length, PA_diameter, Node_nb, SA_nb, SA_average, SA_int, TA_nb, TA_average, TA_int, QA_nb, Sp_nb. GrainsTraits.csv will have the following columns: file_name, Sp_nb, Sp_length, Sp_width,

Table 1
List of (a) panicle structure traits and (b) spikelet traits

Variable name	Trait measured
(a) Panicle structure traits	
pa_length	Length of the rachis
pa_diameter	Diameter of the rachis
sa_po	Position of first primary branch
sa_nb	Number of primary branches
sa_length	Length of primary branch
node_nb	Number of rachis nodes
sa_int	Length of intervals between primary branches
ta_nb	Number of secondary branches
ta_length	Length of secondary branches
ta_po	Position of first secondary branch
ta_int	Length of intervals between secondary branches
qa_nb	Number of tertiary branches
qa_length	Length of tertiary branches
qa_po	Position of first tertiary branch
qa_int	Length of intervals between tertiary branches
(b) Spikelet traits	
sp_nb	Number of spikelets
sp_length	Length of the spikelets
sp_width	Width of the spikelets
sp_area	Area of the spikelets
sp_perimeter	Perimeter of the spikelets
sp_circularity	Circularity of the spikelets
sp_compactness	Compactness of the spikelets
sp_ellipticity	Ellipticity of the spikelets
sp_ar	Aspect ratio of the spikelets

Please *see* Al-Tam et al. [6] for the complete details

Sp_area, Sp_primeter, Sp_circularity, Sp_compactness, Sp_ellipticity, Sp_AR. Please *see* Table 1a, b for the list of traits with short descriptions.

17. Merge both files using the "file_name" column, and then save into a new file. This new file shall contain the final result.

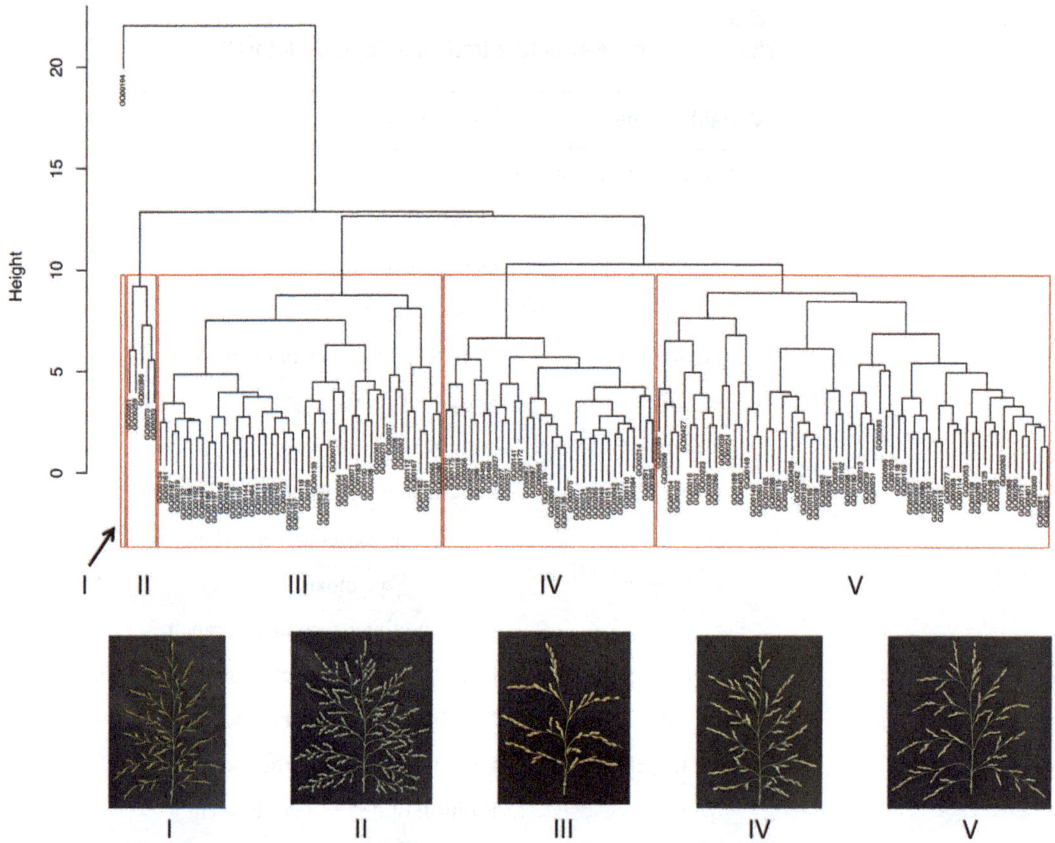

Fig. 2 Hierarchical clustering on 19 traits shows five clusters with approximate representation from each group. The lone accession that is unique from the rest of the panicles analyzed differentiates because of its low values in spikelet ellipticity, compactness, and circularity (*see* Fig. 4). The second smallest cluster (II) shows a combination of spikelet density and unique branching. The rest of the clusters differentiate because of unique combinations of branching and spikelet density, as well as other traits

18. Panicle structure and spikelet related traits measured by P-TRAP are listed in Table 1a, b.

19. An optional but useful step in getting a sense of the data is to do unsupervised statistics such as hierarchical clustering (Fig. 2), principal components analysis (PCA) (Fig. 3a), all-pairs correlation (Fig. 3b), and heatmap generation (Fig. 4). The optimal methods for distance matrix and clustering calculations in generating the dendrograms are "Euclidean distance" and "complete linkage" methods, respectively. Complete linkage ensures good balance between maximum variation between clusters and maximum similarity within a cluster. For the PCA, generate the biplot to see the spatial arrangements in both the sample and variable spaces. The biplot is very useful in making initial inferences about correlations among variables as well as detecting inherent population

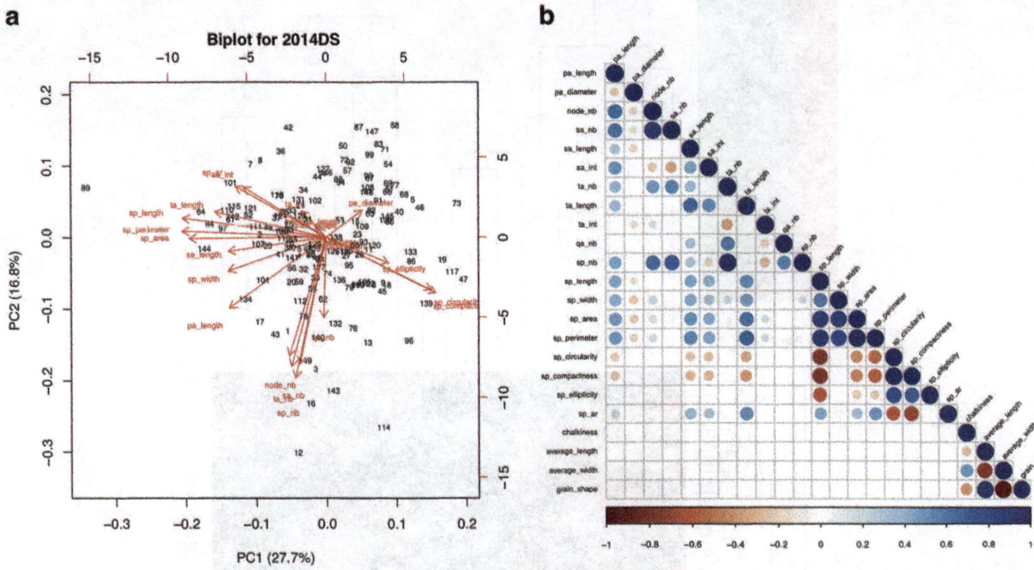

Fig. 3 (**a**) Plot of the first two principal components shows spatial clustering of the variables that suggests their correlations. Variability in the dataset is mostly explained by the spikelet traits where circularity and compactness are the highest positive contributors along the first principal component (PC1), and spikelet length, perimeter and area are negative contributors for PC1. The number of spikelets, nodes, primary and secondary rachis branches are highly correlated and negatively influence the second principal component. These variable correlations are shown in the all-pairs correlation plot (**b**), where the color gradient in the scale bar shows the strength and direction of correlation, while the size of the circles represents the correlation likelihood (*p* value). Blank cells indicate no correlation between the traits

structure. All-pairs correlation among variables is a way of quantifying what is being shown in the biplot's variable space. Hierarchical clustering, PCA, and heatmap can be generated using R. It is important that the distance and clustering functions used in generating both the dendrograms and the heatmap (row and column dendrograms) should be the same: Euclidean and complete linkage, respectively—as mentioned earlier. For PCA, use the prcomp function instead of princomp (both coming from the stats package) because the calculation is done "by a singular value decomposition of the (centered and possibly scaled) data matrix, not by using eigen on the covariance matrix" [17], which is considered more numerically accurate (*see* **Notes 12, 13**).

4 Notes

1. Twenty-four days after fertilization is when the spikelets are of the final size and shape and will not easily shatter when harvested until after a month after mounting.

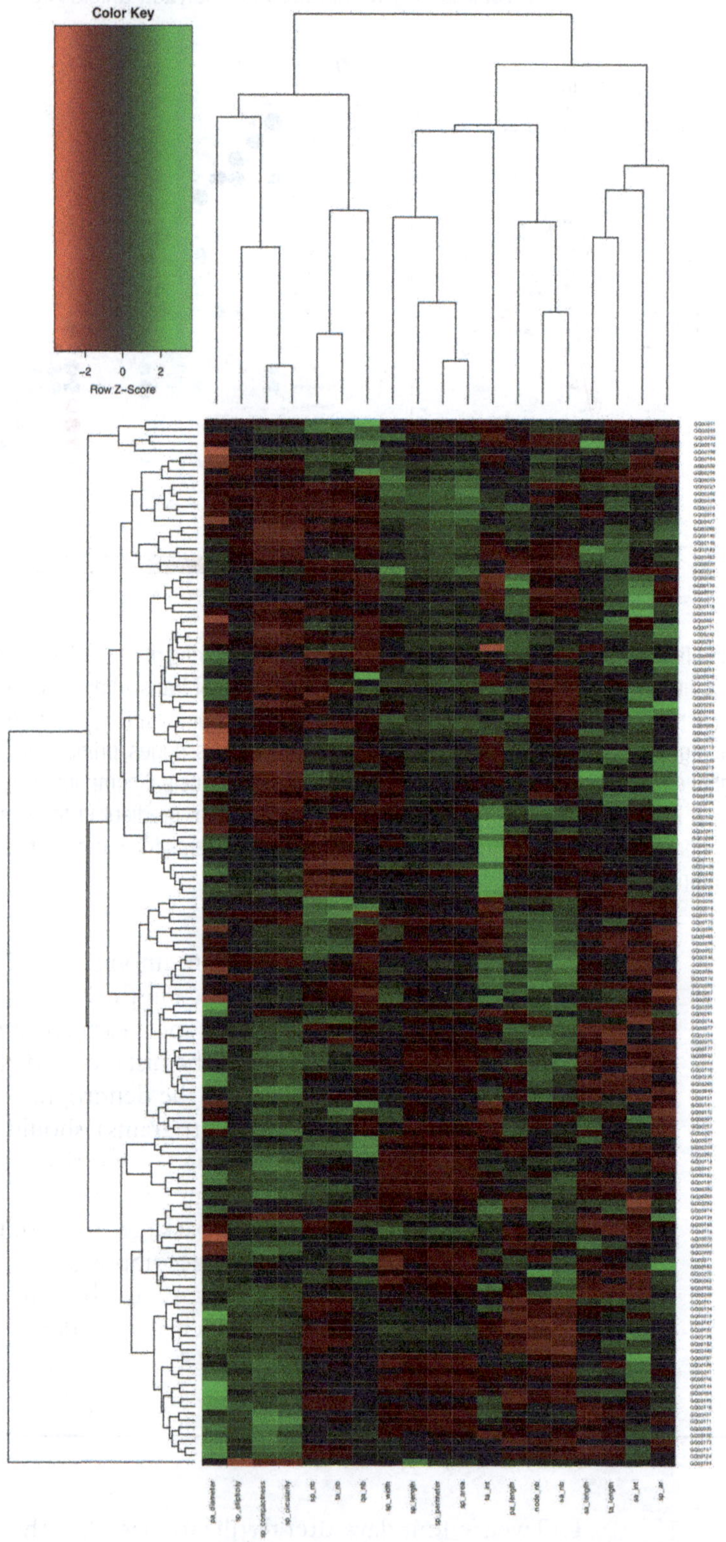

Fig. 4 Heatmap shows both trait and sample clustering. Trait clusters are within these sets: 1. {panicle diameter, spikelet ellipticity, spikelet compactness, spikelet circularity}, 2. {number of spikelets, number of secondary branches, number of tertiary branches}, 3. {spikelet width, spikelet length, spikelet perimeter, spikelet aspect ratio}, 4. {Length of intervals of secondary branches}, 5. {rachis length, number of nodes, number of primary branches}, and 6. {length of primary branch, length of secondary branch, length of intervals of primary branches, spikelet aspect ratio}

2. P-TRAP allows either white or dark colored mounting sheets or board. Just ensure that you match the setting of P-TRAP to whichever you used using the toggle switch displayed in the toolbar labeled by an asterisk. Note, however, that a black background reduces accuracy in measuring the number of spikelets per rachis branch. If this trait is to be considered, a white background should be used instead.
3. For efficiency, a DSLR camera was used and set to 2144 × 1424 pixels resolution. Also, better images may be captured with studio lighting.
4. When taking images of the panicles, it is suggested that you keep a spreadsheet that matches the filename of the panicle's image and its sample ID to prevent confusion later when matching the panicle's image to the data generated.
5. When analyzing multiple panicles, it is more efficient to complete the imaging of all panicles before proceeding with the analysis. It also minimizes batch effects due to variations in the conditions when the images are taken such as camera setting, consistency of the technician, possible variations in lighting, variations in the distance of the mounted panicle to the camera, and other possible confounding factors.
6. When framing the mounted panicle, align as tightly as possible the margins of the mounting paper (e.g., A3 heavy bond paper) with the margins of the view finder. When that is achieved, maintain the position of the camera system and the distance toward the panicle when taking subsequent images.
7. Note to keep an open thumbnail view of captured images to ensure that framing of the images is consistent. This can be done through an open Windows Explorer in thumbnail view.
8. This scaling factor affects the accuracy of the panicle structure and grain dimension measurements.
9. All sheets used in mounting the panicles should be of the same type (dimensions and color).
10. Ensure consistent framing, distance of the mounted panicle to the camera, camera setting for image size and resolution, aperture, film speed, and lighting.
11. In measuring both the panicle and the spikelet traits, of utmost importance is the consistency by which the determinant spots in the panicle are marked. Figure 5 shows an image of the panicle and the resulting graph decomposition. More details about how P-TRAP's algorithm processes this decomposed image are discussed in the paper.
12. Generating the PCA biplot (Fig. 3a) shows the positive correlation and inverse correlation among the variables. Spikelet traits show the highest contributors to data variability, particularly spikelet compactness and circularity. These traits are inversely

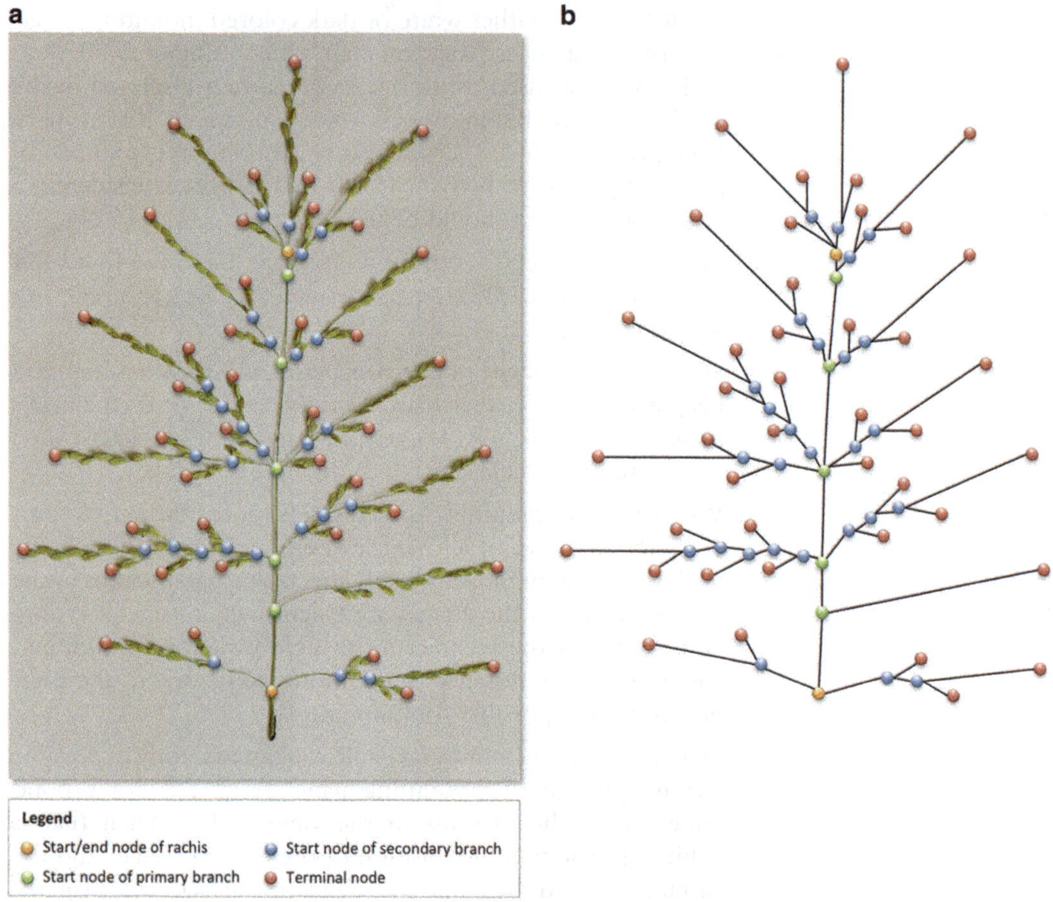

Fig. 5 Graphical decomposition of the panicle images requires precise placement of the start/end nodes of the rachis, the start nodes of the primary, secondary (and tertiary, if present) branches, and the terminal nodes. (**a**) Shows these nodes indicated in the image, and (**b**) shows the resulting graphical decomposition of the panicle's image

correlated to spikelet aspect ratio (calculated as the ratio of spikelet length and width), and to spikelet length, perimeter, and area. Data also shows causal relationship between the intervals between primary branches and the average length of secondary branches. Another interesting cluster of variables are the numbers of nodes, spikelets, branches (although the relative contribution of tertiary branches is minimal) that explains the highest variability along the PC2 axis. Intuitive inverse correlation is shown for the panicle length and diameter. Traits related to milled grains such as chalkiness, dimensions (length and width), and shape (ratio of length and width) do not contribute much to data variability although they show intuitive direct correlations to their counterpart spikelet traits.

13. Panicle architecture analysis can be further improved by generating derived traits based on the measurements from

P-TRAP analysis. Density-related traits may be calculated by getting the ratio of a count trait over the length of another trait. For instance, the density of primary rachis branches may be computed by getting the ratio between the number of primary rachis branches over the length of the rachis. In the same way, the average density of secondary rachis branches may be derived by calculating the mean of all the ratios between the number of secondary rachis branches over the length of a primary rachis branch. P-TRAP results can also be used to partition the panicle into top, middle, or bottom gradients and determine the mean profile for each gradient. To achieve this, the total rachis length must be divided into three partitions and the sum of each internode length starting from the first rachis node until the last one must be calculated. Considering the quotient representing one-third of the length of rachis, the positions of the primary rachis branches can be determined based on the range within which internode they are attached to. Also, panicle shape categories may be created by comparing the means of primary rachis branches' lengths for each panicle gradient. For instance, if primary rachis branches at the lower gradient have a mean length longer than the middle gradient, and the mean length at the middle is also longer than that at the top, then the panicle shape can be described as pyramidal. The inverse of this mean length comparisons results in an upright cone-like shape panicle. If the mean length of primary rachis branches is longest at the middle gradient, the panicle can be described as diamond-like. Otherwise, panicle shape may be categorized as irregular.

Acknowledgment

This work has been supported under the CGIAR thematic area Global Rice Agri-Food System CRP, RICE, Stress-Tolerant Rice for Africa and South Asia (STRASA) Phase III, and Australian Centre for International Agricultural Research (Project ID CIM/2016/046) funding.

References

1. Zeigler RS (2014) Toward food security in 2050: when I think about poverty. CSA News 59:8
2. Vergara BS (1987) Raising the yield potential of rice. Trans Nat Acad Sci Tech 9:397–413
3. Khush GS (2001) Green revolution: the way forward. Nat Rev Genet 2:815–822
4. Khush GS (1995) Breaking the yield frontier of rice. Geo J 35:329–332
5. Crowell S, Falcão AX, Shah A, Wilson Z, Greenberg AJ, McCouch SR (2014) High-resolution inflorescence phenotyping using a novel image-analysis pipeline, PANorama. Plant Physio 166:479–495
6. FM AL-T, Adam H, Anjos A, Lorieux M, Larmande P, Ghesquière A, Jouannic S, Shahbazkia HR (2013) P-TRAP: a panicle trait phenotyping tool. BMC Plant Biol 13:122

7. Xing Y, Zhang Q (2010) Genetic and molecular bases of rice yield. An Rev Plant Biol 61:421–442
8. Zhu Z, Tan L, Fu Y, Liu F, Cai H, Xie D, Wu F, Jianzhong F, Wu J, Matsumoto T, Sun C (2013) Genetic control of inflorescence architecture during rice domestication. Nat Commun 4:2200
9. Mohapatra PK, Panigrahi R, Turner NC (2011) Physiology of spikelet development on the rice panicle: is manipulation of apical dominance crucial for grain yield improvement? Adv Agro 10:333–359
10. Zhang X, Alim N, Lin Z, Liu Z, Li G, Wang Q, Wang S, Ding Y (2013) Analysis of variations in white-belly and white-core rice kernels within a panicle and the effect of panicle type. J Integ Agr Adv Online Pub:1–11
11. Matsue Y, Odahara K, Hiramatsu M (1995) Differences in amylose content, amylographic characteristics and storage proteins of grains on primary and secondary rachis-branches in rice. Japan J Crop Sci 64(3): 601–1606
12. Umemoto T, Nakamura Y, Ishikura N (1994) Effect of grain location on the panicle on activities involved in starch synthesis in rice endosperm. Phytochemistry 36:843–847
13. Iwasaki Y, Mae T, Makino A, Ohira K, Ojima K (1992) Nitrogen accumulation in the inferior spikelet of rice ear during ripening. Soil Sci Plant Nutri 38:517–525
14. Liu ZH, Cheng FM, Cheng WD, Zhang GP (2005) Positional variations in phytic acid and protein content within a panicle of japonica rice. J Cereal Sci 41:297–303
15. Ohsumi A, Takai T, Ida M, Yamamoto T, Arai-Sanoh YM, Ando T, Kondo M (2011) Evaluation of yield performance in rice near-isogenic lines with increased spikelet number. Field Crops Res 120:68–75
16. Peng S, Cassman KG, Virmani SS, Sheehy J, Khush GS (1999) Yield potential trends of tropical rice since the release of IR8 and the challenge of increasing rice yield potential. Crop Sci 39:1552–1559
17. Duan L, Yang W, Huang C, Liu Q (2011) A novel machine-vision-based facility for the automatic evaluation of yield-related traits in rice. Plant Methods 7:44
18. Huang C, Yang W, Duan L, Jiang N, Chen G, Xiong L, Qian L (2013) Rice panicle length measuring system based on dual-camera imaging. Comp Elect Agri 98:158–165
19. Tanabata T, Shibaya T, Hori K, Ebana K, Yano M (2012) SmartGrain: high throughput phenotyping software for measuring seed shape through image analysis. Plant Physiol 160:1871–1880
20. Ikeda M, Hirose Y, Shibata Y, Yamamura T, Komura T, Doi K, Ahikari M, Matsuoka M, Kitano H (2010) Analysis of rice panicle traits and detection of QTLs using an image analyzing method. Breed Sci 60:55–64
21. Zhao S, Gu J, Zhao Y, Hassan M, Li Y, Ding W (2013) A method for estimating spikelet number per panicle: integrating image analysis and a 5-point calibration model. Sci Rep 5:16241
22. Duan L, Huang C, Chen G, Xiong L, Liu Q, Yang W (2015) Determination of rice panicle numbers during heading by multi-angle imaging. The Crop J 3:211–219
23. Jhala T (2015) X-ray computed tomography to study rice (Oryza sativa L.) panicle development. J Expt'l Bot 66(21):6819–6825
24. Xiong X, Duan L, Liu L, Tu H, Yang P, Wu D, Chen G, Xiong L, Yang W, Liu Q (2017) Panicle-SEG: a robust image segmentation method for rice panicles in the field based on deep learning and superpixel optimization. Plant Methods 13:104
25. Rebolledo MC, Peña AL, Duitama J, Cruz DF, Dingkuhn M, Grenier C, Tohme J (2016) Combining image analysis, genome wide association studies and different field trials to reveal stable genetic regions related to panicle architecture and the number of spikelets per panicle in rice. Front Plant Sci 7:1384

Chapter 5

Measuring Head Rice Recovery in Rice

Jennine Rose Lapis, Rosa Paula O. Cuevas, Nese Sreenivasulu, and Lilia Molina

Abstract

Head rice recovery (HRR) is a milling quality attribute that is highly influential toward the market price of rice. It is defined as the proportion of paddy rice that retains 75% of its length after milling. For a new rice variety to be accepted and adopted by farmers, the new variety's HRR should satisfy consumer requirements of at least 55% or above. Hence, HRR is a crucial attribute by which new varieties are selected for release. Although the amount of head rice recovered depends on the genetic background of a rice variety, HRR is also highly affected by postharvest processing conditions that the variety goes through. To determine the maximum HRR, therefore, one must ensure that the processing conditions are as optimal as possible. This book chapter outlines how paddy rice is processed into head rice and how HRR is measured. It also proposes an improved laboratory-scale means for postharvest drying to minimize head rice losses.

Key words Quality evaluation, Dehulling, Milling, Head rice recovery, Drying

1 Introduction

Rice varietal improvement programs aim to develop high-yielding rice varieties. In order to encourage farmers to adopt these improved varieties, the milling quality of these varieties has to be, at least, at par with the varieties that these farmers cultivate regularly. One of the indicators of milling quality is HRR [1, 2], the amount of rice grains that retain at least three-quarters (75%) of their original length after being polished, relative to the amount of paddy rice [3]. Although this attribute can be explained partially by genetics, farming conditions during the grain-filling stage and postharvest conditions also affect HRR [4, 5]. These postharvest factors include moisture content (MC), drying conditions of freshly harvested paddy rice, presence of foreign matter in the sample lot, and grain dimensions. The critical MC values of rice

Nese Sreenivasulu (ed.), *Rice Grain Quality: Methods and Protocols*, Methods in Molecular Biology, vol. 1892, https://doi.org/10.1007/978-1-4939-8914-0_5, © Springer Science+Business Media, LLC, part of Springer Nature 2019

varieties (i.e., MC below the critical MC leads to grain fissuring) vary, but for screening of varietal differences in grain cracking, 14% MC was deemed to be the optimum level [6]. Foreign matter like small stones or soil, and other plant material, on the other hand, are removed prior to milling [7]; too much of these foreign materials can potentially reduce throughput. Long and slender rice grains tend to have greater breakage compared to short and bold rice grains [8].

Although varieties that meet yield expectations, which fail to satisfy the HRR criterion, risk low adoption by farmers because HRR is directly linked to market price of rice set by millers, and therefore projected farmer income may vary. Farmer adoption surveys [9] indicate that modern improved varieties have not seen widespread use, with HRR being an attribute of concern. For instance, rice varieties developed by IRRI's breeding programs were reported to average 45% HRR [10]. It appears, therefore, that there is still room for improving HRR. A two-pronged approach is employed in efforts to increase HRR: (1) decrease the grain's vulnerability to breakage through genetic improvement and (2) improve postharvest conditions to further reduce grain breakage. Chalkiness is recognized as an additional factor in reduced HRR [4] and the potential of eliminating chalkiness in grains through genetic approaches is being explored [8]. On the other hand, drying rough rice is considered the most critical postharvest process that affects milling quality and HRR. Several drying experiments were conducted by Siebenmorgen et al. to understand the response of rice grains to various drying and tempering environments using the glass transition principle and to find ways on how grain fissuring can be minimized [11–14].

The milling process may also contribute to HRR. There are different approaches to mill rice: (1) the traditional method that uses mortar and pestle, (2) the single-stage mill in which the husk and the bran are removed in one pass, and (3) the multistage milling that has different processing steps for dehulling, milling, and separating head rice from the broken grains [15]. In IRRI's Grain Quality and Nutrition Service Laboratory (GQNSL), grains undergo multistage milling.

This book chapter describes the method used in the GQNSL for processing rice from paddy grain to head rice and how HRR is measured. This book chapter also presents a case study exploring various means of drying rice samples, at the laboratory scale, that can possibly contribute to the improvement of HRR measurement through reduced grain breakage and increased throughput. This chapter does not discuss the genetic approach to improving HRR nor does it suggest large-scale modifications that may decrease grain breakage at the industrial level.

2 Materials

1. Drying room with controlled temperature (50 ± 5 °C) and Relative Humidity (63 ± 5%) maintained using an automatic heater and humidifier.
2. Grain moisture tester.
3. Metallic drying racks.
4. Nylon drying net bags.
5. Re-sealable plastic bags used for the tempering stage.
6. Dehuller (Satake, Model THU-35A, Satake Corp., Hiroshima, Japan, Fig. 1).
7. Cleaning brush.
8. Plastic bin for rice husk disposal.
9. Top-loading balance (Model PGW 2502i, Adam Equipment, United Kingdom).
10. Barcode scanner (Model LS2208-SR20007R-UR, Symbol Technologies, Mexico, Fig. 2).
11. Mill (Grainman, Model 60-220-60-DT, Grain Machinery Mfg. Corp., Miami, Florida, USA, Fig. 3).
12. Collecting tray for milled rice.
13. Plastic bin for rice bran disposal.
14. Spatula for Grainman mill head scraping/cleanup.

Fig. 1 Satake blower/dehuller

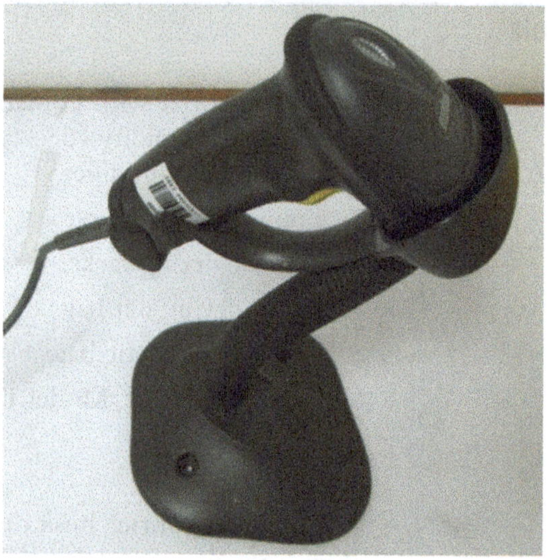

Fig. 2 Barcode scanner

Fig. 3 Grainman mill

15. Head rice separator (Satake rice machine Type TRG with three different trays, Satake Engineering Co., Ltd., Japan, or Rice sizing device, Figs. 4 and 5).
16. Collecting trays for broken and head rice grains.

Fig. 4 Rice sizing device

Fig. 5 Trays (short grain, medium grain, long grain)

3 Methods

1. Drying.
   (a) Determine the moisture content (MC) of the freshly harvested paddy rice sample using a grain moisture tester (*see* **Note 1**).
   (b) Weigh 200 g paddy rice samples and pack in a net drying bag. Arrange samples across drying racks and allow drying inside the controlled drying room (24 h) or until 12–14% MC is obtained.
2. Dehulling.
   (a) Get the moisture content of the paddy rice sample using the grain moisture tester. Clean the paddy rice grain by blowing out the unfilled grains and other impurities in the samples using the Satake dehuller (*see* **Note 2**).
   (b) Weigh 125.00 ± 1.00 g cleaned paddy rice grains.

(c) Dehull the paddy rice grain sample using the Satake dehuller (*see* **Note 3**).

(d) Collect and weigh the unpolished rice grain using the top-loading balance.

3. Milling.

(a) Remove the bran of the unpolished rice grain using the Grainman mill for 45 s (*see* **Note 4**).

(b) Collect and weigh the polished rice grain using the top-loading balance.

4. Determining the head rice recovery.

(a) Remove broken rice grain (<75% of the original grain length after milling) by passing the grains through the vibrating rice sizing device (*see* **Note 5**).

(b) Collect and weigh the head rice grain using the top-loading balance.

5. Data processing.

The proportions of brown, milled, and head rice from paddy rice are calculated based on the equations below (*see* **Note 6**).

$$\%\ \text{Brown rice} = \frac{\text{weight of brown rice}}{\text{weight of paddy rice}\ (\text{g})} \times 100$$

$$\%\ \text{Milled rice} = \frac{\text{weight of milled rice}}{\text{weight of paddy rice}\ (\text{g})} \times 100$$

$$\%\ \text{Head rice} = \frac{\text{weight of head rice}\ (\text{g})}{\text{weight of paddy rice}\ (\text{g})} \times 100$$

4 Applying a Drying Protocol to Improve HRR: A Case Study

In support of IRRI's pursuit on breeding rice with better HRR, a drying protocol that incorporates a tempering step [14] was tested. This additional step to drying supposedly minimizes grain fissuring by preventing abrupt changes in the moisture gradient within the rice kernel. It involved the application of (1) heat while monitoring RH to dry freshly harvested paddy rice and of (2) tempering conditions as a post-drying step.

In this study, four varieties grown at IRRI in the 2016 wet season (Table 1) were obtained right after harvest and were subjected to two laboratory-scale postharvest drying conditions. These varieties were chosen to represent different classes of grain size and shape. NSIC Rc392 is a saline-tolerant variety that was included in the participatory varietal selection (PVS) in 2015 and was favored

Table 1
Description of grain type of four IRRI varieties obtained from IRRI's plant breeding division

Variety	Description
NSIC Rc 392 (IRRI 185)	Long, slender
NSIC Rc 348 (IRRI 175)	Long, bold
NSIC Rc 400(IRRI 186)	Medium, intermediate
NSIC Rc 334 (IRRI 171)	Extra-long, slender

in central Philippines. All other varieties will be introduced in the PVS in the near future.

For samples that were dried in controlled conditions, the freshly harvested samples were threshed and stored in re-sealable plastic bags at 4 °C for 1 week. Grains (200 g) were then transferred into nylon drying bags and were allowed to equilibrate at room temperature overnight. Initial MC was measured using a moisture meter. Then, the samples were placed inside the controlled drying room, in which the temperature range applied was 40–52 °C while RH was monitored (range was 47–63%). When the grains reached the MC range of 12–14%, the drying bags were placed in re-sealable plastic bags for tempering (1 h, 50 ± 5 °C). Samples were removed from the re-sealable bags and then allowed to cool at room temperature until 12–14% moisture content was reached. Samples were refrigerated (4 °C, in resealable plastic bags) until ready for HRR determination.

For samples dried in unmonitored conditions, freshly harvested samples were placed in nylon drying bags and placed in the glasshouse for sun-drying. When the target moisture content range (12–14%) was reached, the samples were hand-threshed, and were stored at 4 °C in resealable plastic bags until ready for HRR determination.

Head rice recovery was determined as described in Subheading 3, in triplicate.

Results (Table 2) suggest that samples subjected to controlled drying conditions had higher HRR compared to those dried in uncontrolled conditions in the glasshouse. Among the varieties tested, IRRI 171 (extra-long and slender grains) appeared to be most susceptible to fissuring among the samples tested under uncontrolled sun-drying, but showed improved HRR when dried under controlled drying condition (Table 2). However, it is noted that only IRRI 185 reached an HRR of 60% under controlled drying conditions. The HRRs of the rest of the samples were around 50% when drying was controlled.

Table 2
Comparison of HRR (%) of four IRRI rice varieties subjected to two drying conditions

Sample	HRR (%)	
	Drying room (controlled)	Glasshouse (unmonitored)
IRRI 185	60.5 ± 0.9	55.0 ± 0.5
IRRI 175	52.1 ± 5.1	43.8 ± 0.4
IRRI 186	51.1 ± 0.6	45.7 ± 0.1
IRRI 171	50.6 ± 2.3	31.6 ± 1.5

The developed controlled drying room protocol, therefore, is recommended as a routine drying method for IRRI varieties requiring HRR analysis

5 Notes

1. Freshly harvested paddy rice will be used for controlled post-harvest drying of paddy samples for HRR analysis. Fresh MC is determined using grain moisture tester by placing an ample amount of the paddy grain to be measured in the sampling tray. The paddy grain must have been presorted to remove unripe or degenerated grains. Insert the sampling tray in the testing chamber at the side of the tester. Rotate the crushing handle counterclockwise until it reaches its limit. Wait until % MC appears on the screen. Moisture content measurements using a moisture meter are based on the conductivity of the paddy rice; which, in turn, is relative to the paddy grain's moisture content [16]. This method is less accurate than the primary (oven-drying) method but produces results within seconds or minutes and is usually used in postharvest management and for trade.
2. For removing the semi-filled and the unfilled grains, the roller distance of the Satake dehuller is adjusted to keep rollers apart. This allows the passage of paddy samples for blowing of unfilled or semi-filled grains without dehulling them. To do this, the machine is switched on, paddy samples are poured on the feed hopper, and the sample passage is controlled to prevent clogging of paddy grains. The filled grains fall into a large container for collection and the semi-filled or unfilled grains will fall into a small container (for discarding). Turn the machine on and off to ensure that no grains are left inside the dehuller. Open the

trap door in the hopper, use a brush or vacuum to clean the inside surface of the dehuller. Clean the machine after each sample.

3. The same Satake dehuller machine is used for dehulling. To do this, the same procedure for cleaning (*see* **Note 2**) is followed, except that the roller distance is adjusted to keep rollers at a distance (i.e., slightly more than half the length of the grain) that allows passage of paddy samples to achieve the optimum huller efficiency at 75–80%, at minimum breakage [3]. Allow two sample passes to ensure maximum paddy grain dehulling. Collect the brown rice samples from the large container. Return samples into their appropriate container (properly labeled envelopes). Discard the undehulled samples.
4. The milling process using Grainman mill removes the outer layer (bran) from brown rice, resulting in a white or polished rice fraction, also known as milled rice. To operate the machine, switch the Grainman mill on and load the brown rice samples into the mill. Lock the milling chamber by pushing the locking handle upward, push the cover downward and hook the weight lever over the cover. Set the timer to 45 s and press the white button to start milling. Once the milling process is complete, remove the weight lever. Clean the milling chamber using a brush to remove the bran on the outside surface of the milling chamber. Place a tray/collection pan under the milling chamber. Gently lift the cover off from the milling chamber. Unlock the milling chamber by pulling the lock handle downward. Turn the milling chamber upside-down in order for the samples to fall into the collection pan. Use spatulas and/or brushes to remove the milled rice off the screen and all sides of the roller, ensuring that all rice grains from the milling chamber are removed. Transfer the milled rice collected on the tray/collection pan into its properly labeled envelope. Clean the mill thoroughly after each use. Use a brush to scrub all accessible parts of the mill (i.e., cover, roller, and chamber). Make sure to remove all the debris on and around the milling chamber, as well as in the collection pans. Once all samples are milled, hold the toggle switch to turn off the mill.
5. The rice sizing device sorts the milled rice to separate the broken grains from head rice. Pass the milled rice sample onto the vibrating or oscillating sieve of the head rice separator. Collect the grains with lengths at least 75% of the unbroken grain (i.e., those grains that do not fit the wells of the sizing device) into a pan. Manually remove broken milled grain from the collected sample.
6. Other terms for "brown rice" and "milled rice" are "unpolished rice" and "polished rice," respectively.

Acknowledgments

The authors thank Marnol Santos, Edgar Amoloza, Anna Carriza Basilio, Dennis Villegas, Teodoro Atienza, Leah Villanueva, for the controlled drying room setup and optimization, and assistance in processing and collecting data for the samples used in the case study at GQNSL. Fresh paddy samples for the drying experiment were donated by Dr. Georgina Vergara and Rafael Julian Panerio while Jun Correa allowed us to avail the rice threshing service at Zeigler Experiment Station in IRRI. This work has been supported under the CGIAR thematic area Global Rice Agri-Food System CRP, RICE, Stress-Tolerant Rice for Africa and South Asia (STRASA) Phase III, and Australian Centre for International Agricultural Research (Project ID CIM/2016/046) funding.

References

1. Zhou L, Liang S, Ponce K, Marundon S, Ye G, Zhao X (2015) Factors affecting head rice yield and chalkiness in *indica* rice. Field Crops Res 172:1–10
2. Patindol J, Wang YJ (2002) Fine structures of starches from long-grain rice cultivars with different functionality. Cereal Chem 79(3):465–469
3. Juliano BO (2007) Rice chemistry and quality. Philippine Rice Research Institute, Munoz, Nueva Ecija, p 402
4. Zhao X, Fitzgerald MA (2013) Climate change: implications for the yield of edible rice. PLoS One 8(6):e66218
5. Pearce MD, Marks BP, Meullenet J-FC (2001) Effects of postharvest parameters on functional changes during rough rice storage. Cereal Chem 78(3):354–357
6. Juliano BO, Perez CM (1993) Critical moisture content for fissures in rough rice. Cereal Chem 70(5):613–615
7. Bhattacharya KR, Ali SZ (2015) An introduction to rice-grain technology, vol 288. CRC Press, Boca Raton, FL
8. Sreenivasulu N, Butardo VM Jr, Misra G, Cuevas RP, Anacleto R, Kavi Kishor PB (2015) Designing climate-resilient rice with ideal grain quality suited for high-temperature stress. J Exp Bot 66(7):1737–1748
9. Laborte AG, Paguirigan NC, Moya PF, Nelson A, Sparks AH, Gregorio GB (2015) Farmers' preference for rice traits: insights from farm surveys in Central Luzon, Philippines, 1966–2012. PLoS ONE 10(8):e0136562
10. Anacleto R, Cuevas RP, Jimenez R, Llorente C, Nissila E, Henry RJ, Sreenivasulu N (2015) Prospects of breeding high-quality rice using post-genomic tools. Theor Appl Genet 128:1449–1466
11. Cnossen AG, Siebenmorgen TJ, Yang W, Bautista RC (2001) An application of glass transition temperature to explain rice kernel fissure occurrence during the drying process. Drying Technol 19(8):1661–1682
12. Cnossen AG, Siebenmorgen TJ (2000) The glass transition temperature concept in rice drying and tempering: effect on milling quality. Trans ASAE 43(6):1661–1667
13. Cnossen AG, Siebenmorgen TJ, Yang W (2002) The glass transition temperature concept in rice drying and tempering: effect on drying rate. Trans ASAE 45(3):759–766
14. Ondier GO, Siebenmorgen TJ, Mauromoustakos A (2012) Drying characteristics and milling quality of rough rice dried in a single pass incorporating glass transition principles. Drying Technol 30:1821–1830
15. Dhankar P (2014) Rice milling. IOSR J Eng 4(5):34–42
16. Janier JB, Maidin MB (2011) Paddy moisture content detector. J Appl Sci 11:1476–1478

Chapter 6

Measurement of Rice Grain Dimensions and Chalkiness, and Rice Grain Elongation Using Image Analysis

Marnol V. Santos, Rosa Paula O. Cuevas, Nese Sreenivasulu, and Lilia Molina

Abstract

Measurements of rice grain dimensions, percent grain chalkiness, and grain elongation used to be tedious and slow due to the manual nature of measurements (e.g., use of calipers to measure grains one at a time) and the subjective nature of scoring based on visual inspection (i.e., chalkiness). Recent developments in imaging technologies have enabled more high-throughput means for measuring physical traits (i.e., grain dimensions and chalkiness) in raw grains and grain elongation by comparing ratio between raw versus cooked rice. The digital images of rice grains are captured through computer scanning and analyzed using software that can calculate area and pixel value statistics of user-defined parameters. The improvements in throughput made possible by the use of imaging technologies will allow faster quality grading of rice varieties. Market quality is usually defined based on the rice grain physical traits (grain size and shape), degree of chalkiness, and the ability of rice to elongate on cooking. In this chapter, the routine methods to measure the physical traits of rice and grain elongation using image analysis are described.

Key words Physical traits, Chalkiness, Size and shape, Grain elongation, Image analyses

1 Introduction

Grain quality can be subdivided into search, experience, and credence attributes [1]. Search attributes are those intrinsic properties that consumers can evaluate and base their purchase decisions on, which includes physical traits for rice grains such as length, width, and degree of chalkiness. Ensuring these intrinsic properties are important market criteria for valuation of milled rice into distinct classes based on shape (shape is expressed as the ratio of grain length to width): slender (>3.1), medium (2.1–3.0), and bold (≤2.0). Traditionally, grain dimensions are determined per grain using calipers or micrometers [2–6], and mean grain dimensions are then reported. However, manual measurements of these properties are slow and tedious. The advent of digital imaging systems has increased the throughput of measurement, with the imaging

Nese Sreenivasulu (ed.), *Rice Grain Quality: Methods and Protocols*, Methods in Molecular Biology, vol. 1892,
https://doi.org/10.1007/978-1-4939-8914-0_6,

systems calibrated using, or compared against, manual measuring techniques using calibrated instruments [7, 8]. Flatbed scanning systems, such as the SeedCount SC5000 Rice Analyser, operate by capturing images of individual seeds within a sample lot using the reflectance mode. Algorithms are used to estimate grain length and width, and calculate the sample lots's mean and its variation [9, 10]. At IRRI, grain length is categorized as short (<5.50 mm), medium or intermediate (5.51–6.60 mm), long (6.61–7.50 mm), and very long (>7.51 mm) [11].

Chalkiness, an attribute that affects the visual quality of the raw rice grain, can also be classified as a search attribute [1] as the percent of opacity ("chalkiness") and translucency of the rice grain factor into purchase decisions of consumers; i.e., in most markets, the higher the degree of chalkiness, the lower is its acceptability in the market [11]. Chalkiness and milling quality (e.g., head rice recovery, HRR) are negatively correlated; i.e., HRR decreases with increased chalkiness [12]. Chalkiness is often expressed as the proportion of opaque areas relative to transluscent areas in a single milled rice grain [13]. Typically, visual observation and scoring based on a scale is used to evaluate the percentage of chalkiness in the grain (reviewed in [14]). However, evaluation of chalkiness in rice by image analysis is more accurate than by human visual inspection [15]. Image analysis also enables an increase in efficiency of measuring chalkiness.

Grain elongation is another visual attribute evaluated in cooked rice rather than in uncooked grains. It is an important attribute for consumers in India and Pakistan, while it is an attribute also observed in sadri varieties (Iran) and in japonica rice [16]. The metrics used to describe this attribute are the grain elongation ratio (the ratio of mean length between cooked grain and raw grain [17] and the grain elongation index (the ratio of mean length-to-width ratios between cooked grain and raw grain [16]. Although grains can be measured one by one directly or through the use of a photo enlarger, digital capture of images of individual cooked and raw grains via an ordinary flatbed scanner and measurement of grain length and width through the ImageJ analysis software and its user-written plug-ins [18] has significantly improved efficiency in characterizing grain elongation.

This book chapter describes the routine sample scanning and image analysis procedures for grain dimensions, chalkiness, and grain elongation measurements being conducted at IRRI's Grain Quality and Nutrition Services Laboratory (GQNSL).

2 Materials

2.1 Sample Preparation

1. Paddy rice grains.
2. Drying oven, Model UFE800, Memmert GmbH + Co. KG, Germany.

3. Dehuller, Satake, Model THU-35A, Satake Corp., Hiroshima, Japan.
4. Mill, depends on the amount of sample to be polished:
   (a) For samples 70–100 g: Grainman Model 60-220-60-DT and 60-230-60-24 T (Grain Machinery Mfg. Corp, Miami, FL, USA).
   (b) For samples 10–15 g: Kett mill (Kett Electric Laboratory, Tokyo, Japan).
5. Head rice separator (Satake rice machine Type TRG, Satake Engineering Co., Ltd., Japan; or Rice sizing device).

2.2 Measuring Physical Traits in Raw Grain

1. SeedCount SC5000 Image Analyzer, Next Instruments, Australia.
2. SC-5000 Image Analysis System Software Version 2.6.3.
3. Image Analyzer sample tray (there are different trays for grains of different lengths and shapes, **Note 1**).
4. Computer running Microsoft Windows Operating System XP or 2007 as scanner controller.
5. Reference rice image dimension template (printed).
6. Reference color chart template (printed).
7. Sample scoop.
8. Sample tweezers/forceps.
9. Soft-bristle paintbrush.

2.3 Measuring Grain Elongation

1. HP Scanjet 8200, Hewlett Packard (for image capture).
2. Computer running Microsoft Windows Operating System Xp/2007 as scanner controller.
3. ImageJ image analysis software, ImageJ 1.47, downloaded from http://imagej.nih.gov/ij
4. 50 mL test tube (for cooking rice sample).
5. Water bath (Temperature set point range: 98.0–100.0 °C).
6. Mild solvent and lint-free tissue for scanner surface cleanup.
7. Reference rice image dimension template (printed).

3 Methods

3.1 Sample Preparation

1. Remove the hull from paddy rice grains using a dehuller.
2. Using a mill suitable for the amount of sample, mill the dehulled grains into white polished rice (*see* **Note 1**).
3. Place the milled rice into bar-code labeled envelopes.

3.2 Measuring Physical Traits in Raw Grain

This procedure is specific for the operations conducted at the GQNSL. Please refer to the SeedCount SC5000 user's guide to customize the settings for other laboratories.

1. Switch on computer and SeedCount unit.
2. Open SeedCount program icon.
3. As quality control, scan the reference rice image dimension template (Fig. 1) before analysis and check if the values are within the calibration standard verification (CSV) limits (Table 1).
4. If the dimension limits are exceeded, clean the scanner surface or restart the scanner and the computer. If these do not help correct the dimension readings, call personnel trained at troubleshooting the equipment (either in-house staff or from the supplier) to recalibrate the system's dimension readings.
5. For measurement of the degree of chalkiness (*see* **Notes 2–4**), presort and label the samples according to the color of milled rice using the reference color chart template (Fig. 2).
6. Using the designated sample scoop, pour at least 100 milled grains onto the tray (medium or long grain tray, depending on the size of the grains, **Note 1**). Make sure that the grains are scattered evenly on the wider wells (bottom part of the tray). All grains must be in the wells.
7. Load the sample tray into the SeedCount unit.

Fig. 1 The SeedCount reference rice image dimension template (not presented to scale), provided by Next Instruments. The dimensions in the template have been validated using a calibrated Vernier caliper prior to use. The template is scanned on the SeedCount unit on a daily basis or as needed to confirm system accuracy and stability

Table 1
Calibration standard verification (CSV) limits for rice dimensions measured using the SeedCount SC5000

	Dimension limits	
Image dimension template1.	**Length (mm)**	**Width (mm)**
CSV1	7.3 ± 0.1	3.4 ± 0.1
CSV2	6.1 ± 0.1	2.8 ± 0.1
CSV3	8.0 ± 0.1	4.0 ± 0.1

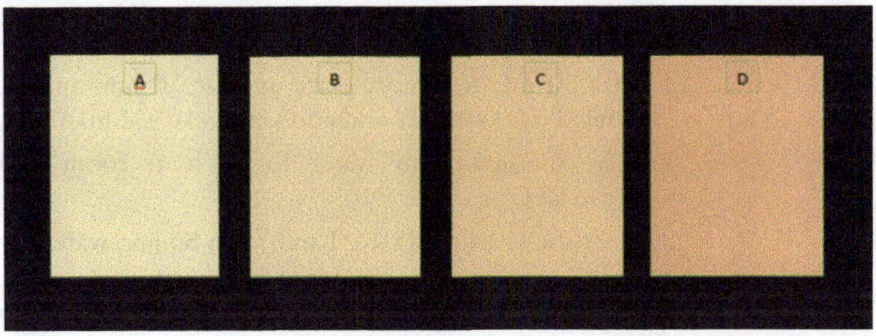

Fig. 2 Reference color chart template, developed in-house. Before scanning, the grain color is compared with the color chart for appropriate adjustment of the reflectance brightness threshold. Color chart values of **a**, **b**, **c**, and **d** correspond to seed count's reflectance slider position values of 121, 122, 123, and 124, respectively

8. In the SeedCount software window, the Module must be set to "White Long Grain Rice" and the Standard must be "ISO."
9. Match the grain color with the color chart template (Fig. 2). Adjust the reflectance brightness threshold slider position from 121 to 124 depending on the corresponding color of the grain on Fig. 2 (*see* **Note 5**).
10. Select "Scan and Analyze."
11. Enter approximate weight of the sample in grams.
12. Wait for the results (grain length, grain width, and % chalkiness) to appear.
13. Select "'Save Data."
14. In the Sample ID window, enter the filename and sample ID or scan the barcode of the sample.
15. Click OK and save.
16. Remove the tray from the SeedCount unit.
17. Return all grain samples into the sample envelope.

18. For succeeding samples, repeat Subheading 3.2, **steps 5–17**.
19. Rescan the reference image dimension template after the analysis. Check the limits as stated on Subheading 3.2, **step 3**.
20. At the end of the analysis, close the SeedCount window.

3.3 Measuring Grain Elongation

3.3.1 Capturing Grain Images

1. Switch on the computer and the image scanner (*see* **Note 6**).
2. Place at least ten whole (unbroken) raw grains directly on the scanner surface for image capture. Ensure that the grains do not touch each other.
3. While leaving the scanner cover open, scan the grains and save the scanned image file.
4. Capture images of unbroken cooked rice grains (Subheading 3.3.2, **steps 1–7**) following Subheading 3.3.1, **steps 1–3**.

3.3.2 Cooking Rice Grains

1. In a 50 mL test tube, place at least 20 raw unbroken grain (milled rice) samples and add water (10 ± 2 mL).
2. Soak the grains in water for 1 h at room temperature (25 ± 5 °C).
3. Heat the test tube (10 ± 1 min) in a boiling water bath.
4. Cool the test tube (5 ± 1 min) in a water bath (20–30 °C) which has a water level at least 1 cm higher than the water height in the test tube.
5. Remove/drain excess water in the test tube.
6. Place the cooked rice on a Petri dish and blot dry using filter paper or tissue.
7. Using tweezers or forceps, place unbroken cooked grains on the scanner surface for image scanning.

3.3.3 Image Capture and Processing Using ImageJ

The methods written below are specific for the operations conducted at the GQNSL. Please refer to the ImageJ user's guide to customise the procedure for other laboratories.

1. Click the ImageJ image analysis software. The system will standardize the pixel-to-mm ratio automatically every time the system starts using the preset or stored reference Vernier caliper image (*see* **Note 7**).
2. Click on the "Plugin" Menu and select the appropriate customized plugin module, then browse and open for the scanned image for measurement.
3. The selected plugin module will measure the lengths and the widths of all objects in the selected file. And will give the following images and result list:

(a) Processed equivalent 8-bit black and white, negative image.

(b) Outlined images with assigned number per object.

(c) Result list of length and width measurement.

4. The analyst may copy the results list onto a spreadsheet file for further data control check (*see* **Note 8**), and for statistical analyses.

5. Selecting the day's scanned reference rice image CSV template (Fig. 3), compare the results with the standards' dimension limits (Table 2).

6. If the dimension limits set are exceeded, clean the scanner surface or restart the scanner and the computer; if the values are still outside the dimension limits, call for technical support (in-house trained staff or from the supplier) to recalibrate the system's dimension readings.

7. For the sample measurement, record the average length, the average width, and the length-to-width ratio of both the raw and the cooked rice grains. Save the file.

8. Report the grain elongation ratio [17] and the grain elongation index [16].

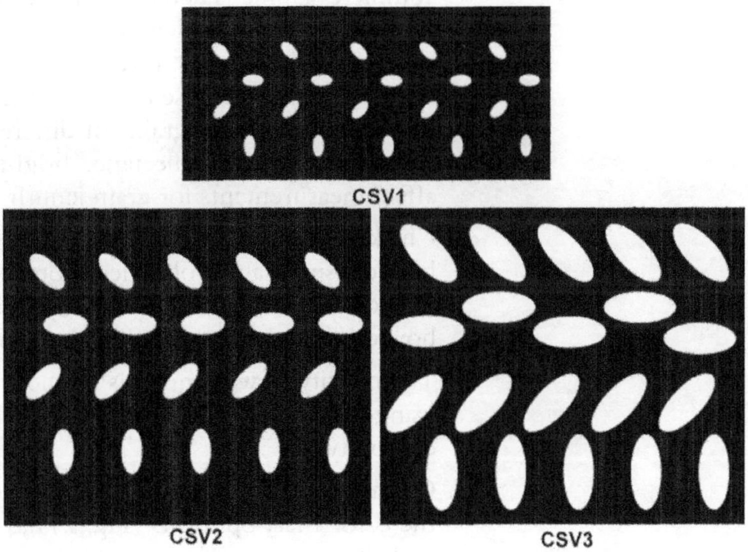

Fig. 3 ImageJ reference rice image dimension template (not shown to scale), developed in-house. Using a certified Vernier caliper, the printed template images dimensions were measured manually and the dimensions were set as the reference values for length and width. The template is scanned on a daily basis or as needed to confirm system accuracy and stability

Table 2
Dimension limits for three image dimension templates used for ImageJ

	Dimension limits	
Image dimension template1.	**Length (mm)**	**Width (mm)**
CSV1	4.1 ± 0.1	2.0 ± 0.1
CSV2	8.9 ± 0.1	4.1 ± 0.1
CSV3	14.7 ± 0.1	6.0 ± 0.1

4 Notes

1. Under-milled and colored rice grains may affect the measurement for % chalkiness. Make sure that grains to be scanned are well milled or polished before analyzing in the SeedCount SC5000. The method is validated for grains with maximum length of 6.5 mm and maximum width of 3.9 mm for medium tray; and maximum length of 8.8 mm and maximum width of 2.5 mm for the long tray.
2. Using SeedCount in reflectance mode, the % chalkiness of the complete sample is determined as the average of the % chalkiness of the individual grains. There is a need to adjust the reflectance brightness threshold (using the reference color chart template) because the default setting provides inaccurate chalkiness values for grains of different shades or translucencies. Adjusting the reflectance brightness threshold does not affect measurements for grain length and for grain width.
3. Chalkiness measurement by the SeedCount SC5000 was calibrated using values obtained from processing raw rice grain images using the ImageJ software. Figure 4 is an example of how chalkiness measurements were obtained using ImageJ.
4. The opacity of waxy grain is neither determined nor validated using the color chart and is not covered by the machine capability.
5. Grain sample scanned with a darker color shade than that of the color chart option D (Fig. 2) has been provided improper % chalkiness values for rice samples.
6. Always clean the scanner surface by using any mild solvent such as ethyl or isopropyl alcohol, or water and lint-free tissue prior to and after use.
7. Scan the reference rice image dimension template at least twice, as the initial and the last sample for the day's quality control verification run. Then using the ImageJ software, compare results with the standards' set values.

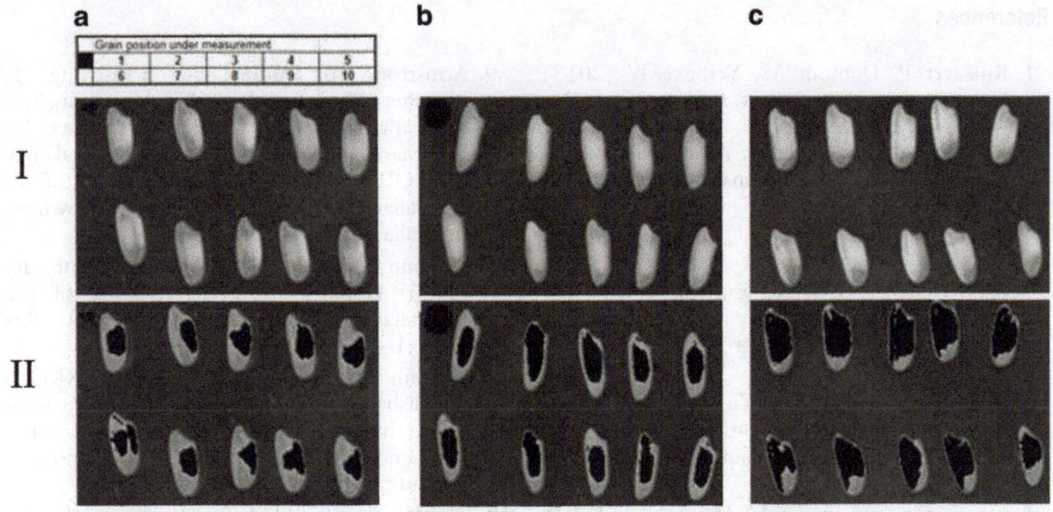

(a)

grain	area pixel whole	chalk	% chalk
1	818	307	38
2	866	249	29
3	827	303	37
4	899	321	36
5	969	297	31
6	902	329	36
7	957	267	28
8	870	222	26
9	931	302	32
10	885	234	26
		ave	32

(b)

grain	area pixel whole	chalk	% chalk
1	1001	382	38
2	992	626	63
3	990	556	56
4	966	382	40
5	923	411	45
6	903	406	45
7	828	429	52
8	968	375	39
9	890	402	45
10	925	405	44
		ave	47

(c)

grain	area pixel whole	chalk	% chalk
1	796	589	74
2	869	624	72
3	851	630	74
4	802	602	75
5	749	547	73
6	739	465	63
7	777	560	72
8	720	490	68
9	746	563	75
10	633	336	53
		ave	70

Fig. 4 Different degrees of chalkiness in grains, with chalkiness of the complete sample reported as the average of % chalkiness of the individual grains. These images were captured using a scanner (Subheading 3.3). There are ten grains per image, positioned into two rows with five grains each row (I). ImageJ generates grain images in which the chalky portions are blackened or shaded (II). The % chalkiness is then based on the area pixel of the chalky area in relation to the whole grain (III). Sets (**a**), (**b**), and (**c**) illustrate samples whose calculated degrees of chalkiness are 32%, 47%, and 70%, respectively

8. The resulting value for the scanned working standards should be compared to the set value and should be within ±0.1 mm of the validated record or from the control chart, where available. If these requirements are met, start processing the sample images; otherwise, recalibrate the pixel-to-mm value based on manual measurements using a Vernier caliper.

Acknowledgments

The authors thank Teodoro Atienza (who operates the SeedCount SC5000), and Leah Villanueva (who measures grain elongation metrics) for providing technical assistance during method development and optimization of the image analysis protocols. This work has been supported under the CGIAR thematic area Global Rice Agri-Food System CRP, RICE, Stress-Tolerant Rice for Africa and South Asia (STRASA) Phase III, and Australian Centre for International Agricultural Research (Project ID CIM/2016/046) funding.

References

1. Rutsaert P, Demont M, Verbeke W (2013) Consumer preferences for rice in Africa. In: Wopereis MCS, Johnson D, Ahmadi N, Tollens E, Jalloh A (eds) Realizing Africa's rice promise. CAB International, Wallingford, pp 293–301
2. Fan C, Xing Y, Mao H, Lu T, Han B, Xu C, Li X, Zhang Q (2006) *GS3*, a major QTL for grain length and weight and minor QTL for grain width and thickness in rice, encodes a putative transmembrane protein. Theor Appl Genet 112(6):1164–1171
3. Li M-m XL, J-f R, G-l C, Yu L-q, He H-h, L-z H, H-j K (2010) Identification of quantitative trait loci for grain traits in japonica rice. Agr Sci China 9(7):929–936
4. Xie X, Song M-H, Jin F, Ahn S-N, Suh J-P, Hwang H-G, McCouch SR (2006) Fine mapping of a grain weight quantitative trait locus on rice chromosome 8 using near-isogenic lines derived from a cross between *Oryza sativa* and *Oryza rufipogon*. Theor Appl Genet 113(5):885–894
5. Govindaraj P, Vinod K, Arumugachamy S, Maheswaran M (2009) Analysing genetic control of cooked grain traits and gelatinization temperature in a double haploid population of rice by quantitative trait loci mapping. Euphytica 166(2):165–176
6. Fan C, Yu S, Wang C, Xing Y (2009) A causal C-A mutation in the second exon of *GS3* highly associated with rice grain length and validated as a functional marker. Theor Appl Genet 118(3):465–472
7. Koutroubas SD, Mazzini F, Pons B, Ntanos DA (2004) Grain quality variation and relationships with morpho-physiological traits in rice (*Oryza sativa* L.) genetic resources in Europe. Field Crops Res 86(2–3):115–130
8. van Dalen G (2005) Characterisation of rice using flatbed scanning and image analysis. In: Riley AP (ed) Food policy, control and research. Nova Biomed, pp 149–186
9. Armstrong B, Aldred G, Armstrong T, Blakeney AB, Lewin L (2005) Measuring rice grain dimensions with an image analyser. In: Blanchard C, Truong H, Allen H, Blakeney AB, O'Brien L (eds) Cereals 2005: 55th Australian cereal chemistry conference, Sydney, Australia pp 326–329
10. Tikapunya T, Fox G, Furtado A, Henry RJ (2016) Grain physical characteristic of the Australian wild rices. Plant Genet Res 15(5):1–12
11. Graham R (2002) A proposal for IRRI to establish a Grain Quality and Nutrition Research Center. In: IRRI discussion paper series no. 44, International Rice Research Institute, Los Banos p 15
12. Zhao X, Fitzgerald MA (2013) Climate change: implications for the yield of edible rice. PLoS One 8(6):e66218
13. Lisle AJ, Martin M, Fitzgerald MA (2000) Chalky and translucent rice grains differ in starch composition and structure and cooking properties. Cereal Chem 77(5):627–632
14. Raju GN, Manjunath N, Srinivas T (1991) Grain chalkiness in cereals. Trop Sci 31:407–415
15. Yoshioka Y, Iwata H, Tabata M, Ninomiya S, Ohsawa R (2007) Chalkiness in rice: potential for evaluation with image analysis. Crop Sci 47(5):2113–2120
16. Juliano BO, Perez CM (1984) Results of a collaborative test on the measurement of grain elongation of milled rice during cooking. J Cereal Sci 2(4):281–292
17. Dela Cruz NM, Khush GS (2002) Rice grain quality evaluation procedures. In: Singh RK, Singh US, Khush GS (eds) Aromatic rices. Mohan Primlani, New Delhi, pp 15–28
18. Schneider CA, Rasband WS, Eliceiri KW (2012) NIH image to ImageJ: 25 years of image analysis. Nat Methods 9(7):671–675

Chapter 7

Method Development of Near-Infrared Spectroscopy Approaches for Nondestructive and Rapid Estimation of Total Protein in Brown Rice Flour

Rosario Jimenez, Lilia Molina, Iman Zarei, Jennine Rose Lapis, Ruben Chavez, Rosa Paula O. Cuevas, and Nese Sreenivasulu

Abstract

Rice varietal development and improvement programs are constantly seeking means to shorten the breeding cycle in order to deliver new, consumer-acceptable rice varieties to farmers and to consumers. Advances in molecular biology technologies have enabled breeders to use high-throughput genotyping to screen breeding lines. However, current phenotyping technologies, particularly for rice cooking and eating properties, have yet to match the efficiency of genotyping methodologies. A high-throughput and cost-effective phenotyping suite is essential because without phenotype, the value of genotypic information cannot be maximized. In this book chapter, we explore the application of near-infrared spectroscopy (NIRS), a high-throughput and nondestructive approach in characterizing rice grains, primarily describing method development and validation, instrument calibration, upgrading, and maintenance. We then focus on estimating protein content (PC) in brown rice as a case study because (1) PC is an attribute that contributes to the cooking behavior and the eating properties of cooked rice; and (2) proteins contain chemical bonds that can easily be detected by NIRS.

Key words Protein content, Near-infrared spectroscopy (NIRS)

1 Introduction

Measurement of indicators of rice cooking and processing characteristics is considered important in varietal improvement programs and thus has always been an essential component in selecting materials in rice breeding programs [1].The most commonly quantified metrics for predicting grain quality include amylose content (AC), gelatinization temperature (GT), and gel consistency (GC) [2–4]. The methodologies that have been developed for determining these attributes are relatively high-throughput (e.g., differential scanning calorimetry, alkali spreading test, iodine colorimetry, and gel consistency test) but are often labor- and resource-intensive.

Nese Sreenivasulu (ed.), *Rice Grain Quality: Methods and Protocols*, Methods in Molecular Biology, vol. 1892,
https://doi.org/10.1007/978-1-4939-8914-0_7,

Because current improvements in rice breeding programs attempt to shorten delivery times of new rice varieties, efforts have been made to increase efficiency of breeding line selection through the use of high-throughput genotyping approaches [5–8]. However, current routine grain quality phenotyping methodologies, albeit being high-throughput, are neither fast nor efficient enough to match the rapid pace of genotyping pipelines and the increasingly efficient modern breeding approaches. It is, therefore, important to develop easy-to-implement, high-throughput, multi-trait, and cost-effective grain quality phenotyping tools to ensure the timely delivery of new rice varieties with desirable quality attributes.

A cost-effective, high-throughput, and nondestructive method with the potential to screen for phenotypic variability in rice is NIRS. In this technique, the wavelengths within the NIR range (λ = 750–2500 nm) are absorbed differently by XH_n functional groups (O–H, N–H, C–H), due to specific vibrations of the chemical bonds of the atoms in these functional groups [9]. These chemical bonds (like in O–H, N–H, and C–H) differ in strength, leading to differences in the amounts of energy required to change the chemical bonds' normal vibration patterns (e.g., transitioning from a ground state to the second excited state). These energy differences can then be visualized as spectra of varying absorptions bands (e.g., overtones and combination bands) at different wavelengths [10, 11]. The absorbed energy can then be associated with the concentration of the functional groups [12].

Since the 1970s, NIR instrumentation has been used for a wide variety of applications in the grain industry as well as in proximate analysis of other foods and feeds, and in on-line process control in pharmaceutical manufacturing [11, 13–16]. However, NIRS was slow to gain wide acceptance as an analytical tool [17] because it is an indirect method whose accuracy is influenced by the sum of errors of both the reference method and the NIR calibration system, and by interferents or other variables not necessarily related to the analyte of interest (e.g., particle size distribution and moisture content [18]). With the advent of more sophisticated NIR instrumentation, more powerful microprocessors, advanced chemometrics software, and computational algorithms, more robust and reliable NIRS prediction models for various analytes of interest have become possible. These prediction models are typically regressions taking on the general form:

$$Y = a + b_1X_1 + b_2X_2 + \ldots + b_kX_k,$$

where Y is the response variable; X_1, X_2,…,X_k are k independent variables such that each variable is a combination of spectral values; b_1, b_2, …, b_k are the regression coefficients; and a is the intercept. Chemometrics is used to determine the regression coefficients during prediction model development using known X and Y values, and is used to predict Y given a set of X_k and b_k [16].

The improved application of NIRS is evident in the increasing number of NIRS publications for predicting AC and other quality components in whole-grain brown and milled rice [19–23], rice flour [24–26], as well as in individual milled and bulk milled kernels [27]. The best standard error of prediction (SEP) for AC by NIRS obtained in earlier studies made by Villareal [19], Delwiche [20, 21], and Bao [24] ranged from 1.0 to 1.6%. Besides AC, a more comprehensive NIR prediction of rice starch quality parameters was reported by Bao [24] for GT, GC, pasting parameters, and gel texture. NIRS was found reasonably accurate in predicting pasting parameters of setback (SB) (SEP = 13.6 RVU, R^2 = 0.92), and breakdown (BD) (SEP = 10.2 RVU, R^2 = 0.88); and gelatinization peak temperature (T_p) (SEP = 1.33 8C, R^2 = 0.89). Although the various studies reported SEP and/or standard error of cross-validation (SECV) that were reasonably acceptable for the range of samples analyzed, most of the calibration models developed were restricted to specific types of rice varieties that do not cover the entire compositional range. Rice samples used in these studies were sometimes specific to the region where they were grown or were selected from commercially released rice varieties and advanced breeding lines. The accuracy of NIR global prediction models for rice with widely diverse physical and compositional traits still has to be validated.

Protein content (PC), although not routinely measured during early selection breeding, likewise plays a substantial role in affecting the cooking behavior of rice and the texture of cooked rice because of protein's ability to imbibe water during the cooking process [28]. Although protein contributes to the nutritional value of rice, high-protein rice may be unacceptable to some consumers because it increases hardness, reduces the stickiness, and affects milled rice color [29]. Much of the variability in the PC of rice is believed to be caused by environmental factors [30] or by G × E interactions more than by general heritability [31]. Moreover, a negative correlation between PC and grain yield was reported in earlier studies done at IRRI [30]. Therefore, a method that can estimate PC in a high-throughput manner is important. The NIRS method for protein estimation has been established with good precision for years. In fact, the first commercial NIRS instrument was designed to analyze PC in grain and the success and importance of the technique was demonstrated in the determination of protein in wheat [32–34], using partial least squares (PLS), reported that the calibration models obtained from near-infrared transmission (NIRT) spectroscopy had an SEP of 0.24 for brown rice protein content and 0.22 for milled rice PC. The derivative method for NIR calibration to determine both PC and AC in milled rice flour, resulting in a SECV of 0.23% for PC using four factors and an SECV of 1.0% for AC using ten factors [35].

While NIRS is viewed as a fast and simple method, the complexity lies in the development and maintenance of reliable and robust calibration models for the analytes or traits of interest. Most commercial near-infrared spectrophotometers can now be purchased with built-in calibrations. Some employ more complex but efficient methods such as Artificial Neural Networks (ANN) for the development of global NIR calibration models on large databases [36]. However, since instrument manufacturers have limited access to all possible variations in the types of samples the client may encounter, pre-calibrated instruments are often not satisfactory or reliable enough to be applied to a particular client's samples. This could be due to a number of reasons, mainly:

1. The range in composition of the samples used in the calibration does not cover all possible variation;
2. The sampling technique or presentation used for NIR spectra collection is not applicable to the client's limited amounts of rice grains available for analysis;
3. The reference data is unreliable or the reference method is highly susceptible to artifacts, and therefore problematic;
4. The calibrations provided with the instrument need to be validated and updated with client's samples or entirely new calibration models need to be developed that fit the specific purpose of the assay.

This book chapter aims to present a general guideline for the development, calibration, validation, as well as the upgrading and maintenance of NIRS procedures. Figure 1 outlines the general workflow for an NIRS analytical procedure. This workflow is divided into two main phases: the initial calibration development phase and the subsequent management phase of the NIRS procedure lifecycle which includes constant validation and monitoring, and updating and maintenance of the NIRS library or database.

Phase 1 involves initial NIRS calibration development for the analyte or trait of interest using selected and representative samples. The spectral pattern of a set of diverse samples are first explored by scanning all samples in the spectrophotometer using the vis-NIR wavelength range (400–2500 nm) and interpreted by the use of chemometric techniques [37]. A standard and well-established reference method is required to develop a reliable calibration model. Ideally, the reference analysis should be performed at the same time or immediately after NIR scanning. Information on the performance (linearity, accuracy, and precision) of the reference method should be known since it will influence the performance of the NIR prediction model that is calibrated against it. A calibration model is developed by relating spectral data to laboratory values using multivariate statistical algorithms and then validated by testing the model with an independent set of samples (usually around 10%, but not found in the calibration set) [18].

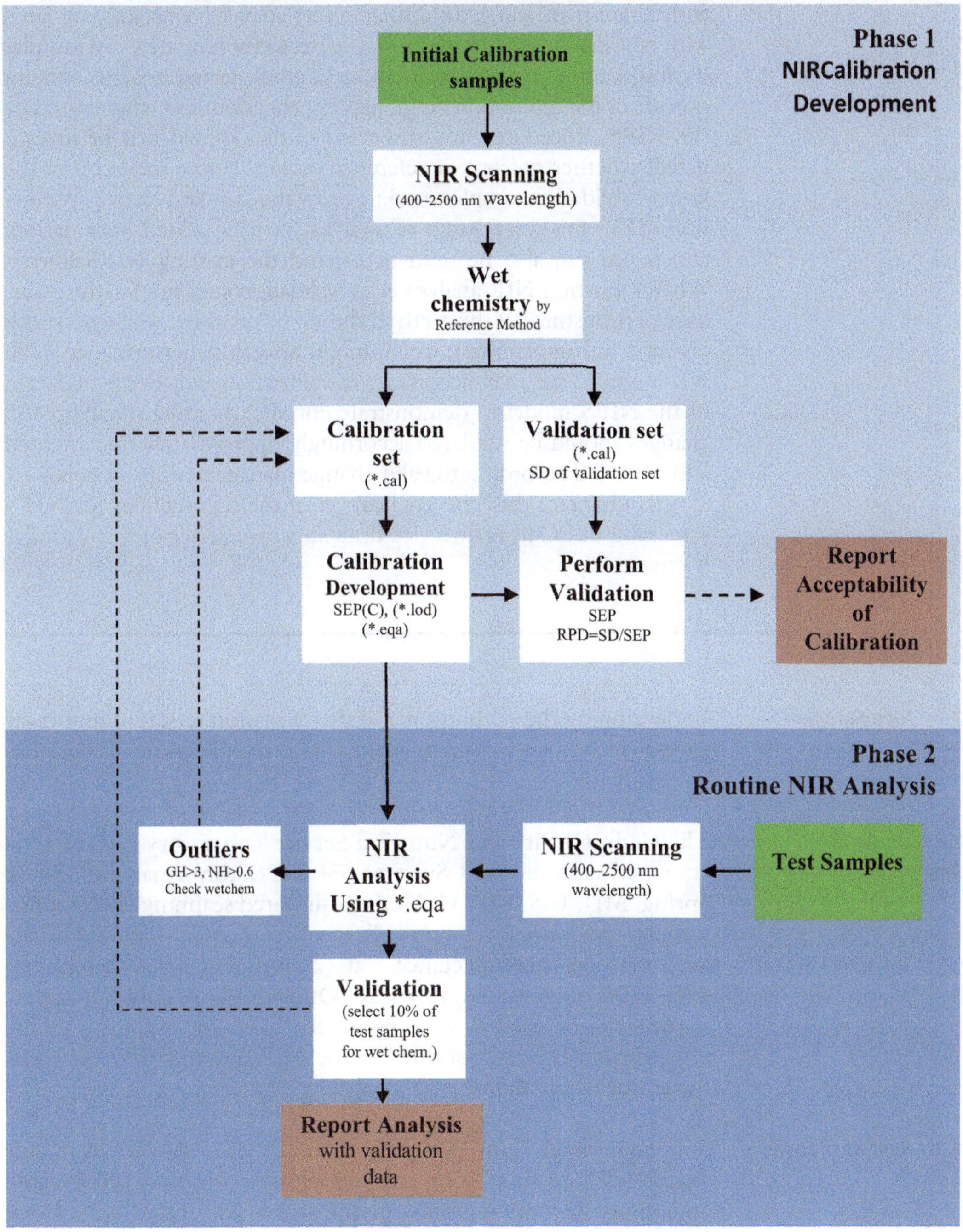

Fig. 1 General workflow for method development, validation, and routine analysis by NIRS

Phase 2 involves calibration maintenance and updating while performing routine assay of succeeding batches of the same type of test samples using the prediction model developed in Phase 1. The development and implementation of a NIRS calibration is dynamic and iterative, meaning the NIRS library may be constantly updated with new data, usually with outliers detected among new test samples during routine analysis. Validation samples having analyte content outside of the calibration range may appear as outliers when tested by the NIRS procedure. Any suspected outlier should first be investigated, whether spectral or reference outlier, before inclusion in the library. Validation samples used to verify the standard error of prediction (SEP) for every batch of analysis, provide added information, and hence can also be used to expand the existing NIRS library. When reporting NIR analysis data, validation statistics for the assurance of robustness of the method should be provided. Changes (both planned and unplanned), which might affect the performance of an NIRS procedure, may necessitate re-calibration and/or re-validation of the NIRS model to demonstrate continued model suitability. All changes should be validated accordingly, appropriately documented and recorded according to valid change management protocols.

To illustrate this, the application of these guidelines for NIRS screening of PC in brown rice flour in a diverse set of rice varieties is reported.

2 Materials

2.1 Rice Samples

In developing the calibration and the validation sets, it is important to use rice samples coming from a diverse collection of materials (*see* **Note 1**).

2.2 Vis-NIRS Spectrophotometer

The Grain Quality and Nutrition Service Laboratory (GQNSL) is equipped with the NIR Systems 6500 (Perstorp Analytical, Silver Spring, MD, U.S.A.), a visible/near-infrared scanning spectrophotometer (*see* **Note 2**) able to collect reflectance readings and calculate for log (1/reflectance) at 2 nm intervals within the 400–2500 nm wavelength range. Other NIR spectrophotometer models can be used as well; however, this book chapter presents settings that may be specific for the NIR Systems 6500 as optimized for rice samples.

2.3 Sample Module

Spinning sample cup (for reflectance mode), 36 mm diameter (depth = 9 mm) black ring cups with quartz windows on one side and disposable white backs on the other side, fitted with black inserts with 12 mm diameter windows.

2.4 Standardized Reference (Check Cell)

The standardized "Check Cell" is provided with the instrument to evaluate and keep track of overall instrument performance over time.

2.5 Sample Preparation Devices

1. Dehuller.
2. Retsch MM400 mixer mill (Retsch GmbH, Retsch-Allee 1-5, 42781 Haan, Germany) for fine-grinding (up to ~5 μm) small amounts of grains.

2.6 Chemometrics Software

The chemometrics software available with the NIRSystem6500 is WinISI v1.02a (Infrasoft International, LLC) although other third-party multivariate analysis software (e.g., Unscrambler X 10.3, CAMO Software, Oslo, Norway) may also be utilized for this purpose (*see* **Note 3**).

2.7 Apparatus and Materials for Micro-Kjeldahl N Analysis (Reference Method)

1. Skalar San++ Continuous Flow Analyzer unit consisting of sampler, manifold, proportioning pump, heating bath, colorimeter (with a 660 nm filter).
2. Analytical balance, 1000 mg capacity.
3. Aluminum block digestor with automatic temperature control.
4. Digestion tubes (2.5× 20 cm) calibrated to 20.000 ± 0.100 g of the same solution matrix as the calibration standard (K_2SO_4-H_2SO_4).
5. Dispensers, 10 and 50 mL capacities.
6. Vortex mixer.
7. Calibrated scoops (for catalyst mixture).
8. Polyethylene bottles.
9. Volumetric flasks.
10. Funnel.
11. Autosampler cups.
12. Concentrated sulfuric acid (Analytical Grade, 18 M).
13. Ammonium sulfate (AR grade).
14. Potassium sulfate (AR grade).
15. Ammonium chloride (AR grade).
16. Anhydrous Potassium sulfate/Selenium mixture, 100 g:2 g (*see* **Note 4**).
17. Calibration standards (0, 10, 30, 50, 70, 90, 120, 150 mg N/kg from 1000 ppm N ammonium sulfate).
18. Sodium hydroxide solution, 20% (Stock).
19. Sodium potassium tartrate solution, 20% (Stock).
20. 0.5 M Buffer Solution (Stock) (*see* **Note 5**).
21. Working Buffer, a solution made up of 20% sodium potassium tartrate (250 mL), 0.5 M stock buffer solution (200 mL), and 20% sodium hydroxide (250 mL); the volume is then made up with deionized water to 1 L.
22. Sulfuric acid (0.135 M)/sodium chloride (1.71 M) solution.
23. Sodium salicylate (0.94 M)/sodium nitroprusside (0.00026 M) solution.
24. Sodium hypochlorite solution, 0.525%.

25. Internal rice reference standard, 99G (rice flour).
26. 30% Brij35 (AR grade).

3 Method Development

3.1 Sample Preparation

This section describes method development for brown rice samples. However, the approach should be applicable to milled rice as well.

1. Dehull the paddy rice samples using the rice dehuller to obtain brown rice.
2. Grind and homogenize the brown rice grains (2–3 g) using the Retsch MM400 mixer mill to get very fine flour (*see* **Note 6**).
3. Prior to NIR scanning and wet chemical analysis, the samples are oven-dried (80 °C, 30 min), and then kept in a desiccator cabinet until ready to be loaded into NIR sample cups or weighed for chemical analysis.

3.2 NIRS Scanning of Brown Rice Flour Samples

Since NIRS is a nondestructive method, scanning is usually done prior to destructive wet chemical analysis, especially for samples with limited available grains. The method described here is specific for the NIRSystems 6500. For more details, refer to the manufacturer's manual.

1. Power on and warm up the (NIRSystems6500) spectrophotometer (1 h).
2. Perform the following instrument diagnostics according to instructions from the manufacturer [18]:
   (a) *Instrument Response* refers to the absolute reflectance from the ceramic reference tile (supplied by the manufacturer).The maximum value for absolute reflectance in the NIR Systems 6500 is 65,535. The auto-gain function of the instrument allows the reflectance at the visible range to be above 50,000 and that of the NIR range to be above 35,000.The dark period values should be less than 1600 (*see* **Note 7**).
   (b) *Wavelength Accuracy* is measured using internal standards: didymium (for the visible range) and polystyrene (for the NIR). In aligning the wavelengths, the error should be within ±0.30 nm; i.e., any peak found within the 0.30 nm margin of the wavelength is acceptable. The reported wavelength location and suggested or nominal locations for the wavelengths should not have day-to-day deviations of more than 0.10 nm. Wavelength accuracy can be tested by comparing the *current* *K* and Phi values, and its associated error value. There is no need to change the *K* and Phi values if the current error does not exceed twice the suggested error (*see* **Note 8**).

(c) *NIR Repeatability* indicates the magnitude of deviation (noise) in optical data at each wavelength. Testing for repeatability requires the use of the ceramic tile, scanned as a reference, as an unknown sample, and then as a reference. Repeatability is measured using the statistic average root mean square corrected for bias (RMS(C)), calculated in the WinISI software. The RMS(C) of five scans should be <20 in a room with stable temperature (*see* **Note 9**).

(d) Scan "Check cell" before measurements are made on test samples to evaluate overall day to day performance of the instrument (*see* **Note 10**). This step may also be included within the sample sequence. The spectra of the "Check cell" are stored in the Chk.NIR file (*see* **Note 11**).

3. Record details in the NIR sample logbook.

(a) List the sample codes beside the corresponding sample cup numbers. Sample cups are numbered 1–100 on the disposable backs or covers.

(b) List down other relevant information such as the sample designation, the source of sample, the harvest year and season, the date of scanning, and the filename where the NIR spectral data will be stored. Also indicate the insert used (*see* **Note 12**).

4. Pack the flour into sample cups

Standard small ring cups (IH-0307, i.d. = 36 mm, depth = 9 mm) have to be uniformly and completely filled with flour samples. However, if not enough sample is available, the ring cups can be fitted with black ring inserts to reduce the volume of the cups that have to be filled with smaller amounts of sample (*see* **Note 12**). It is important to decide from the start what type or size of cups to use based on expected amount of samples available for the entire application or intended use of the procedure.

(a) Arrange the disposable white back covers on piles according to number (1–100, 1 on top).

(b) Clean the sample cup and the insert using a soft camel hair brush and/or a small vacuum cleaner. Additional cleaning may be done by wiping the quartz glass with lint-free tissue.

(c) Mix the flour sample in the envelope thoroughly using the stainless steel spatula.

(d) Scoop portions from the envelope with the spatula and pour into the sample cup (with or without the insert). Overfill the cup a little to ensure that there are no empty air spaces inside.

Fig. 2 NIRSystems 6500

(e) Use the white disposable back to cover the cup. Make sure to use the correct sample cup numbers corresponding to the sample codes (refer to the NIR sample logbook).

(f) Lightly push the cover to secure it and to compress the sample so that there will be no loose or empty spaces inside.

(g) Vacuum or wipe again the outer quartz window of the sample cup to ensure it is clean or free of dust.

5. Collect NIR spectra of samples

(a) Load the sample cup into the Autocup magazine connected to the Spinning Sample Cup module (for reflectance mode) of the NIRSystems 6500 (Fig. 2).

(b) Create a file where all sample spectra are to be stored, for example, "BRF-DS15.NIR" for brown rice flour sampled from 2015 dry season field plots. "NIR" is the file extension for files containing spectra.

(c) Scan the sample over the entire visible-NIR range (λ = 400–2500 nm) in 2 nm increments. Each sample spectrum recorded is the average of 32 scans.

(d) Check the sample cups for any air spaces after scanning. Note the sample number and repeat **step 4**.

6. Cleaning sample cups
   (a) Collect the sample and return to the designated envelope.
   (b) Clean the sample cup, white disposable cover, and insert (if used) by using a vacuum, soft camel hair brush, and lint-free tissue.

3.3 Reference Method (Wet Chemical Analysis)

The GQNSL performs protein content determination by Kjeldahl method developed and modified based on the procedure reported by Yoshida et al. [38].

1. Convert protein N to NH_4^+ by chemical digestion
   (a) Weigh 50.0 ± 0.1 mg of oven-dried sample in a 2.5 × 20 cm pre-calibrated digestion tube.
   (b) Add 1 scoop (approximately 1.0–1.3 g) of potassium sulfate/selenium mixture into the digestion tube.
   (c) Add 2.0 mL concentrated sulfuric acid from a calibrated dispenser and mix using a Vortex mixer to completely wet the brown rice flour material.
   (d) Place the tube in the digestion block preheated to 380 ± 10 °C to digest the sample (1 h), shaking the tube after 30 min to ensure complete digestion (to wash down flour that may cling to the side of the tube into the digestion mixture).
   (e) Remove the tube from the digestor and let it stand to cool at room temperature before adding deionized water to final volume.
   (f) Transfer the diluted digest into a 20 mL test tube and stand overnight to allow for sedimentation before analysis.
2. Measure the NH_4^+ concentration based on its colorimetric reaction (an emerald green complex) with sodium salicylate in the presence of sodium hypochlorite and nitroprusside (modified Berthelot reaction [39], *see* **Note 13**).
   (a) In the software of the Skalar San++, create the "Sample Sequence Run," which includes:
      i. Primer, the standard with the highest concentration (150 ppm).
      ii. Wash Solution (deionized water).
      iii. Calibration curve standards: blank, 10 ppm, 30 ppm, 50 ppm, 70 ppm, 90 ppm, 120 ppm, 150 ppm $(NH_4)_2SO_4$.
      iv. Unknown samples in batches of ten.
      v. The 90-ppm standard, treated as quality check sample, is analyzed for every ten unknown samples.
      vi. Calibration standards as unknowns.
      vii. Wash solution.

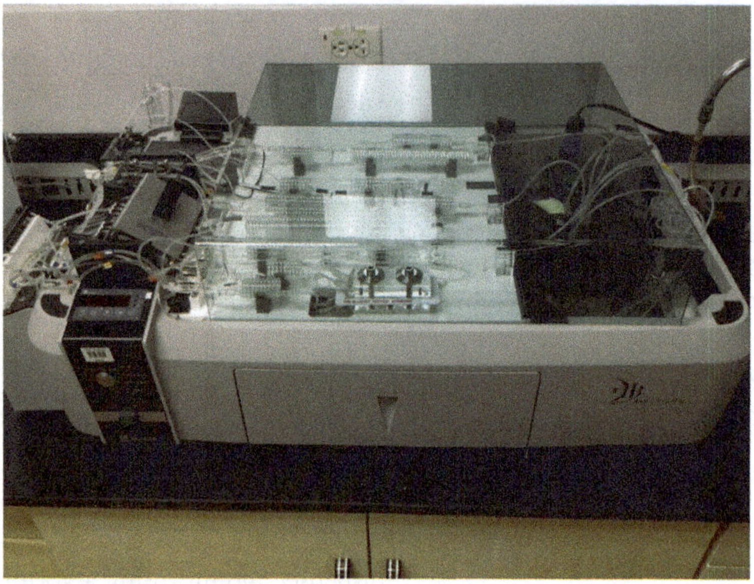

Fig. 3 Skalar San++ Continuous Flow Analyzer

(b) Prepare the reagents
   i. Sodium salicylate/sodium nitroprusside solution (*see* **Note 14**).
   ii. Sodium hypochlorite solution (*see* **Note 15**).

(c) Pour the wash solution, the standards, and the diluted sample digests into autosampler cups and arrange according to the "Sample Sequence Run" order.

(d) Start the sequence run of the autoanalyzer (Fig. 3) with the following settings:
   i. Absorbance is measured at $\lambda = 660$ nm;
   ii. Sample flowrate is set at 0.16 mL/min;
   iii. Sampler settings at 30 s (sample time; the time allowed for the sample to pass through the chemistry manifold), 60 s (wash time) and 6 s (air time);
   iv. Peak detection based on peak height and using a Calibration I Order ISO 8466.

(e) Construct a standard curve based on the absorbance values of the calibration standards.

(f) Calculate the % Kjeldahl N based on the standard curve, which is based on the Beer-Lambert's Law [40].

(g) Protein content (%) is calculated by multiplying % Kjeldahl N by 5.95 [40].

(h) The expanded uncertainty for percent Kjeldahl N is ±0.072 with coverage factor = 2, which gives a level of confidence of approximately 95%.

3.4 Initial NIRS Calibration Development

1. Sample Selection for the Calibration Set

   There are no hard and fast rules to determine the optimum number of selected samples to include in the initial or "starter" calibration set. The calibration set should be composed of samples that are sufficient to provide all of the likely sources of variance (Infrasoft International 1995). A NIRS method development procedure can make use of large sets of data already available, such as from a database in a rice quality evaluation service laboratory or from a screening experiment. There are three different approaches to select the calibration set. Choose one approach depending on the availability of samples and reference values:

   (a) Selection from a large data pool based on reference data

   A large database can provide valuable source of all kinds of variation. Taking into consideration the range and distribution of reference data, samples with uniform distribution across the expected range may be assembled to compose the calibration set. A large number of samples with nonuniform distribution in the calibration set may, however, lead to poor prediction models due to increased risk of involving noise in the calibration and bias toward the mean or toward the range with dense population [41]. Representative and evenly distributed samples should be selected to cover the entire range of component of interest.

   (b) Selection from the spectra files without available data from the reference method

   It is possible that spectra of a large set of samples are available but reference data and the range of variability are still unknown. Collection of NIR spectra of all samples prior to wet chemical analysis proves to be a more simple and cost-saving approach since only samples with selected representative spectra that are unique are included in the calibration set and are sent to the laboratory for wet chemical analysis.

   (c) Selection from a large sample spectra file with reference data

   When there is no prior knowledge about the diversity and composition of all available samples, NIR scanning and wet analysis are performed simultaneously on all samples even before selection is done. This is usually the case when a laboratory starts to explore the applicability of an NIR method as an alternative for routine wet chemical analysis. Although the process may be labor-intensive at the start, this is considered as the best approach since the SELECT algorithm in WinISI fine-tunes the scores to maximize the differences among the spectra as well as the reference value.

2. Processing of NIRS spectra from the calibration set

(a) Select the wavelength segment of the spectra to be used for creating scores

For the protein calibration, the factory default segment or the normal recommended wavelength for the instrument in the NIR range (1108–2492, 8) is used. This means that the wavelengths used for creating scores by multivariate statistical algorithm are every 8 nm from 1100 to 2500 nm (NIR region) where the first (1100 nm) and last (2500 nm) wavelengths are not included.

(b) Mathematical treatment of spectral data

The "SNV and Detrend" option is applied on all spectra for scatter correction. The SNV (Standard Normal Variate) function scales each spectrum to have a standard deviation of 1.0 to help reduce particle size effects. "Detrend" removes the linear and quadratic curvature of each spectrum.

A good exploratory math treatment is a (1, 4, 4, 1) where:

1 = first derivative

4 = gap over which the derivative is calculated

4 = smoothing of points

1 = no second smoothing

An example of "raw" spectra of 3148 samples is shown in Fig. 4a and the resulting normalized, or mathematically treated, spectra in Fig. 4b. Reducing the particle size effects through normalizing the data set has visibly reduced the noise (Fig. 4b).

(c) Create the score file from the spectra file, done using the WinISI software

A "score file" is created wherein the mathematically treated spectral data are first reduced to independent sources of variation and converted to scores by principal component analysis (PCA). This option is important in understanding the structure and distribution of a population of samples based on spectra; no reference values are used. However, it places a lot of emphasis on particle size, moisture [18], and other variables that influence the spectral pattern more than the sample chemical composition.

(d) Use SELECT algorithm for pattern recognition

The method uses an algorithm called SELECT in the chemometrics software. This algorithm identifies patterns in a group of normalized or math-treated spectra that

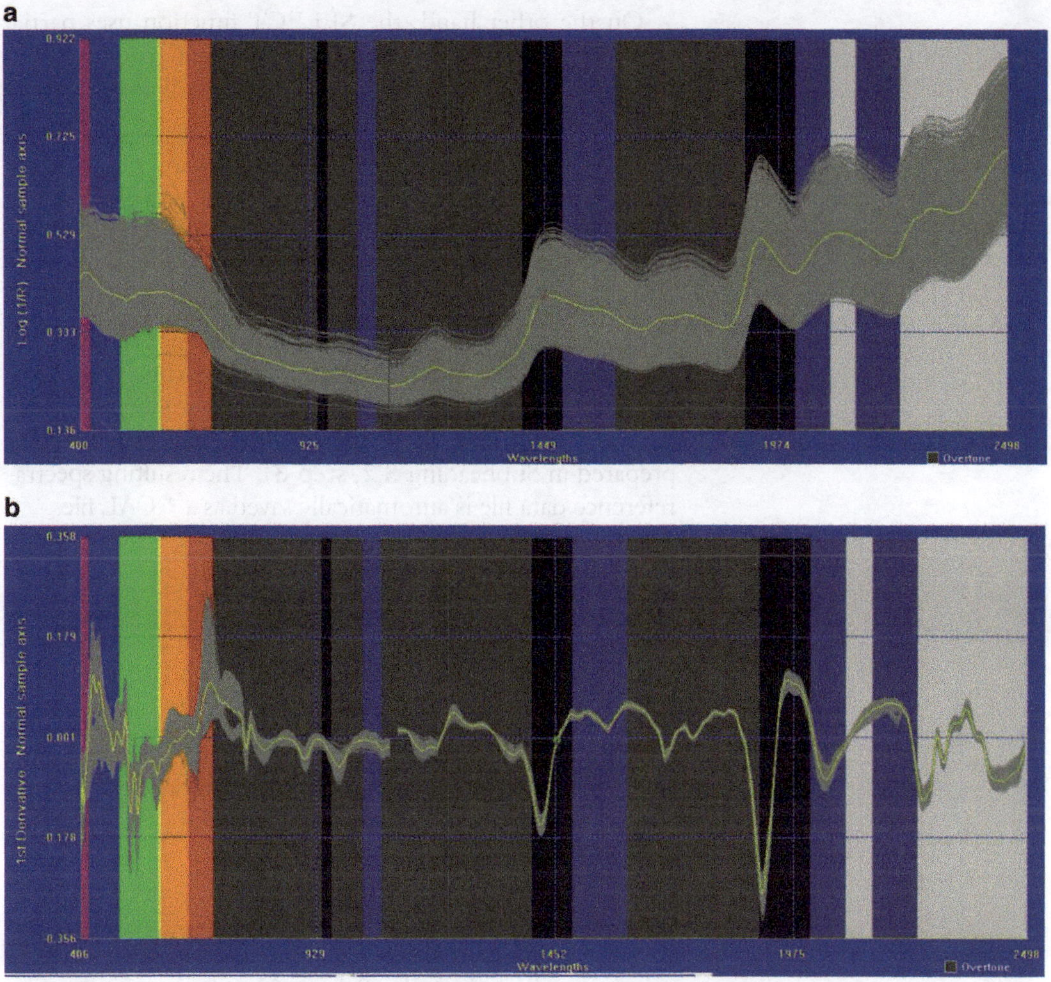

Fig. 4 NIR spectra of 3148 brown rice flour samples showing the average spectrum (in yellow) for (**a**) raw data and (**b**) normalized data

contribute most to the variation among the spectra. Each sample score is ranked by its distance from the center of a spectra file (global H or GH) and by its distance from every other sample score in the file (neighborhood H or NH). Redundant samples or samples whose scores are close to selected neighbors (NH value) are excluded in the calibration set. The cutoff NH is set according to the spectral range of the data (0.6–1.0 for monochromators collecting data at λ = 1100–2498 nm). If too many samples are automatically selected by the SELECT algorithm in WinISI based on the cutoff NH value, the number can be reduced to whatever is deemed enough by WinISI to represent the entire population.

On the other hand, the SELECT function uses partial least squares (PLS) to identify patterns in a group of spectra that contribute most to the variation within spectra-reference data.

3. Calibration development in the WinISI software

   The following steps represent a good place to start to make an initial calibration:

   (a) Create the calibration file

      Use the *.NIR file or the file containing the spectra of selected calibration samples (from **step 1**). Encode or import the reference data into the file, taking care to properly match the reference data (PC by Kjeldahl N, Subheading 3.3) with the correct sample number or code (Refer to logbook prepared in Subheading 3.2, **step 3**). The resulting spectra-reference data file is automatically saved as a *.CAL file.

   (b) Wavelength selection

      There are two wavelength segments that can be considered for the calibration:

      i. Visible range: 408–1092, 8

      ii. NIR range: 1108–2492, 8

   (c) Mathematical pretreatment of spectral data

      The same scatter correction and math treatment method applied to create a score file from the spectra file should be used (as in **step 2b**).

   (d) Open/Create the required files

      i. Input files

         - *Calibration file*: Select the file containing the spectra and reference data of samples in calibration set from **step 3a** (*.CAL).
         - *Repeatability file*: Select the file containing spectra of the "Check C ell" repeatedly scanned on different days and at specific intervals during routine analysis of test samples (e.g., "Chk.NIR" file from Subheading 3.2, **step 2d**).

      ii. Output files

         - *Equation file*: Create the resulting filename where the calibration equation or the prediction model will be saved. This file usually has the same name as calibration file (sample set) but with the default extension (*.EQA).
         - *Loadings file*: Create the resulting filename that will contain the loadings or the patterns that explain the spectral variation in the calibration file. This file usually has the same name as the calibration file but with the default extension (*.LOD).

iii. Choose and perform the regression method

The recommended "Regression Method" is modified PLS, which reduces the spectral data to a few combinations of absorptions that account for both the spectral information and reference values. The modified PLS procedure algorithm [42] copes more effectively with non-analyte interference and provides an estimate of the interference spectrum that can be used to correct the measured spectrum before determination of analyte concentration.

4. Sample Selection for the Validation Set

   Validation of calibration models is normally performed on an independent set of samples or samples not found in the calibration set. However, in the course of calibration development, simultaneous validation is also performed using internal samples within the calibration set (cross-validation).

   (a) External or independent validation set

   To demonstrate linearity, it is likewise required that samples in the validation set are distributed across the specified range of interest. From among the redundant samples excluded from the calibration set in **step 1**, around 10% of the number of samples in calibration set may be chosen either randomly (with uniform distribution across the protein content range) or by using the SELECT algorithm on the score file of external samples.

   (b) Internal validation set

   The internal validation set is composed of samples in the calibration set where each sample is cross-validated against the resulting calibration model. The algorithm usually performs this automatically during the calibration procedure to determine the optimum number of terms (or factors) that will result in the least possible SECV.

5. Validation

   NIRS validation presents evidence for the accuracy and robustness of the calibration model in predicting the composition of samples for which the calibration was made. The range should be confirmed by use of a suitable validation set (**step 2**) which matches in extent the proposed range for the intended use of the procedure. The NIRS predicted analyte concentration of samples in the validation set should be compared with reference values.

   (a) Create the validation file

   Use the spectra file (*.NIR) of the validation set containing the selected independent samples or samples not included in the calibration set (selected from **step 2a**). Send the samples for laboratory analysis if reference data are still unknown (or if samples are selected from a score file). Enter or import the reference data into the spectra file (*.NIR), taking care to properly match the reference data (PC by Kjeldahl N method) with the correct sample number or code. The resulting file containing the spectra and reference data is likewise saved with the default extension (*.CAL) but use another character in the file name to differentiate it from the calibration set (ex. *v.CAL).

(b) Use the output calibration equation file (*.EQA developed from **step 3dii**) to predict the PC analyte concentration of samples in the validation set (*v.CAL).

(c) Compare NIR predicted and reference values of samples in validation file and report validation statistics (SEP, R2, range and standard deviation of analyte concentration).

(d) Rate the performance of the calibration model by the ratio of the standard deviation of analyte concentration in the validation set to the SEP.

4 Application of the NIRS Method Development Pipeline for Protein Content Estimation in Brown Rice

A diverse set of rice samples ($N = 3148$) belonging to indica, tropical and temperate japonica, and aus varietal groups were grown and harvested at IRRI's experiment station during the 2013 dry season. These samples were selected based on the diversity of their cooking and texture characteristics (e.g., AC, GT, GC) as well as their physical characteristics (e.g., chalkiness, grain length, grain width, grain shape). The paddy grains were dehulled to obtain brown rice and were subjected to quantification of PC by Kjeldahl N method. The brown rice was ground using the Retsch MM400 mixer mill to get very fine flour (2–3 g). The flour was oven-dried (80 °C, 30 min) then packed into the NIRS sample cups. NIR spectra were obtained using the NIRSystems 6500, scanning over the entire visible-NIR range ($\lambda = 400$–2500 nm in 2 nm increments).

Results showed that the range of PC was from 5.36 to 13.98% ($\mu = 9.50\%$). The frequency distribution of PC, more or less, follows the bell shape of a normal distribution and shows that majority of the samples have PC at around 8.1–9.0% (Fig. 5), consistent with reported literature [34].

Determination of the calibration set was then conducted using the three approaches specified (Subheading 3.4, **step 1**).

In the first approach (selection based on reference data), 10% of the population was randomly selected from within each bin in the frequency distribution (Fig. 5) such that each bin, except for the tails, was equally represented (Fig. 6).

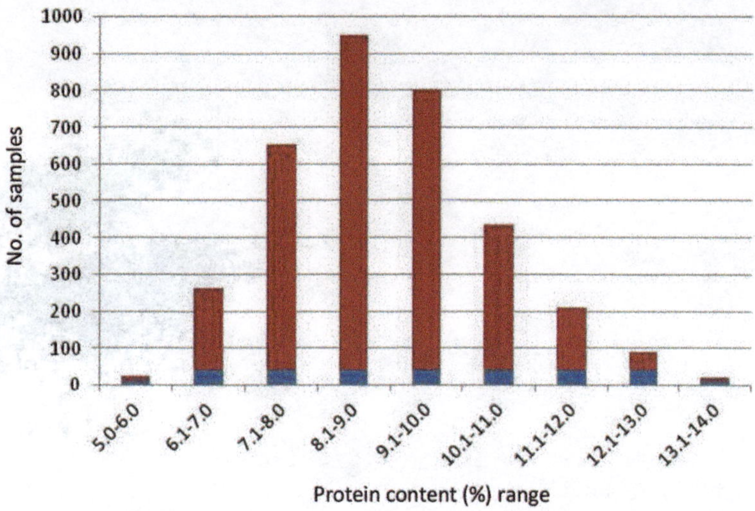

Fig. 5 Histogram of PC in a diverse set of 3148 brown rice samples. The blue shaded portions represent the number of samples selected from each range for the calibration set

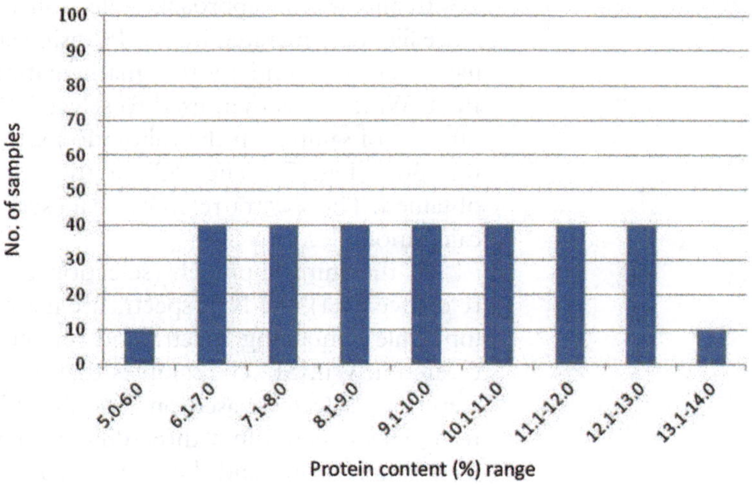

Fig. 6 Uniformly distributed samples selected from each PC range bin for the calibration set

For the population of 3148 samples, around 300 samples were selected for the calibration set such that the second to the eighth range bins were represented by 40 samples; the first and last bins were each represented by 10 samples since there were fewer samples in those bins (Fig. 6). The NIR spectra of these samples were obtained and subjected to further mathematical treatment. The spectra-reference data sets were used to create a calibration file.

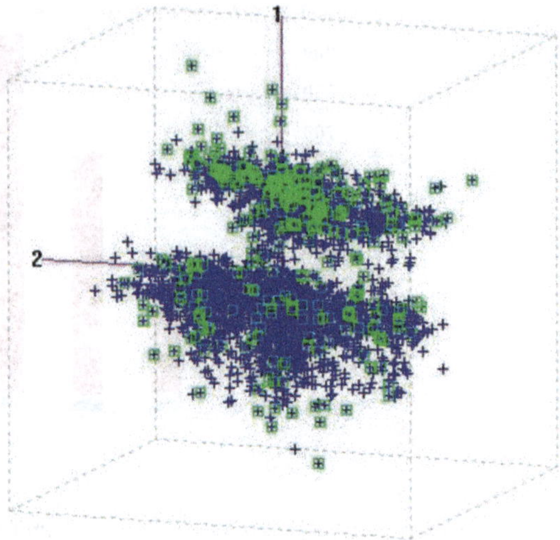

Fig. 7 Three-dimensional scores plot of 3148 sample spectra (blue) generated by PCA showing distances from the center of the spectral hypersphere and the 301 representative samples (green) selected by WinISI for the calibration set

In the second approach (selection based on NIR spectra), a score file was generated by WinISI using all available spectra (which have already undergone mathematical treatment) through PCA. WinISI was then used to select 10% of the population. The number of samples in the calibration set (n_{CS}), selected by WinISI was 301 (Fig. 7). The PCs of the selected samples were then obtained. The spectra-reference data sets were then combined in a calibration file.

In the third approach (selection based on both spectra and reference data), the NIR spectra are mathematically treated and an input file containing spectra and reference data (N = 3148) was created in WinISI. The samples that went into the calibration file were then selected based on modified PLS on the entire population. This resulted in a three-dimensional scores plot that visualized sample scores and the samples (n_{CS} = 307) that were selected for inclusion in the calibration set (Fig. 8).

Calibration models were developed for each approach, the number of terms (i.e., wavelengths used) for each equation is found in Table 1. The coefficient of determination (R^2) for the calibration models ranged from 0.951 to 0.970, indicating that the models were highly linear.

The standard error of calibration (SEC) ranged from 0.303 to 0.373. As the number of wavelengths (terms) used in the regression model increased, the SEC and the R^2 decreased. On the other hand, samples composing ~10% of the calibration set were randomly selected to be part of the independent validation set (Table 1). For the first approach, the samples were selected randomly based on the PC frequency distribution (Fig. 6). For the

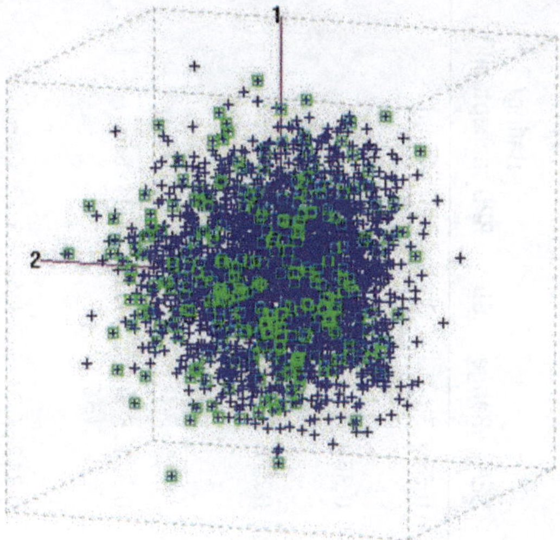

Fig. 8 Three-dimensional scores plot of 3148 spectra-reference data (blue) generated by PLS showing distances from the center of the spectral hypersphere and the 307 samples (green) selected by WinISI for the calibration set to represent the whole population

second and third approaches, the samples were selected by WinISI. The spectral data of the independent validation set were fed into each of the calibration models to predict PC. Results showed that the R^2 ranged from 0.919 to 0.977, indicating that the models predicted PC with high accuracy; while the SEP ranged from 0.312 to 0.376.

To select the best calibration model for predicting PC routinely, the balance between SEC and SEP must be achieved (or at least be within 20% of each other). The data (Table 1) shows that the SEC and the SEP ranges overlap and are well within 20% of each other, hence it was difficult to select the best calibration model. Therefore, the RPD values were reviewed. The highest RPD values were from the first approach where calibration set defined based on reference data (Table 1). According to Baillères et al. [43] and Agelet and Hurburgh [44], at this RPD, the calibration model can be used for quality control purposes. The other models were good enough for screening purposes, according to the standards previously reported [44, 43]. Hence, for predicting PC by NIR, it is recommended to use the first calibration model (based on the calibration set with uniformly distributed PC reference data) (*see* **Note 16**).

It is recommended that the calibration models developed for brown rice in the population ($N = 3148$) be used for routine PC prediction and to expand the NIRS by including outlier samples (both spectral and reference data) to the calibration model.

Table 1
Validation of calibration models developed for protein in brown rice flour

Calibration sample set				Calibration models developed					Validation test with independent samples						
Selection method	n_{CS}	Range	Mean	Filename	No. of terms	R^2	SEC	SECV	No. of samples	R^2	SEP	Range	SD	RPD	Rating/ acceptability
1. Uniform distribution	300	5.36–13.98	9.50	C123s-1.eqa	8	0.970	0.373	0.401	30	0.977	0.356	5.64–13.86	2.35	6.6	Good
									2848	0.919	0.359	5.64–13.86	1.24	3.45	Fair
2a. Spectra scores (PCA) Sample cutoff = 300	301	5.72–13.74	9.03	C123s-2.eqa	9	0.954	0.338	0.366	33	0.938	0.376	5.60–11.66	1.46	3.88	Fair
2b. Spectra scores (PCA) NH cutoff = 0.6	1245	5.36–13.98	9.03	C123s-22.eqa	10	0.951	0.323	0.334	1903	0.94	0.321	5.70–13.74	1.305	4.06	Fair
3a. Spectra scores-Reference data Sample cutoff = 300	307	5.70–13.98	9.12	C123s-3.eqa	11	0.967	0.303	0.355	30	0.966	0.312	5.74–13.68	1.63	5.22	Good
3b. Spectra scores-Reference data NH cutoff = 0.6	299	5.73–13.51	9.08	C123s-32.eqa	11	0.959	0.322	0.369	2849	0.937	0.343	5.35–13.98	1.34	3.91	Fair

Where: SEC = Standard Error of Calibration, SECV = Standard Error of Cross-validation, SEP = Standard Error of Prediction, an indicator of overfitting of the model such that the minimum SEP value indicates the maximum number of terms that can be put in the model without overfitting, SD = Standard Deviation of the range of PC (data based on the reference method) in the validation set, RPD = Ratio of Performance to Deviation, the ratio of standard deviation of the PC data in the validation set to the SEP (SD/SEP). Ratings of RPD are as follows: Excellent (>10); Very good (7–10); Good (5–7); Fair (3–5); Not satisfactory (<3)

Note: Range and SD are derived from data derived from the reference method

5 Notes

1. The diversity and selection of samples used to build the database is crucial to the robustness and reliability of the calibration or the prediction model for the substance or trait being analyzed [45]. The samples ideally cover the whole range of known variations in concentration, rice genotype, particle size of ground material, growing season, location, and other expected and unexpected sources of variation. Subsets may also be maintained for rice samples prepared and analyzed in different forms (e.g., whole grain vs. flour, or brown vs. milled) and those that used other NIRS data collection mode (e.g., reflectance vs. transmittance mode).
2. The NIR spectrophotometer is a means for collecting and measuring the intensity of the transmitted or reflected light by the sample in the near-infrared region ($\lambda \sim 780–2500$ nm) that are mainly due to overtone and combination bands primarily of O–H, C–H, N–H, and C–O groups (Infrasoft International 1995). Essentially, NIRS takes a "fingerprint" or spectral pattern of a sample based on its organic constituents. Most NIR instruments also include the visible region ($\lambda = 400–780$ nm) that permit visible color difference analysis.
3. An appropriate chemometrics procedure is required to be able to extract meaningful information from the NIR spectra. Analyte concentrations are not directly determined from the NIR spectra because the spectra at the NIR range feature overtone and represent combination bands primarily of O–H, C–H, N–H, and C–O groups [10, 11]. Thus, an equation or a mathematical model relating spectral data (treated as the independent variables) to wet chemistry-based laboratory values (the response variables) is necessary. This calibration process normally uses multivariate statistical procedures and computer algorithms on a large number of sample spectra with their corresponding reference data to arrive at an appropriate calibration model for predicting analyte concentrations. It is important for an analyst who uses NIR to understand the performance and suitability of the procedure. The calibration model then needs to be tested or validated either on independent samples or cross-validated on randomly selected samples from the calibration set.
4. Mix well each time before use as the mixture tends to be heterogeneous. Use this mixture within 1 year.
5. Dissolve 134 g of sodium phosphate dibasic heptahydrate in deionized water (800 mL). Add sodium hydroxide pellets (20 g), and then add deionized water to make 1 L volume. Mix thoroughly. Use this solution within 3 months.

6. Store the flour in properly labeled coin envelopes at −20 °C until ready for wet chemical analysis and NIR scanning.
7. These values may vary from one sample attachment module to another but these values are under the control of the instrument manufacturer. If the instrument response is not within the specified range, it may be time to replace lamp.
8. If current error is equal to or more than twice the suggested error, change *K* and Phi values. The *K* and the Phi values are coefficients for linearizing the spectra and for aligning the wavelengths.
9. If the average RMS(C) is not within the specified range, room conditions (i.e., temperature and humidity) may not be stable, record and monitor room conditions to establish their specific ranges where the average RMS(C) is less than 20.
10. The current analysis is compared to previous values stored in the ISISTDSF.CHK file. The "*T*" value is computed based on how far the current values are from the mean of the previous values. If check cell analysis fails (e.g., $T > 3$), an error message will appear to warn that one or more analyses have changed. This could be due to one of the following:
    (a) a module has been changed;
    (b) a repair has been done;
    (c) the check cell may be out of date.

    If it is due to the any of the first two, there is a need to re-standardize the NIRS. If check cell is out of date, replace the current check cell with a new one.
11. The Chk.NIR file is helpful in the calibration procedure to minimize the effects of variation in temperature and instrument optics and to make the calibration model insensitive to these changes during analysis.
12. Different sizes of inserts can be used if the amount of flour sample is not enough to fill the standard sample cup. For flour samples
    (a) >2 g: use sample cup with no insert
    (b) >500 mg–2 g: use sample cup with size 1 insert (18 mm hole, SI1)
    (c) >200–500 mg: use sample cup with size 2 insert* (12 mm hole, SI2)
    (d) >40–200 mg: use sample cup with size 3 insert (6 mm hole, SI3)

    *This is the recommended size of insert to use for scanning of brown rice flour prepared from 2 to 3 g paddy rice.

    Separate files have to be created for samples that were characterized using samples cups with different inserts. The amount of sample available and the size of the "window" of the insert will affect the resulting NIR spectrum. Therefore separate calibrations are performed for samples characterized in samples cups equipped with different inserts.

13. For the NIRS calibration for protein, the accepted and recognized Kjeldahl method for the wet chemical analysis of protein in brown rice flour samples is reported by Bietz [46].
14. Dissolve of sodium salicylate (150 g) in about 600 mL of deionized water and add 0.3 g of sodium nitroprusside to the solution. Filter the solution through a fast filter paper into a 1 L volumetric flask and make the volume to 1 L. Add about 1 mL of 30% Brij-35 and mix thoroughly. Use this solution within 1 month.
15. Dilute 10.0 mL of any commercial bleach solution (containing at least 5.25% available chlorine) to 100 mL with deionized water. Add two drops of 30% Brij-35 and mix thoroughly. Prepare fresh daily.
16. A large number of samples with nonuniform distribution (i.e., the PC bins are not equally represented) in the calibration set may lead to poor prediction models due to increased risk of involving noise in the calibration and bias toward the mean or toward the range with dense population [41]. Representative and evenly-distributed samples should be selected to cover the entire range of component of interest, such as PC in this case.

Acknowledgments

The authors thank the following GQNSL staff for the technical assistance in generating the protein data: Jose Rosales, Edgar Amoloza, Leonel Borebor, Anna Carissa Basilio, and Ruben Chavez. This work has been supported under the CGIAR thematic area Global Rice Agri-Food System CRP, RICE, Stress-Tolerant Rice for Africa and South Asia (STRASA) Phase III, and Australian Centre for International Agricultural Research (Project ID CIM/2016/046) funding.

References

1. Cuevas RP, Pede VO, McKinley J, Velarde O, Demont M (2016) Rice grain quality and consumer preferences: a case study of two rural towns in the Philippines. PLoS One 11(3):e0150345
2. Cuevas RP, Daygon VD, Corpuz HM, Reinke RF, Waters DLE, Fitzgerald MA (2010) Melting the secrets of gelatinisation temperature in rice. Func Plant Biol 37:439–447
3. Cagampang GB, Perez CM, Juliano BO (1973) A gel consistency test for eating quality of rice. J Sci Food Agric 24:1589–1594
4. Juliano BO, Perez CM, Resurreccion AP (2009) Apparent amylose content and gelatini-

zation temperature types of Philippine rice accessions in the IRRI Gene Bank. Phil Agric Sci 92(1):106–109

5. Tanaka J, Hayashi T, Iwata H (2016) A practical, rapid generation-advancement system for rice breeding using simplified biotron breeding system. Breed Sci 66(4):542–551
6. Janwan M, Sreewongchai T, Sripichitt P (2013) Rice breeding for high yield by advanced single seed descent method of selection. J Plant Sci 8(1):24–30
7. Spindel J, Begum H, Akdemir D, Virk P, Collard BCY, Redoña E, Atlin G, Jannink J-L, McCouch SR (2015) Genomic selection and association mapping in rice (Oryza sativa): effect of trait genetic architecture, training population composition, marker number and statistical model on accuracy of rice genomic selection in elite, tropical rice breeding lines. PLoS Genet 11(2):e1004982
8. Varshney RK, Terauchi R, McCouch SR (2014) Harvesting the promising fruits of genomics: applying genome sequencing technologies to crop breeding. PLoS Biol 12(6):e1001883
9. Rahim HA, Ibrahim S (2013) Near infrared spectroscopy measurement: the assessement of amylose content in rice grain. Sensors Transducers 156:251–258
10. Bokobza L (1998) Near infrared spectroscopy. J Near Infrared Spectrosc 6:3–17
11. Pasquini C (2003) Near infrared spectroscopy: fundamentals, practical aspects and analytical applications. J Brazilian Chem Soc 14(2): 198–219
12. Agelet LE, Hurburgh CR Jr (2014) Limitations and current applications of Near Infrared Spectroscopy for single seed analysis. Talanta 121:288–299
13. Roggo Y, Chalus P, Maurer L, Lema-Martinez C, Edmond A, Jent N (2007) A review of near infrared spectroscopy and chemometrics in pharmaceutical technologies. J Pharma Biomed Anal 44(3):683–700
14. Luypaert J, Massart DL, Heyden YV (2007) Near-infrared spectroscopy applications in pharmaceutical analysis. Talanta 72(3): 865–883
15. Butz P, Hofmann C, Tauscher B (2005) Recent developments in noninvasive techniques for fresh fruit and vegetable internal quality analysis. J Food Sci 70:R131–R141
16. Givens DI, De Boever JL, Deaville ER (1997) The principles, practices and some future applications of near infrared spectroscopy for predicting the nutritive value of foods for animals and humans. Nutr Res Rev 10:83–114
17. Martens H, Martens M (2001) What makes NIR data so information-rich? In: Martens H, Martens M (eds) Multivariate analysis of quality: an introduction. Wiley, Chichester, pp 415–417
18. Infrasoft International (1995) NIRS 2 Version 3.10 Routine Operation and Calibration Development Software for Near Infrared Instruments, State College PA, USA
19. Villareal CP, Dela Cruz NM, Juliano BO (1994) Rice amylose analysis by near-infared transmittance spectroscopy. Cereal Chem 71(3):292–296
20. Delwiche SR, Bean MM, Miller RE, Webb BD, Williams PC (1995) Apparent amylose content of milled rice by near-infrared reflectance spectroscopy. Cereal Chem 72(2): 182–187
21. Delwiche SR, McKenzie KS, Webb BD (1996) Quality characteristics in rice by near-infrared reflectance analysis of whole-grain milled samples. Cereal Chem 73(2):257–263
22. Rahim HA, Ibrahim S (2013) Using near-infrared spectroscopy to investigate the amylose content in rice. Appl Mech Mater 239–240:163–166
23. Lee H-S, Choi Y-M, Lee Y-Y, Ma K-H, Gwag J-G, Lee JR, Yoon Y-T, Cho Y-G, Lee S-Y (2014) Selecting high amylose rice germplasm combined with NIR spectroscopy at the RDA genebank conserved. Plant Breed Biotechnol 2(4):380–385
24. Bao J-S, Cai YZ, Corke H (2001) Prediction of rice starch quality parameters by near-infrared reflectance spectroscopy. J Food Sci 66(7): 936–939
25. Shu Q, Wu D, Xia Y, Gao M, McClung AM (1999) Apparent amylose content of rice by near infrared reflectance analysis of ground milled samples. Chinese J Rice Sci 13(3): 189–192
26. Wu JG, Shi CH (2007) Calibration model optimization for rice cooking characteristics by near infrared reflectance spectroscopy (NIRS). Food Chem 103(3):1054–1061
27. Rash J, Meullenet J-FC (2010) Apparent amylose content prediction using near infrared spectroscopy of individual and bulk rice kernels. AAES Res Ser:312–321
28. Martin M, Fitzgerald MA (2002) Proteins in rice grains influence cooking properties. J Cereal Sci 36(3):285–294
29. Juliano BO (2003) Rice chemistry and quality. Philippine Rice Research Institute, Munoz, p 492
30. Gomez KA (1979) Effect of environment on protein and amylose content of rice. In:

Workshop on chemical aspects of rice grain quality. Rice Grain Res, pp 59–68

31. Shi C, Zhu J, Yang X, Yu Y, Wu J (1999) Genetic analysis for protein content in indica rice. Euphytica 107:135–140
32. Osborne BG (2006) Applications of near infrared spectroscopy in quality screening of early-generation material in cereal breeding programmes. J Near Infrared Spectrosc 14: 93–101
33. Williams PC (1975) Application of near infrared reflectance spectroscopy to analysis of cereal grains and oilseeds. Cereal Chem 52: 561–576
34. Kawamura S, Natsuga M, Itoh K (1999) Determination of undried rough rice constituent content using near-infrared transmission spectroscopy. Trans Am Soc Agric Eng 42(3): 813–818
35. Sohn M, Barton FE, McClung AM, Champagne ET (2004) Near-infrared spectroscopy for determination of protein and amylose in rice flour through use of derivatives. Cereal Chem 81(3):341–344
36. Nørgaard L, Lagerholm M, Westerhaus MO (2013) Artificial neural networks and near infrared spectroscopy: a case study on protein content in whole wheat grain. FOSS, Hillerød, Denmark
37. Wang H, Peng J, Xie C, Bao Y, He Y (2015) Fruit quality evaluation using spectroscopy technology: a review. Sensors 15: 11889–11927
38. Yoshida S, Forno DA, Cock JH, Gomez KA (1976) Laboratory manual for physiological studies of rice, 3rd edn. International Rice Research Institute, Los Banos
39. Searle PL (1984) The berthelot or indophenol reaction and its use in the analytical chemistry of nitrogen. A review. Analyst 109:549–568
40. Holtzhauer M (2006) Basic methods for the biochemical lab, 1st edn. Springer, Berlin
41. Cao N (2013) Calibration optimization and efficiency in near infrared spectroscopy. Iowa State University, Ames, p 184
42. Shenk JS, Westerhaus MO (1991) Population structuring of near infrared spectra and modified partial least squares regression. Crop Sci 31(6):1548–1555
43. Baillères H, Davrieux F, Ham-Pichavant F (2002) Near infrared analysis as a tool for rapid screening of some major wood characteristics in a eucalyptus breeding program. Ann For Sci 59:479–490
44. Agelet LE, Hurburgh CR Jr (2010) A tutorial on near infrared spectroscopy and its calibration. Crit Rev Anal Chem 40:246–260
45. Kays SE, Barton FE II, Windham WR (2000) Predicting protein content by near infrared reflectance spectroscopy in diverse cereal food products. J Near Infrared Spectrosc 8:35–43
46. Bietz JA (1974) Micro-Kjeldahl analysis by an improved automated ammonia determination following manual digestion. Anal Chem 46: 1617–1618

Chapter 8

Multi-Dimensional Cooking Quality Classification Using Routine Quality Evaluation Methods

Lilia Molina, Rosario Jimenez, Nese Sreenivasulu, and Rosa Paula O. Cuevas

Abstract

A battery of assays to characterize the cooking and eating attributes of rice have been in routine use for several decades. The classification system to group rice varieties into different quality types are often based on cooking and eating attributes defined based on amylose content, rather than being considered a set of attributes contributing to an overall quality type based on multi-dimensional approach. In this chapter, the methods developed to measure the cooking quality attributes of rice are described. Instead of considering each attribute on its own, the authors employ multidimensional data generated from the estimation of amylose content, gel consistency, gelatinization temperature, Rapid Visco-Analyzer parameters to classify rice into distinct cooking quality ideotypes. If used universally, such an approach can improve prediction of cooking quality classifications of rice varieties in the breeding programs.

Key words Quality evaluation, Amylose content, Gelatinization temperature, Rapid Visco-Analyzer, Multivariate analyses

1 Introduction

Rice grain quality evaluation programs have been dependent largely on three indicators: amylose content (AC), gel consistency (GC), and gelatinization temperature (GT). Eventually, the Rapid Visco-Analyzer (RVA) also became a routine part of grain quality analyses [1] as it can provide information on samples' cooking behaviors in conditions simulating real-life processing conditions [2]. Through these indicators, rice varieties could be grouped into distinct cooking quality classes on a per-trait approach. This approach toward classification has always been based on ranges for each indicator: AC classes, for example, are as follows: waxy (0–2%), very low (3–9%), low (10–19%), intermediate (20–25%), and high (> 25%); GC of cooked rice pastes, on the other hand, can be used to group high-AC varieties into hard (gel length < 40 mm), medium (gel length 41–60 mm), and soft (gel length > 61 mm);

Nese Sreenivasulu (ed.), *Rice Grain Quality: Methods and Protocols*, Methods in Molecular Biology, vol. 1892,
https://doi.org/10.1007/978-1-4939-8914-0_8,

rice samples may have high (>74 °C), intermediate (70–74 °C), or low (<70 °C) GT [3–7]. Moreover, the different classes of AC, GT, and GC have been associated with different alleles. These alleles can potentially be developed further into molecular markers that can be used for routine marker-assisted screening for predicting cooking quality [8, 9]. Just using established AC, GT, and GC classes, it is possible to define 45 quality classes, assuming that each AC-GT combination has three classes of GC. However, considering that GC is primarily designed as a secondary screening tool for high-AC samples, there are possibly 21 quality classes. RVA parameters [10] are not even considered yet; if RVA metrics are added to the classification, the number of quality classes will inevitably increase into impractical numbers, with many of these quality classes not represented by samples. The per-trait approach, therefore, is not appropriate because (1) it paints a rather disjointed picture of cooking quality; (2) it makes classification of rice samples and development of product profiles complex; and (3) the advent of post-genomics tools [11] and the growing dependence of breeding programs on genotyping technologies [12, 13] requires a deep understanding of overall cooking quality, particularly at the phenotypic level.

An alternative approach is to characterize cooking quality attributes using all available quantitative values. In this multivariate approach, the traditional quality classes, based on individual attributes, are eschewed; rather, the metrics that predict the cooking quality of a sample are considered simultaneously. Through this means, it is predicted that defining quality classes will become more grounded in statistics and quality classes that reflect upon real sample behavior will be formulated.

In this book chapter, methods routinely used to evaluate grain quality of cooked rice are described. For each method, appropriate rice reference materials and intermediate checks are incorporated. The reference materials are traceable to SI units of measurement or to certified reference materials, against which internal reference samples and working standards are checked. Meanwhile, to demonstrate the value of the holistic approach of characterizing rice varieties using all available metrics simultaneously, a case study involving a multivariate approach has been presented to distinguish cooking quality ideotypes.

2 Materials

2.1 Sample Preparation

1. Paddy rice grains.
2. Drying oven.
3. Dehuller: Satake, Model THU-35A, Satake Corp., Hiroshima, Japan.

4. Mill (depends on sample size): Grainman Model 60-220-60-DT and 60-230-60-24 T (Grain Machinery Mfg. Corp, Miami, FL, USA) or Kett mill (Kett Electric Laboratory, Tokyo, Japan).
5. Grinder: Udy cyclone mill, Model 3010-019 with sieve mesh size 60 or 0.25 mm (Udy Corporation, Fort Collins, CO, USA).

2.2 Apparent Amylose Content (AC) Determination

1. Rice flour (prepared as per Subheading 3.1).
2. Milled rice reference materials (flour of IR65, mixed IR65/IR24, IR24, IR64, and IR8; preparation from Subheading 3.1).
3. Spatula.
4. Weighing paper.
5. Analytical balance (0.1 mg accuracy): Mettler Toledo Model ML 204/02, Switzerland.
6. Volumetric flask (capacities: 100 mL and 1 L).
7. Aluminum foil.
8. Magnetic stirrer: Cole Parmer Model 4810, USA.
9. Plastic carboy container, 10 L.
10. Pipettes (capacities: 5 mL, 10 mL).
11. Test tubes, 20 mL.
12. Water bath 95°C: Blue M Model MW-1140C-1, USA.
13. Skalar San++ System Segmented Flow Analyser (SFA): Model 3000 (Skalar Analytical B.V. 4800 De Breda, The Netherlands). It consists of an autosampler, an amylose chemistry unit (manifold, proportioning pump, and colorimeter (with 620 nm filter)), an interface, and a computer for data processing (*see* **Note 1**, Fig. 1).

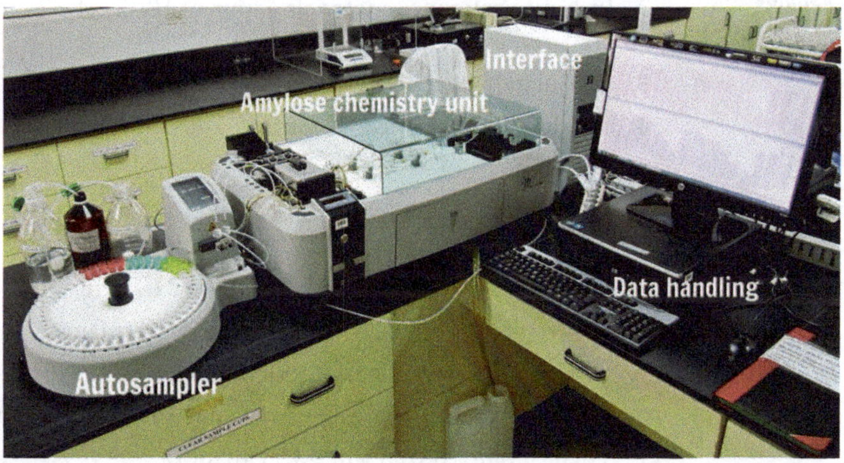

Fig. 1 The San++ SFA set-up with three components: (**a**) autosampler; (**b**) amylose chemistry unit; (**c**) interface, and (**d**) computer for data processing

14. Deionized water.
15. Sodium hydroxide (NaOH), AR Grade.
16. Glacial acetic acid (CH_3COOH), AR Grade.
17. Potassium iodide (KI), AR Grade.
18. Iodine (I_2), AR Grade.
19. Ethanol (EtOH), technical grade.

2.3 Measurement of Gelatinization Temperature by Differential Scanning Calorimetry

1. Rice flour (prepared as per Subheading 3.1).
2. In-house quality control checks (e.g., flour of IR 36, and IR 42, prepared as per Subheading 3.1).
3. Deionized water.
4. Differential scanning calorimeter (DSC): TA Instruments Q100, USA (*see* **Note 2**).
5. TA Instruments Universal Analysis software.
6. Analytical balance: Mettler Toledo Model AG 204, Switzerland.
7. Encapsulating press: TA Instruments.
8. Hermetic seal pre-forming tool/crimping die: TA Instruments, USA.
9. Pipettor, 10 μL.
10. Spatula.
11. Hermetic aluminum pans: TA Instruments, USA (*see* **Note 3**).
12. Hermetic aluminum lids: TA Instruments, USA (*see* **Note 4**).
13. Nitrogen gas, HP and industrial grade.
14. Indium standard.

2.4 Measurement of Gel Consistency

1. Rice flour (prepared as per Subheading 3.1).
2. In-house reference materials (flour of IR42, IR32, and IR48, prepared as per Subheading 3.1).
3. Deionized water.
4. Top-loading balance: Mettler Toledo Model AG 2014, Switzerland.
5. Adjustable-volume dispenser bottle.
6. Culture tubes (13-mm OD × 100-mm length).
7. Vortex mixer: Cole Parmer Model 4721-00, USA.
8. Hot-water bath: Julabo Model SW23, Germany.
9. Cold-water bath: Julabo Model 33, Germany.
10. Leveled reading board lined with millimeter graphing paper.
11. Thymol blue (0.025% in 95% Ethanol).
12. 0.2 M KOH.

2.5 Characterization of Rice Viscosity Profiles Using RVA

1. Rice flour (from Subheading 3.1).
2. In-house reference material (IR32 flour, prepared as per Subheading 3.1).
3. Deionized water.
4. Top-loading balance: Mettler Toledo Model PB1502-S, Switzerland.
5. Adjustable-volume dispenser bottle.
6. Pasteur pipette.
7. Rapid Visco-Analyzer Series 4 (RVA-4), Newport Scientific, USA.
8. RVA-4 canisters and paddles.
9. Thermocline for Windows software, Analysis Tool for RVA, Newport Scientific v. 2.6.
10. Cooling water bath: Julabo Model F12, Germany.
11. Viscosity Reference Standard, Product Code S2000S, Paragon Scientific Ltd., UK.

3 Methods

3.1 Flour Preparation

1. Gradually air-dry the paddy grains in the laboratory to bring down moisture content to 14%.
2. Remove the hull from paddy rice grains using a dehuller.
3. Mill the unpolished rice to remove the bran (*see* **Note 5**).
4. Select a test portion of grains (20–50 grains) (*see* **Note 6**).
5. Grind the rice grains to flour using the Udy mill to pass a 0.25-mm mesh.
6. Allow the rice flour to equilibrate to room temperature before the assays.

3.2 Determination of Apparent Amylose Content Via Iodine Colorimetry

This procedure is based on published ISO methods [14, 15]. It has been modified and adapted for routine operations using the Skalar San++ SFA system.

3.2.1 Preparation of Reagents

1. *1 N NaOH.* Weigh NaOH pellets (400 g) into a plastic carboy containing about 9 L deionized water and mix using a magnetic stirrer for 5 min. Remove the magnetic bar and make up the volume with deionized water to 10 L.
2. *1 N CH_3COOH.* In a 1-L volumetric flask, add 57.2 mL (60 g) of glacial CH_3COOH to 500 mL deionized water. Make up the volume to 1-L using deionized water.
3. *(2%:0.2%) KI-I_2 stock solution.* Dissolve KI (20 g) in 500 mL deionized water, by stirring, in a 1-L volumetric flask. Add I_2

(2 g) and mix until it is completely dissolved. Make up the volume to 1L with deionized water (*see* **Note 7**).

4. *KI-I_2 + CH_3COOH solution.* Mix 1 N CH_3COOH (100 mL) with KI-I_2 stock solution (300 mL) in a 1L volumetric flask. Make up the volume to 1-L with deionized water.
5. *0.09 N NaOH.* Add of 1 N NaOH (9 mL) to deionized water and make up the volume to 100mL with deionized water. Use a 100-mL volumetric flask.

3.2.2 Gelatinization of Rice Flour

This method is applicable for both sample and reference flour materials. *See* **Note 8** for special instructions.

1. Turn the water bath on and heat to 95 °C.
2. Add flour (100 ± 0.5 mg) in a 100-mL volumetric flask.
3. Add 95% ethanol (1 mL) into the flask.
4. Add of 1 N NaOH (9 mL) into the flask.
5. Heat the flask for 10 min in the water bath to gelatinize the starch.
6. Remove the flask from heat and cool at room temperature for at least 2 h (or up to 16 h or overnight).
7. Make up the volume to 100 mL with deionized water.

3.2.3 Preparing a Standard Curve Using Rice Flour Standards

1. Selected IRRI-released rice varieties are used as reference materials in constructing the standard curve for AC determination: IR 65 (waxy), IR 65/24 (very low AC), IR 24 (low AC), IR 64 (intermediate AC), and IR 8 (high AC). The AC of these samples were determined following the ISO 6647-1: 2015 Reference method [14]. The Skalar San++ Segmented Flow Analyzer (wavelength = 620 nm) was used to determine the absorbance of the reference materials (*see* Subheading 3.2.4).

3.2.4 Iodine Absorbance Measurements of Samples by SFA

Detailed instructions on how to operate the SFA can be found in the instrument's manual.

1. Set up the SFA (Fig. 1) and warm up the colorimeter for at least 30 min. Use the following conditions:
   (a) Filter wavelength: 620 nm.
   (b) Sample injection rate: 0.42 mL/min.
   (c) I_2-CH_3COOH injection rate: 0.60 mL/min.
   (d) Flow cell: 10 mm.
2. Dip all the reagent tubings in deionized water.
3. Set up the pump in the amylose chemistry unit.
4. Place assigned reagent tubings into their corresponding reagents:

(a) KI-I_2-CH_3COOH.

(b) Deionized water.

(c) 0.09 N NaOH.

5. Let the system run for 15 min and make sure that bubbles are discretely segmented.
6. Monitor the baseline until a good reagent baseline is achieved.
7. Pour samples (from the overnight standing, Subheading 3.2.2) and internal reference materials into individual sample cups and arrange them as per the pre-prepared sample sequence on the autosampler.
8. Start sample injections into the SFA.
9. After the sequence run, flush the system with deionized water for 15 min.
10. Switch off the interface, the autosampler, and the chemistry unit.

3.2.5 Data Processing

The Flow Access Software automatically calculates the ACs of the samples based on their absorbance values (indicated by peak height) and on the standard curve generated based on the reference materials.

1. Using the Flow Access Software of the Skalar San++ Analyzer, in the data processing window, click on the file for processing.
2. Click on the "Realtime" tab to view the chart. In this screen, manually correct the baseline as necessary. Click on "Results" tab to see corrected peak height values and ACs.
3. Click on "Options" and "Export" to "Excel format" and save the file.

3.3 Measurement of Gelatinization Temperature by Differential Scanning Calorimetry

3.3.1 Rice Sample Preparation

This method is applicable for both sample and reference flour materials. In preparing the samples for DSC, use metal forceps to handle the pans and the lids. These items may be too small for hands to handle. Also, the forceps are used for handling to avoid contaminating the pans and the lids. Instructions on operating the sealing equipment can be found in the manual provided by the manufacturer.

1. Weigh flour (4.0 ± 0.1 mg) onto the center of the aluminum pan.
2. Add Milli-Q-filtered water (8 μL) onto the flour (*see* **Note 9**).
3. Put the lid on the pan with the concave side up.
4. Seal the pan and lid combination hermetically using the forming tool and the encapsulating press.
5. Place the hermetically sealed pans onto the DSC autosampler. Add the sample identifier (name or code) into the DSC's

software as soon as the sample is positioned in the autosampler. A typical sequence run will have the following sequence:

Pan No. 1 = Indium Standard, use as calibrant.

Pan No. 2–3 = Reference Materials (IR 36 and IR 42 rice flour).

Pan No. 4 onwards = samples.

6. Allow the first sample to equilibrate for 30 min (to allow the water to be fully absorbed by the flour) before starting the experiment.

3.3.2 Differential Scanning Calorimetry

Instructions on how to use the software are found in the instrument's manual. The DSC is kept switched on to maintain the temperature within the DSC cell above ambient temperature. It is **NOT** switched off, unless it will not be used for an extended length of time. Ensure that the instrument is calibrated prior to sample analyses (*see* **Note 10**). The method used is based on TA- Q100 Differential Scanning Calorimetry Manual- TA Instruments [16].

1. Allow N_2 gas to flow through the system.
2. Turn on the refrigerated cooling system (RCS).
3. Set up sequence runs with the following conditions using the TA Thermal Advantage Instrument Control software (*see* **Note 11**):

   For indium only:

   Sample size: 3.000–5.000 mg.

   Heating Ramp Rate: 10.0 °C/min.

   Final Temperature: 180.0 °C.

   Load and Unload Temperature: 25.0–35.0 °C.

   For rice flour samples only:

   Sample size: 4.0 mg.

   Heating Ramp Rate: 10 °C/min.

   Final Temperature: 120.0 °C.

   Load and Unload Temperature: 25.0–35.0 °C
4. Start the sequence run.
5. At the end of the sequence run, turn off the RCS and wait until the flange temperature has returned to ambient.
6. Stop the flow of the N_2 gas.

3.3.3 Data Processing

Instructions on how to use the software for data processing are found in the instrument's manual.

1. Load the TA Universal Analysis software.
2. To view the results, open the sample file and set the y-axis to heat flow and the x-axis to temperature.

3. Set the left and right baselines of the melting curve and determine, using the software, the following parameters: onset temperature (T_o), peak temperature (T_p), and enthalpy (ΔH).

3.4 Measurement of Gel Consistency

This method is based on Cagampang et al. [3]. *See* **Note 12** for limitations to its application.

3.4.1 Preparation of Reagents

1. Thymol blue (0.025% in 95% ethanol): Dissolve thymol blue indicator (250 ± 10 mg) in 95% ethanol in a 1-L volumetric flask. Mix then add ethanol to make up the volume to 1 L.
2. *0.2 M KOH*: Dissolve KOH (12.76 ± 0.01 g, ACS, JT Baker, 87.7%) in deionized water in a 1-L volumetric flask. Mix and add deionized water to make up the volume to 1L.

3.4.2 Procedure

1. Turn water bath ON and heat to ≥90 °C. Turn ON cooling bath to ≤10 °C.
2. Weigh flour of the sample (100.0 ± 0.1 mg) into a culture tube and place the tube in a test tube rack. Each test tube rack is a sample batch. Prepare duplicate batch.
3. Weigh flour of (100.0 ± 0.1 mg) in a culture tube for each of the reference materials. Assign three different reference material samples for each batch. Prepare the reference tubes in duplicate.
4. Add the 0.025% thymol blue solution (200 ± 8 μL) to each of the tubes.
5. Add the 0.2 M KOH (2.0 ± 0.1 mL) to each of the tubes and mix thoroughly.
6. Place an empty steel test tube rack (with side handles) in the boiling water bath.
7. Place the freshly mixed samples (from one batch) in the steel test tube rack (in the boiling water bath) and cover the tubes with glass marbles. The water should not reach half the height of the culture tubes to prevent the boiling water from spilling into the samples.
8. Heat the tubes for 8 min. Immediately start the timer as soon as tubes (i.e., usually a set of four tubes) are in the boiling water bath.
9. Occasionally lift the rack out of the water to prevent the contents of the tubes from boiling over.
10. Allow the samples to cool at room temperature for 5 min. Immediately start the timer once the first tube or set of tubes are placed at room temperature.
11. Cool the tubes further in an ice-water bath (15 min), again starting the timer as soon as the samples or set of samples are placed in the ice-water bath.

12. Lay the tubes horizontally on the leveled reading board lined with millimeter graphing paper and allow the gel to solidify (1 h).
13. Measure the length of the blue gel in the horizontally positioned tubes, using the millimeter graphing paper as basis.

3.4.3 Quality Control

Included in the analysis are three different reference materials with each sample batch. The reference materials are IR42, IR32, and IR48.

3.5 Viscosity Profiling by RVA

This method is based on the AACC Method 61-02 [17]. Instructions on how to use the software and the instrument are in the instrument's manual. *See* **Note 13** for instructions about sample handling.

1. Switch on the water bath (set at 10 °C) and the RVA-4.
2. Load the Thermocline software for Windows (TCW.exe).
3. Use the following conditions for the RVA run:
   (a) Temperature ramp up range: 50–95 °C.
   (b) Temperature ramp down range: 95–50 °C.
   (c) Temperature ramp rate: 11.8 °C/min.
   (d) Holding time at 95 °C: 2 min 30 s.
   (e) Rotational speed during RVA run: 160 rpm.
4. Weigh a test portion of rice flour (3.00 ± 0.01 g) into a canister and add deionized water (25 ± 0.01 g). This applies for samples and for reference materials. The reference materials must be included in every sample batch.
5. Place a paddle in the canister and use it to suspend the samples and to disperse lumps.
6. Place the canister onto the sample platform under the tower of the RVA and insert the paddle on top of the coupling.
7. Lower the tower and start the analysis.
8. Remove the canister and the paddle from the tower after the run has completed.
9. Data processing: When a test run has been completed, the TCW software will display the results of the analysis based on predefined formulas. The results include the following parameters: Peak1 (PV), Trough1 (TV), Breakdown (BD), Final Viscosity (FV), Setback (SB), Peak Time (Pk_Time), Pasting temperature (PT), and Retrogradation (also known as lift-off, LO).

3.5.1 Quality Control

Aside from the inclusion of in-house reference material per sample batch, the calibration of the RVA unit is checked at least once per month using a reference standard oil of known viscosity.

4 Applying a Multi-Dimensional Approach in Characterizing Rice Samples: A Case Study

Ten indica rice lines were grown and harvested in IRRI in the dry season of 2014. Samples were dehulled, milled, and then ground to flour. The ground samples were characterized based on the methodologies described for AC (by iodine colorimetry), GT (by DSC), GC, and RVA attributes. In total, there are 10 parameters that were used to characterize the rice lines. Pasting temperature was excluded from the parameters in the analyses because it is also an indicator of cooking time, like GT. Instead of putting the rice lines into quality classes per attribute, the quantitative nature of the dataset was viewed holistically. This allows categorization of rice samples beyond the classic system, possibly into more comprehensive quality classes because all available phenotypic data points were considered. Because the parameters used in this study had different ranges, the data was pre-processed through scaling; i.e., the data set was scaled so that for each attribute, the mean = 0 and the SD = 1. The data was then plotted into spider plots. The shapes of these plots were then compared.

Results showed that the ten indica rice lines represented the five AC classes, the low- and the intermediate-GT classes, while the two high-AC samples belonged to the soft GC class (Table 1). The rest of the samples had soft GC as well.

The RVA plot points PV, TV, and FV suggest that the samples' properties are in a continuum, rather than classifiable into discrete quality classes; however, the derived attributes (SB, LO, and BD) have indications of groupings. Samples 1–3 had low BD, high LO, and positive SB. Samples 4–10, however, have negative SB. These samples can further be grouped based on BD and LO: Samples 4–6 had higher BD and LO than Samples 7–10. Also, it is noted that Samples 1–6 attained PV at around 5–6 min while Samples 7–10 attained PV at around 4 min.

Plotting scaled values for all these attributes shows that the indica rice lines could be clustered into three groups based on the similarities of the spider plots (Fig. 2). The advantage of the spider plot over other graph types is that spider plots allow simultaneous multivariate presentation in two dimensions and comparison of the samples in the dataset, such as in this case study.

It is important to consider the overall shape of the plots and not just the values of individual attributes, especially in attempts to understand cooking and eating quality beyond the typical quality classes. This can then be used as a first step in developing prediction models for cooking and eating quality using quantitative parameters.

Table 1
Cooking and eating quality metrics for 10 indica rice lines

Designation	AC (%)	GT (°C)	GC (mm)	PV (cP)	Peak Time (min)	TV (cP)	FV (cP)	BD (cP)	LO (cP)	SB (cP)
Sample 1	26.3	68.68	84	2628	6.1	2088	3894	540	1806	1266
Sample 2	22.9	68.73	70	2727	6.1	2133	4234	594	2101	1507
Sample 3	26.4	68.85	70	3708	6.2	2903	4750	805	1847	1042
Sample 4	9.8	70.25	70	3689	5.3	1417	2076	2272	659	−1613
Sample 5	11.6	70.54	76	3911	5.9	2064	3189	1847	1125	−722
Sample 6	14.1	71.03	84	3756	5.9	1736	2718	2020	982	−1038
Sample 7	2.7	69.97	90	2487	3.7	1217	1616	1270	399	−871
Sample 8	1.3	70.17	100	1911	3.7	981	1200	930	219	−711
Sample 9	3.3	70.25	81	2405	4.0	1472	2087	933	615	−318
Sample 10	1.3	70.33	100	2482	3.9	1326	1733	1156	407	−749

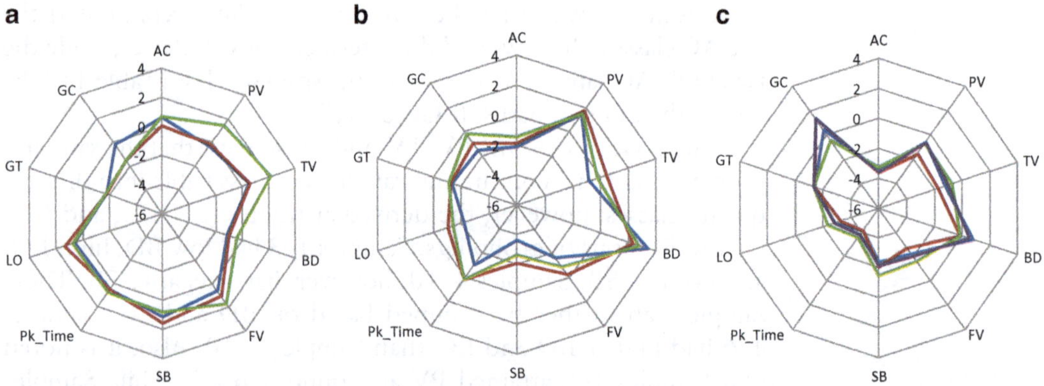

Fig. 2 Spider plots indicating the differences among the 10 indica lines categorized into three classes (**a**, **b**, **c**) based on 11 quantitative grain quality attributes (*AC* amylose content, *PV* peak viscosity, *TV* trough viscosity, *BD* breakdown, *FV* final viscosity, *SB* setback, *Pk_Time* time of PV, *LO* lift-off, *GT* gelatinization temperature, and *GC* gel consistency). Each plot represents one rice line

5 Notes

1. The Skalar San++ SFA Analyzer has been set–up with Amylose Module 1 and 2 for amylose determination in rice flour by iodine method.
2. The TA Q100 DSC is composed of a calorimeter and a RCS. Other DSC models can be used. The method defined here is specifically targeted for using the Q100 and its accessories.

3. TA hermetic aluminum pans: P/N 900793.901.
4. TA hermetic aluminum lids: P/N 900794.901.
5. The 20–50 grains selected for grinding have to be well-milled (i.e., no bran left on the grain surface) and must not be discolored.
6. The type of mill to use to polish the unpolished samples depends on the amount of sample available. For small amounts (<70 g), the Kett mill is typically used. However, for larger sample amounts (i.e., ≥70 g), the Grainman mill can be used. The type of mill used has the potential to change the results of the tests due to the differences in the degree of milling.
7. Iodine (I_2) is light-sensitive. Cover the flask used for preparing the KI-I_2 solution with aluminum foil during preparation of the solution. Dissolution via stirring may take a few minutes.
8. For **steps 3–4**, while adding the solutions, tilt the volumetric flask and slowly put the solutions in such that the solutions do not fall directly into the flour and cause it to splatter inside the flask.
9. Make sure that the flour that gets displaced during the process is scraped back into the middle of the pan. Also, ensure that the water has been absorbed by the flour and that there is no water on the side of the pan. Gentle agitation may be necessary to put all displaced material in the middle of the pan.
10. Calibration of the DSC can be conducted each time the N_2 tank is replaced or at a pre-set frequency. Refer to the User's Manual for the instructions on how to conduct DSC calibrations.
11. The conditions of DSC runs may be modified to fit one's research objectives. Changes in temperature ranges and ramp rates may affect endotherm formation and must be taken into account when comparing sample profiles across different experiments. The Q100 offers a "modulated DSC" option, a setting in which the temperature ramp is sinusoidal instead of linear, but still following the ramp rate set for the run.
12. The gel consistency test differentiates varieties of high- and intermediate-AC in terms of tenderness of cooked rice. It classifies high-amylose rice into three categories based on gel length.
13. Right after the sample has been suspended in water, it must be analyzed in the RVA-4. Therefore, only add water when the sample is about to be analyzed. Otherwise, keep the canister covered to minimize changes in moisture content of the flour, which is hygroscopic.

Acknowledgments

The authors thank Dennis Villegas, Teodoro Atienza, Leah Villanueva, Jennine Rose Lapis, Reah Gonzales, Eric Jhon Cruz, Marnol Santos, Ruben Chavez, and Mitzi Alodia Asih for assistance in processing and collecting data for the samples used in the case study and in the validation and optimization of the methodologies in GQNSL. This work has been supported under the CGIAR thematic area Global Rice Agri-Food System CRP, RICE, Stress-Tolerant Rice for Africa and South Asia (STRASA) Phase III funding.

References

1. Blakeney AB, Lewin L, Reinke RF (2001) Quality rice for North Asia. Rural industries research and development corporation. RIRDC, Kingston, ACT, p 34
2. Deffenbaugh LB, Walker CE (1989) Comparison of starch pasting properties in the brabender viscoamylograph and the rapid visco-analyser. Cereal Chem 66:493–499
3. Cagampang GB, Perez CM, Juliano BO (1973) A gel consistency test for eating quality in rice. J Sci of Food and Agr 24:1589–1594
4. Kumar I, Khush GS, Juliano BO (1987) Genetic analysis of *waxy* locus in rice (*Oryza sativa L.*). Theor Appl Genet 73: 481–488
5. Graham R (2002) A proposal for IRRI to establish a grain quality and nutrition research center. IRRI Discussion Paper Series No. 44. International Rice Research Institute, Los Banos, p 15
6. Cuevas RP, Daygon VD, Corpuz HM, Reinke RF, Waters DLE, Fitzgerald MA (2010) Melting the secrets of gelatinisation temperature in rice. Funct Plant Biol 37:439–447
7. Cuevas RP, Fitzgerald MA (2012) Genetic diversity of rice grain quality. In: Caliskan M (ed) Genetic diversity in plants. InTech, Rijeka, pp 285–310
8. Tuaño AP, Perez LM, Padolina TF, Juliano BO (2015) Survey of grain quality of Philippine farmers' specialty rices. Phil Agric Sci 98:446–456
9. Zhao X, Zhou L, Ponce K, Ye G (2015) The usefulness of known genes/QTLs for grain quality traits in an *indica* population of diverse breeding lines tested using association analysis. Rice 8:29
10. Fitzgerald MA, Martin M, Ward RM, Park WD, Shead HJ (2003) Viscosity of rice flour: a rheological and biological study. J Agri Food Chem 51:2295–2299
11. Anacleto R, Cuevas RP, Jimenez R, Llorente C, Nissila E, Henry RJ, Sreenivasulu N (2015) Prospects of breeding high-quality rice using post-genomic tools. Theor Appl Genet 128:1449–1466
12. Collard BCY, Septiningsih EM, Das SR, Carandang J, Pamplona AM, Sanchez DL, Kato Y, Ye G, Reddy JN, Singh US, Iftekharuddaula KM, Venuprasad R, Vera Cruz CN, Mackill DJ, Ismail AM (2013) Developing new flood-tolerant varieties at the International Rice Research Institute (IRRI). SABRAO J Breed Genet 45:42–56
13. Gregorio GB, Islam MR, Vergara GV, Thirumeni S (2013) Recent advances in rice science to design salinity and other abiotic stress tolerant rice varieties. SABRAO J Breed Genet 45:31–41
14. International Organization for Standardization (2015) ISO 6647-1: 2015-Rice—Determination of amylose content—Part 1: Reference method p 1–4
15. International Organization for Standardization (2015) ISO 6647-2: 2015–Rice—Determination of amylose content—Part 2: Routine methods p 1–4
16. TA Instruments (2003) Differential scanning calorimeter q series getting started guide. TA Instruments –Waters LLC, New Castle
17. American Association of Cereal Chemists Inc (2000) Approved methods of the American Association of Cereal Chemists. American Association of Cereal Chemists International, St. Paul, MN

Chapter 9

Characterization of Mechanical Texture Attributes of Cooked Milled Rice by Texture Profile Analyses and Unraveling Viscoelasticity Properties Through Rheometry

Rosa Paula O. Cuevas, Pawan S. Takhar, and Nese Sreenivasulu

Abstract

The mechanical attributes of cooked rice grains reflected by textural characteristics capture consumers preferences. Two of these attributes such as hardness and stickiness are typically indicated in grain quality evaluation programs by the amylose content of rice. However, the association of amylose content with two other textural attributes such as cohesiveness and springiness remains unknown. Hence, texture profile analyses play a role in quantifying these mechanical parameters of texture. Rheometry on the other hand can be utilized to characterize both viscous and elastic properties of rice during cooking in a water-rice proportion closer to what consumers typically use for cooking. In this chapter, methods for texture profiling and rheometry are presented to capture inferences on cooking quality modeling. Data extracted from rice texture and viscoelastic properties that go beyond what amylose content can predict, has been deciphered through mathematical modeling, which can help predict cooking quality of new rice breeding lines to improve textural and cooking quality specifications.

Key words Rice, Rheometry, Texture, Texture profile analyses, Viscoelastic properties

1 Introduction

In rice varietal development the cooked grain's eating quality of breeding lines is typically assessed by measuring amylose content, gel consistency, and gelatinization temperature [1]. These three tests are used extensively all along breeding line selection processes. Amylose content determination by iodine colorimetry has long been used routinely as an indicator of hardness and stickiness [2]. These two mechanical attributes of textural characteristics are reported to heavily influence Asian consumer preferences [3, 4].

Mechanical textural attributes are those that describe deformation of food caused by biting and chewing forces [5]. However, other mechanical textural attributes may not be modeled to amylose

Nese Sreenivasulu (ed.), *Rice Grain Quality: Methods and Protocols*, Methods in Molecular Biology, vol. 1892, https://doi.org/10.1007/978-1-4939-8914-0_9, © Springer Science+Business Media, LLC, part of Springer Nature 2019

content. For instance, premium and second-best rice varieties often of the same amylose content class were reported to be distinguishable by a trained group of descriptive sensory profilers [6], implying that amylose content is not the sole determinant of rice eating quality. Other components of the rice grain, such as amylose versus amylopectin proportions, protein [3, 7], non-starch polysaccharides [4, 8, 9], and lipids [10], are reported to contribute to viscoelastic and textural properties of cooked grains. Thus, by relying heavily on amylose content and its supplementary test gel consistency [11], there is a probability that current screening methods provide incomplete information about a rice variety's textural quality when cooked. Such incomplete information poses a challenge in breeding programs to capture the premium cooking quality (e.g., [12, 13]). Texture is multi-faceted trait and must be understood by considering different attributes such as hardness, adhesiveness, cohesiveness, and springiness. Expanding phenotyping capacity by measuring other facets of mechanical texture may prove useful in further understanding rice eating quality and their contributions to the overall sensory experience to support breeding for improved cooking quality in rice. Two ways of doing so are by characterizing the viscoelastic properties of cooked rice flour by rheometry (static testing), and by measuring compression-dependent attributes of cooked rice grains by texture profile analyses (TPA).

TPA is a two-compression-cycle test often performed with a texture analyzer. It provides a sense of the chewing behavior of food products. Hence, it is often called a "two-bite test." The TPA procedure provides information about four parameters [14, 15]:

- *Hardness (HRD)*: The maximum force reached during the first compression (positive value).
- *Adhesiveness (ADH)*: It indicates the force required to separate the probe from the base plate after the sample has been compressed.
- *Cohesiveness (COH)*: It indicates the capacity of a material to withstand two consecutive deformations.
- *Springiness (SPR)*: It measures how much a material has returned to its original shape after it has been previously compressed and allowed to recover before the next compression.

Rheology is the study of the deformation properties of materials upon application of stress [16]. Rheometry can provide insight into the textural properties of food materials, which exhibit both viscous and elastic properties [17]. This is particularly important in product development for food purposes since consumer satisfaction is strongly linked with food texture. The viscoelastic properties can be measured using static or dynamic testing. The static testing involves measuring stress-relaxation or creep compliance functions by applying constant-strain or -stress to a sample. During dynamic testing, the samples are oscillated sinusoidally at a constant frequency and often the temperature of the samples increased during cooking [16–18]. Some of these viscoelastic properties are:

- *Storage modulus (G′)*: A measure of a material's elastic (i.e., solid-like, rigid) behavior under shear stress. It is an indication of how much energy is stored in a sample as the components return to their initial configurations.
- *Loss modulus (G″)*: A measure of a material's viscous (i.e., liquid-like) behavior under shear stress. It indicates mechanical energy lost as heat as the structure (e.g., molecular) of the material is damaged, allowing it to flow.
- *Phase (loss) angle δ (°):* The phase shift between the oscillatory deformation and the material's response. The angle range is $0° < \delta < 90°$.
- *Tan (δ)*. The ratio of G'' to G'. Its value decreases with an increase in solid-like, elastic behavior.
- *Storage compliance (J′)*: Like G', J' is also a measure of energy storage in a material. It is the ratio of strain in phase with the applied sinusoidal stress to the stress [19].
- *Loss compliance (J″)*: It is a measure of a material's energy dissipation ability or its liquid like behavior. It is the ratio of strain, which is 90° out of phase with the applied sinusoidal stress to the stress.
- *Stress Relaxation Function [G(t)]*: Obtained by compressing a sample at a constant strain, while recording the decay in stress with time. $G(t)$ is the ratio of stress decay with constant applied strain.
- *Creep Compliance Function [J(t)]*: Obtained by compressing a sample at a constant stress, while recording the change in strain with time. $J(t)$ is the ratio of strain as a function of time with constant applied stress.

The static and dynamic testing provides complementary information on viscoelastic characteristics of biopolymers at different time scales. The static testing measures the viscoelastic characteristics of biopolymers at long-time scales are generally performed using texture analyzer. The dynamic tests are performed using a rheometer for gels or liquid foods, and dynamic mechanical analyzer for solid foods. Some newer dynamic testing equipment can perform static testing as well.

The dynamic properties are usually more sensitive in detecting transitions such as glass-transition, gelatinization, gelation, etc. The static properties are easier to correlate with textural attributes such as stiffness, hardness, sogginess etc., and can be directly utilized in mathematical models describing mechanical texture change during imbibition of water into rice during cooking. To obtain complete understanding of polymers behavior in a material, the data from tests performed at a wide range of time scales need to be combined [19].

In this chapter, the methodologies for rheometry and texture profile analyses are described as phenotyping tools to understand the mechanical attributes of cooked rice and case studies are presented. Here we demonstrate the value of texture profiles and rheometric information, which add value as selection tools in rice breeding pipelines to capture premium cooking properties. Furthermore, cooking quality modeling will be inferred with the previously mentioned phenotyping measures. Since it may be tedious to perform all complementary rheological measurements; often the mathematical relations converting data from one type of test to the other type can be utilized. As an example, case study 2 has been presented, where the methodology for converting storage, and loss moduli to the corresponding compliance moduli or vice-versa; and the stress relaxation function to the creep compliance function or vice-versa has been demonstrated.

2 Materials

2.1 Texture Profile Analyses (TPA)

1. Uncooked polished (milled) rice grains (*see* **Note 1**).
2. Spatula (metal).
3. 50 mL glass test tubes.
4. Marbles or foil (to seal the test tubes during cooking).
5. Water bath set at 50 °C (*see* **Note 2**).
6. Water bath set at 100 °C (*see* **Note 3**).
7. Milli-Q water.
8. Texture Analyzer (TA-XT Plus, Stable Micro Systems) fitted with a 5 kg load cell and equipped with a 35 mm diameter cylinder flat-end probe.
9. 2 kg weight for calibration.

2.2 Rheometry

1. Rice flour.
2. Milli-Q water.
3. Rheometer (Advanced Rheometer 2000, TA Instruments) equipped with a peltier plate system, and a stainless steel parallel plate geometry (diameter = 40 mm). A water circulation system is used to control the cooling rate of the rheometer.
4. Beaker.

5. Metal spatula.
6. Balance.

3 Methods

The objective in performing TPA (a two-compression test) is to quantify four mechanical attributes of texture: hardness, adhesiveness (or stickiness), cohesiveness, and springiness. It is important that the cooked rice grains chosen for TPA are not broken or split and are of the same height.

3.1 Texture Profile Analyses (TPA)

In this procedure, three setups per sample (i.e., three test tubes) are prepared. Hence, there are three cooking replicates.

3.1.1 Rice Sample Preparation

1. Milled rice grains that are not broken (25 grains) are placed in one test tube.
2. Wash grains thrice with water. The wash water is decanted.
3. Put water with 1:2 (w/v) ratio with rice into the tube.
4. Seal the test tube with a marble or foil, and tie with rubber band.
5. Place the test tube in the 100 °C water bath and heat for 15 min. During this cooking process the water has been absorbed by the grains.
6. Transfer the test tube to the 50 °C water bath. Keep it there until ready to conduct TPA on the cooked grains.

3.1.2 TPA

For each cooking replicate, three measurements are performed. Hence, nine cooked rice grains are selected for TPA per test tube. In total, nine measurements are performed for each sample (3 measurements per tube × 3 tubes). The Texture Analyzer is calibrated once daily before samples are tested (*see* **Note 4**). For instructions on how to use the software controlling the Texture Analyzer, refer to the instrument manual.

1. Calibrate the Texture Analyzer as per instructions in the instrument manual.
2. Set the Texture Analyzer parameters to the following values (*see* **Note 5**):
   (a) Pretest speed is set to 1 mm s^{-1}.
   (b) Test speed and post-test speed are both set to 5 mm s^{-1}.
   (c) Target mode is set to Strain.
   (d) Strain is set to 90% (this means that the cooked grains are compressed to 10% of their heights; the software detects

the initial heights of the grains and automatically calculates from these values).

(e) Time is set to 5 s.

(f) Trigger type is set to Autoforce.

(g) Trigger force is set to 3 g (this is the force that is being recognized by the instrument to indicate that probe has made contact with the samples).

(h) Tare mode is set to Auto.

3. Select three cooked rice grains of the similar heights (from the same tube kept at 50 °C) and place them in parallel at the center of the base plate such that the grains will all be compressed in their entirety by the probe.
4. Using the Exponent Lite software to control the instrument, compress the grains twice. The output graph is a profile composed of two positive peaks and two negative peaks (Fig. 1).
5. Clean the probe and the base plate with a moist cloth. Make sure that residues from the compressed rice grains have been thoroughly removed. And then repeat **steps 3** and **4**.
6. Data preprocessing can be conducted using the Exponent Lite software. Export the data as .csv files, as per the software manufacturer's instructions. The pre-processing will provide hardness (HRD, Force 2) and adhesiveness (ADH) values along with other parameters.
7. To calculate cohesiveness (COH) and springiness (SPR), use the following formulae (this can be performed as an additional macro in the software or on a spreadsheet after the data has been exported) based on Fig. 1:

$$\mathrm{COH} = \frac{A_{4:6}}{A_{1:3}} \tag{1}$$

$$\mathrm{SPR} = \frac{t_{4:5}}{t_{1:2}} \tag{2}$$

3.2 Rheometry

3.2.1 Rice Sample Preparation

In this procedure, three replicates are tested. Hence, at least 3 g of flour is needed per sample. Each flour-water suspension has to be prepared just before the rheometry test as to avoid settling of the flour to the bottom of the beaker.

1. Put 1 g flour in a glass beaker.
2. Add 2 g water.
3. Stir the mixture until flour is thoroughly wetted and suspended; i.e., there are no clumps of flour at the bottom of the beaker or streaks of dry flour on the side of the beaker.

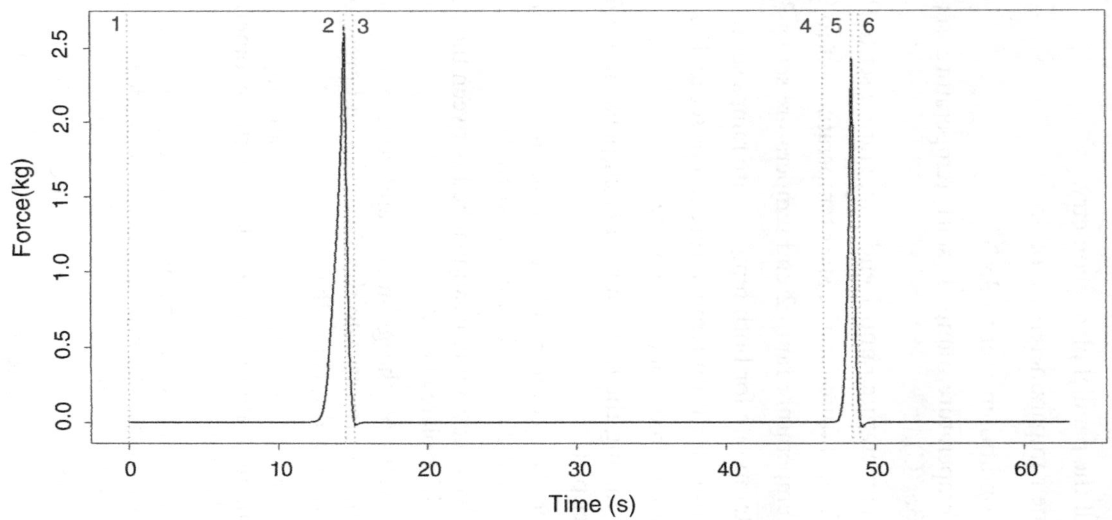

Fig. 1 An example of a TPA profile for cooked rice. Six dashed numbered lines divide the profile into different regions. These lines indicate the boundaries of the different components of the TPA profile. These features of the curve are: (F_2) peak force during the first compression (hardness); ($A_{1:3}$) area of work during the first compression; ($A_{1:3}$) area of work during the first compression; ($A_{3:4}$) are of work during the upstroke after the first compression (adhesiveness); ($A_{4:6}$) area of work during the second compression; ($t_{1:2}$) travel time on the downstroke of the first compression; ($t_{4:5}$) travel time on the downstroke of the second compression. These variables are used in calculating the derived values, cohesiveness, and springiness

3.2.2 Rheometry

For instructions on how to use the software, please refer to the instrument's manual. A rheometry procedure (a "run") consists of a heating ramp and a cooling ramp (*see* **Note 6**).

1. Start up the Advanced Rheometer 2000 according to the instrument manufacturer's instructions. This includes turning on the air compressor, the circulating water bath, the rheometer, and the computer.
2. Install the parallel plate geometry.
3. Set the run parameters as follows:
   (a) Conditioning set to 35 °C.
   (b) Temperature ramp 1 start temperature set to current (35 °C).
   (c) Temperature ramp 1 end temperature set to 95 °C.
   (d) Temperature ramp 2 start temperature set to 95 °C.
   (e) Temperature ramp 2 end temperature set to 35 °C.
   (f) Ramp rate for both temperature ramps set to 4 °C min^{-1}.
   (g) Strain for both temperature ramps set to 1%.
   (h) Single frequency set at 1 Hz.
4. After mixing the rice flour-water suspension, pour it onto the Peltier plate.
5. Lower the geometry up to the geometry gap using the TA Advantage 2003 software (*see* **Note 7**).
6. Remove excess sample sandwiched between the geometry and the Peltier plate (Fig. 2a).
7. Place water on the geometry and cover the system with the solvent trap to saturate the air within and prevent the sample from drying out (Fig. 2b).
8. Using the TA Advantage 2003 software, supply the necessary information about the sample and the geometry, and then start the run.
9. After each run, clean the geometry and the Peltier plate using a moist cloth. Do not scrape or make deep scratches while cleaning the geometry and the Peltier plate.
10. Repeat **steps 1–6** for each replicate of each sample.
11. For each run, values for storage moduli (G'), loss moduli (G''), and *tan* (δ), among other parameters are recorded by the software during the temperature ramps. Export these into a .csv file as needed. The graph of the viscoelastic profiles is in Fig. 3.
12. Data processing steps and correlation analysis have to be outlined.

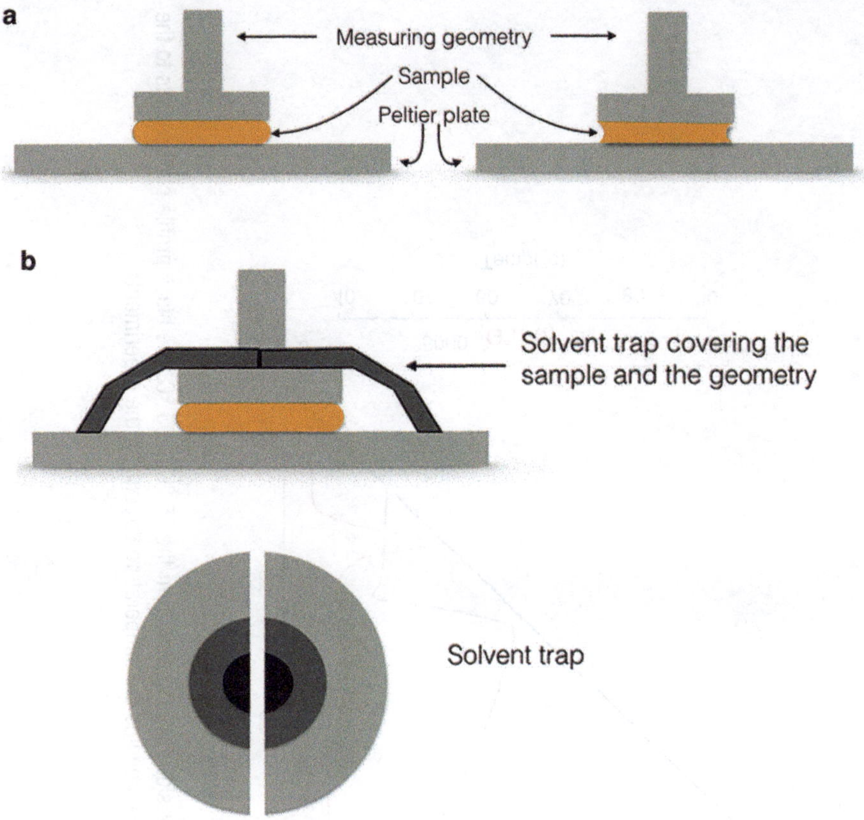

Fig. 2 Proper sample preparation for rheometry using a parallel plate geometry and a peltier plate. (**a**) How an adequate amount of sample looks like when viewed from the side; (**b**) Installation of a solvent trap to prevent evaporation of moisture from the sample during heating

3.3 Case Study 1: Correlations of TPA and Rheometry Properties from Diverse Lines

The relationships among the routine grain quality indicators (i.e., amylose content, gel consistency, and gelatinization temperature) and more direct measurements of mechanical textural attributes (TPA) and viscoelastic properties (rheometry) were studied using a subset of diverse collection of indica rice varieties ($n = 61$). The contrasting lines depict extreme differences in TPA and rheometric properties (Fig. 4). Characterizing the diversity panels documented with huge phenotyping variation for TPA and rheometric properties might be useful to define cooking quality characteristic features.

The correlation results for the diversity samples indicate that elevated amylose content is positively correlated with hardness and also negatively correlated with adhesiveness (Fig. 5). These results agree with previously published literature (e.g., [20]). Amylose content is also correlated with springiness and cohesiveness, though these correlations are weaker than with adhesiveness and with hard-

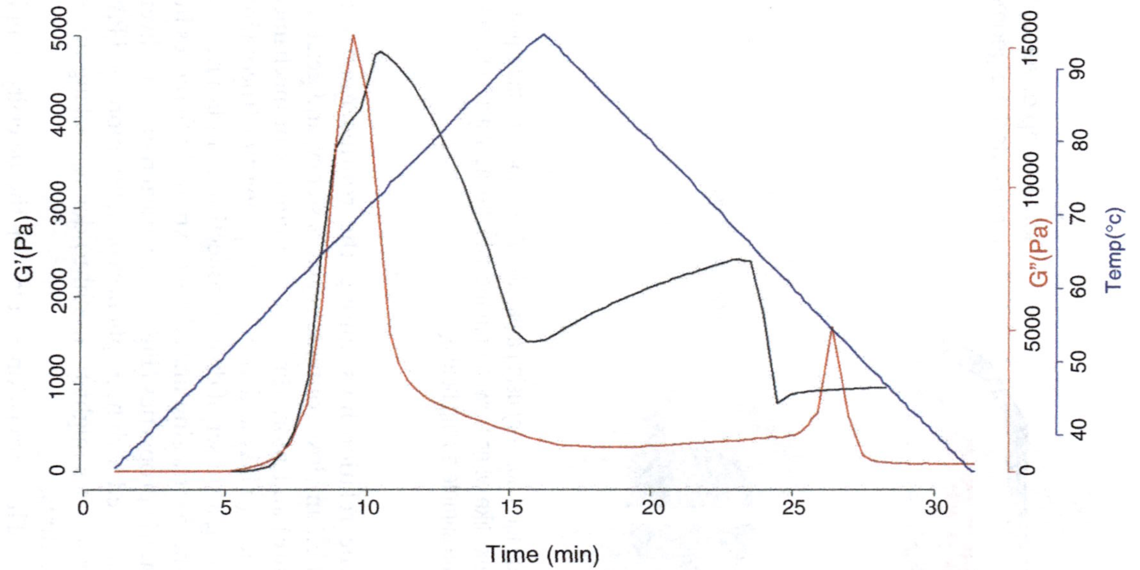

Fig. 3 A profile of rheometric properties obtained froma 1:2 (w/v) flour-to-water slurry was heated from the 35 °C to 95 °C. The black profile corresponds to the storage modulus (*G'*); the red profile is the loss modulus (*G"*): and the blue profile denotes the temperature throughout the experiment

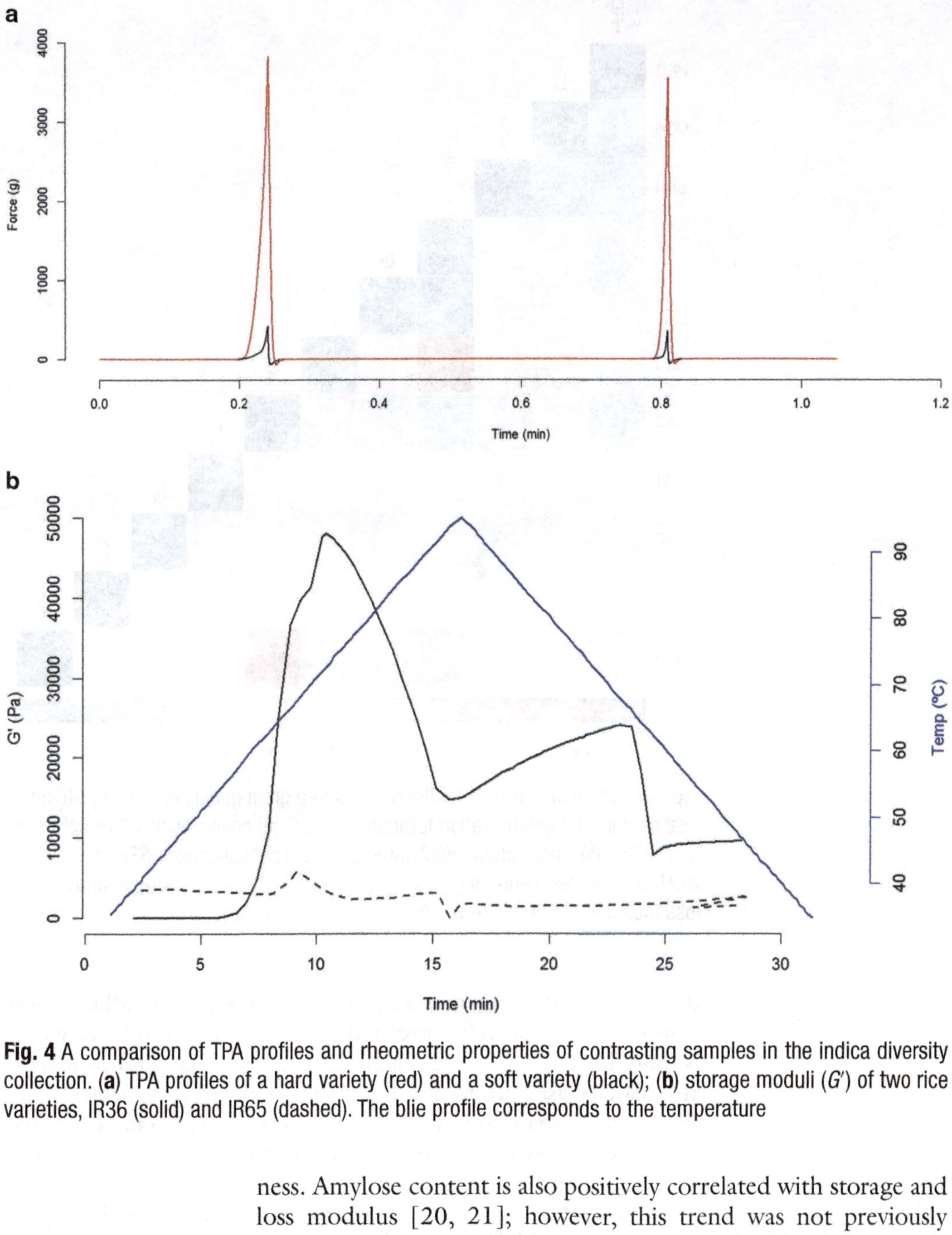

Fig. 4 A comparison of TPA profiles and rheometric properties of contrasting samples in the indica diversity collection. (**a**) TPA profiles of a hard variety (red) and a soft variety (black); (**b**) storage moduli (*G'*) of two rice varieties, IR36 (solid) and IR65 (dashed). The blie profile corresponds to the temperature

ness. Amylose content is also positively correlated with storage and loss modulus [20, 21]; however, this trend was not previously observed in slurries of high concentration and with inherent amylose [22]. These trends indicate that AC remains to be an interesting parameter in the determination of the cooking and eating quality of the rice grain. Consumers may negatively associate hardness with adhesiveness. However, Fig. 5 suggests that the relationship may not be strong. This indicates that there may be varieties

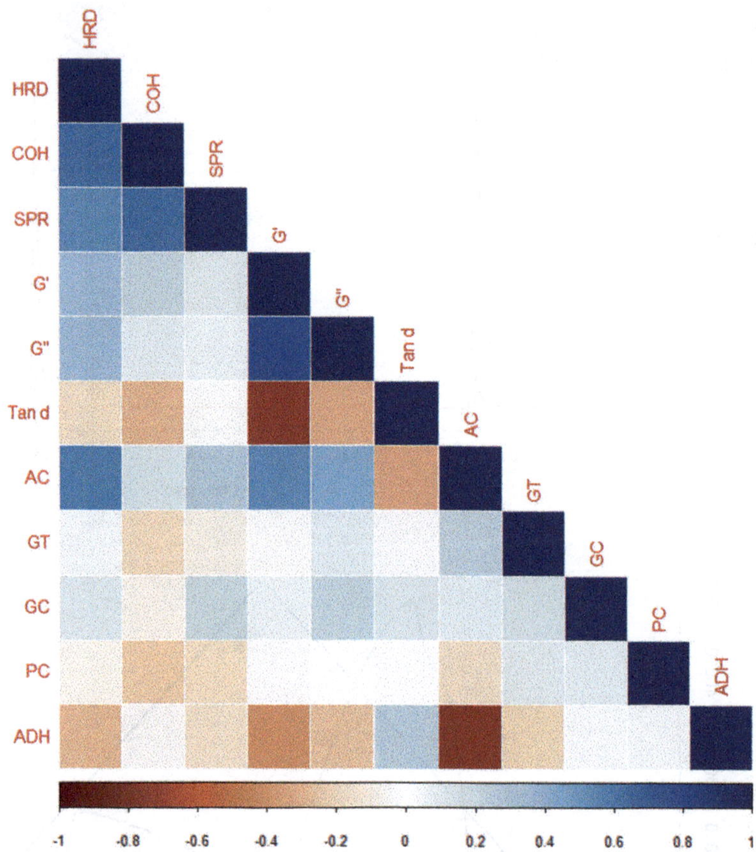

Fig. 5 Correlation matrix of routinely measured grain quality properties (*AC* amylose content, *GT* gelatinization temperature, *GC* gel consistency, *PC* protein content) with TPA parameters (*HRD* hardness, *COH* cohesiveness, *SPR* springiness, *ADH* adhesiveness) and rheometric attributes (*G′* maximum storage modulus, *G″* loss modulus at maximum *G′*, *Tan d tan (δ)* at maximum *G′*)

that are soft but are not necessarily sticky, and vice versa. Adhesiveness appears to be negatively correlated with the storage modulus, which is quite intuitive, because materials that are sticky tend to exhibit attenuated elastic properties. It is rather expected that stickier materials exhibit enhanced flow properties; however Fig. 5 indicates that the adhesiveness is negatively correlated with the loss modulus. Clearly, starchy materials such as rice-water slurries are complex materials and thus it remains important to capture not simply viscosity properties but also elastic properties.

Protein content negatively correlated with TPA and rheometric properties with moderate significance. Most notable among these correlations are protein content correlation with cohesiveness and springiness, two textural properties of the cooked rice grain that have not gained as much attention as hardness. Despite the weaker contributions of protein to texture, it still warrants a closer look since varieties with similar amylose contents may have differences in textures (i.e., [6]) with proteins as a discriminating factor. It is also interesting to note that interactions between amylose versus amylopectin and amylose to protein ratio may also contribute to unique textural and rheological behaviors that make indica varieties distinctive among each other.

These new insights gleaned from diversity panel of 61 samples show deeper understanding of cooking and eating quality. Although amylose content is an important attribute and is still worth exploring, other measures such as rheology and texture profile analyses may comprehend better knowledge of cooking and eating quality, and provide breeders with new and efficient tools that will allow them to expand selection criteria to distinguish distinct cooking quality classes.

3.4 Case Study 2: Conversion of Storage (G′) and Loss Moduli (G″) to Storage (J′) and Loss Compliance (J″)

It is straightforward to convert the storage and loss moduli to the corresponding compliance properties using the complex functions (involving the use of imaginary number, *i*) [19]. The storage and loss moduli can be denoted using the complex modulus function (G^*):

$$G^* = G' + iG'' \tag{3}$$

Similarly J' and J'' can be denoted using the complex compliance function (J^*):

$$J^* = J' - iJ'' \tag{4}$$

The conversion relation between the two complex functions is [19]:

$$G^* = \frac{1}{J^*} \tag{5}$$

or

$$G' + iG'' = \frac{1}{J' - iJ''} = \frac{J' + iJ''}{J'^2 + J''^2} \tag{6}$$

By relating the real and imaginary components on left and right hand sides of (Eq. 6) we get

$$G' = \frac{J'}{J'^2 + J''^2}$$
$$G'' = \frac{J''}{J'^2 + J''^2} \tag{7}$$

3.5 Conversion of Stress Relaxation Function to Creep Compliance Function

The stress relaxation data collected using a texture analyzer can be described using a mathematical relation. Often the generalized Maxwell relation is used to describe the stress relaxation data [17]. The stress relaxation function ($G(t)$ is related to stress ($\sigma(t)$) in the material by:

$$\sigma(t)=\int_0^t G(t-\tau)\frac{d\varepsilon(\tau)}{d\tau}d\tau \tag{8}$$

where $\varepsilon(t)$is the strain in the material. The above equation represents a convolution. Assuming that the food is free from stress and strain before the start of an experiment, the Laplace transformation of (Eq. 8) gives

$$\bar{\sigma}(s)=s\bar{G}(s)\bar{\varepsilon}(s) \tag{9}$$

where the bars over a function denote the corresponding Laplace transformed function, and s denotes the frequency in Laplace space.

The creep compliance function ($J(t)$) is related to strain using the following relation:

$$\varepsilon(t)=\int_0^t J(t-\tau)\frac{d\sigma(\tau)}{d\tau}d\tau \tag{10}$$

Laplace transformation of (Eq. 10) gives

$$\bar{\varepsilon}(s)=s\bar{J}(s)\bar{\sigma}(s) \tag{11}$$

By substituting $\bar{\varepsilon}(s)$ from (Eq. 11) in (Eq. 9) and simplifying, we get

$$J(s)=\frac{1}{s^2G(s)} \tag{12}$$

The inverse Laplace transformation of (Eq. 12) would give the creep compliance function, $J(t)$. Similarly, if the creep compliance function ($J(t)$) were measured using a mechanical test, it can be converted to the corresponding stress relaxation function ($G(t)$) by first performing Laplace transformation of $J(t)$, and then using (Eq. 12) to obtain ($G(s)$) and finally performing the inverse Laplace transformation of $G(s)$ to obtain $G(t)$.

4 Notes

1. Milled, or polished, rice are eaten by most rice consumers. In preparing milled rice, paddy rice grains are dehulled and then milled. The milling process removes the outer layers of

the rice grain [23]. The methodologies outlined in this chapter can also be applied to brown (unpolished) rice bran. The rheological and textural profiling data will, expectedly, be different because the layers removed during polishing may affect transfer of heat and/or water absorption into the rice grains during cooking [24].

2. The water bath (50 °C) will keep the cooked rice grains from retrograding. Retrogradation is the development of firmness of the cooked grain brought about by the formation of networks of amylose molecules [25]. With the temperature set below gelatinization temperatures (based on a survey of highly diverse samples reported by [26]), rice grains will not get overcooked. It is important to keep the water bath covered to keep the air saturated with water vapor; this will prevent the rice grains from becoming dehydrated before they are subjected to TPA.
3. The water bath (100 °C) is used to cook rice grains in the test tubes. If a water bath is not available, a sand bath set at 100 °C or a beaker with water being heated to 100 °C on a hot plate can be used instead.
4. To calibrate the Texture Analyzer, follow the instrument manufacturer's instructions before compressing any samples. With the 5 kg load cell, use the 2 kg weight.
5. The settings of the TPA compression test can be preprogrammed in a Project. See the instrument's manual for instructions on how to do this. Alternatively, there already is a template for TPA in the Texture Analyzer. Parameters may be customized as well.
6. The procedure and the geometry can be further customized by changing the temperature ramp rates, the geometry (e.g., use of the starch cell or of the steel cone geometry), the oscillation stress, among other parameters. These changes should depend on the objective of the study.
7. The software currently being used to operate the Advanced Rheometer 2000 is the TA Advantage Software 2003 (v.4.0.0). Other rheometer brands may have other proprietary software while more recent models of the Advanced Rheometer from TA Instruments may have later versions of the software.

Acknowledgments

We thank Cyril John Domingo and Lucena Samadio for assistance in obtaining the data used in this chapter. The authors acknowledge the Office of International Programs in College of ACES at

University of Illinois, Urbana-Champaign for helping initiate this collaborative work. This work has been supported under the CGIAR thematic area Global Rice Agri-Food System CRP, RICE, Stress-Tolerant Rice for Africa and South Asia (STRASA) Phase III funding.

References

1. Cuevas RP, Fitzgerald MA (2012) Genetic diversity of rice grain quality. In: Genetic diversity in plants. InTech, Rijeka, pp 285–310
2. Juliano BO, Perez CM, Barber S, Blakeney AB, Iwasaki TA, Shibuya N, Keneaster KK, Chung S, Laignelet B, Launay B, del Mundo AM, Suzuki H, Shiki JI, Tsuji S, Yama JT, Tatsumi K, Webb BD (1981) International cooperative comparison of instrument methods for cooked rice texture. J Texture Stud 12:17–38
3. Zhou Z, Robards K, Helliwell S, Blanchard C (2002) Ageing of stored rice: changes in chemical and physical attributes. J Cereal Sci 35:65–78
4. Mestres C, Ribeyre F, Pons B, Fallet V, Matencio F (2011) Sensory texture of cooked rice is rather linked to chemical than to physical characteristics of raw grain. J Cereal Sci 53:81–89
5. Szczesniak AS, Brandt MA, Friedman HH (1963) Development of standard rating scales for mechanical parameters of texture and correlation between the objective and the sensory methods of texture evaluation. J Food Sci 28:397–403
6. Champagne ET, Bett-Garber KL, Fitzgerald MA, Grimm CC, Lea J, Ohtsubo KI, Jongdee S, Xie L, Bassinello P, Resurreccion AP, Ahmad R, Habibi F, Reinke RF (2010) Important sensory properties differentiating premium rice varieties. Rice 3:270–281
7. Martin M, Fitzgerald MA (2002) Proteins in rice grains influence cooking properties. J Cereal Sci 36:285–294
8. Das M, Gupta S, Kapoor V, Banerjee R, Bal S (2008) Enzymatic polishing of rice–a new processing technology. Food Sci Tech 41: 2079–2084
9. Shibuya N, Iwasaki T (1984) Effect of cell wall degrading enzymes on the cooking properties of milled rice and the texture of cooked rice. J Jap Soc Food Sci Tech 31:656–660
10. Biliaderis CG, Tonogai JR (1991) Influence of lipids on the thermal and mechanical properties of concentrated starch gels. J Agri Food Chem 39:833–840
11. Cagampang GB, Perez CM, Juliano BO (1973) A gel consistency test for eating quality in rice. J Sci Food Agri 24:1589–1594
12. Unnevehr LJ (1986) Consumer demand for rice grain quality and returns to research for quality improvement in Southeast Asia. Am J Agri Econ 68:634–641
13. Vandeplas A, Minten B (2015) Food quality in domestic markets of developing economies: a comparative study of two countries. Agri Econ
14. Chen L, Opara UL (2013) Texture measurement approaches in fresh and processed foods: a review. Food Res Int 51:823–835
15. Champagne ET, Lyon BG, Min BK, Vinyard BT, Bett KL, Barton FE, Webb BD, McClung AM, Moldenhauer KA, Linscombe S, McKenzie KS, Kohlwey DE (1998) Effects of postharvest processing on texture profile analysis of cooked rice. Cereal Chem 75:181–186
16. Schramm G (2000) A practical approach to Rheology and Rheometry. Gebrueder HAAKE GmbH, Dieselstrasse, p 291
17. Steffe JF (1996) Rheological methods in food process engineering. Freeman Press, East Lansing, p 418
18. Mezger TG (2006) The Rheology handbook: for users of rotational and oscillatory rheometers, vol 299. Vincentz Network GmbH, Hannover
19. Ferry JD (1980) Viscoeslastic properties of polymers. John Wiley & Sons, New York, p 672
20. Biliaderis CG, Juliano BO (1993) Thermal and mechanical properties of concentrated rice starch gels of varying composition. Food Chem 48:243–250
21. Li JY, Yeh AI (2001) Relationships between thermal, rheological characteristics and swelling power for various starches. J Food Eng 50:141–148
22. Lii CY, Tsai ML, Tseng KH (1996) Effect of amylose content on the rheological property of rice starch. Cereal Chem 73:415–420
23. Bryant RJ, Dorsch JA, Peterson KL, Rutger JN, Raboy V (2005) Phosphorus and mineral concentrations in whole grain and milled low phytic acid (lpa) 1-1 rice. Cereal Chem 82:517–522

24. Juliano BO, Bechtel DB (1985) The rice grain and its gross composition in rice: chemistry and technology. Am Assoc Cereal Chem:17–57

25. Tang MC, Copeland L (2007) Investigation of starch retrogradation using atomic force microscopy. Carb Pol 70:1–7

26. Cuevas RP, Daygon VD, Corpuz HM, Reinke RF, Waters DLE, Fitzgerald MA (2010) Melting the secrets of gelatinisation temperature in rice. Func Plant Biol 37: 439–447

Chapter 10

Characterizing Starch Molecular Structure of Rice

Cheng Li, Hongyan Li, and Robert G. Gilbert

Abstract

A better understanding of the nutritional properties of rice starch is important because of the rapid rise of diet-related health complications, particularly obesity, type 2 diabetes, and colorectal cancers. Rice starch that is slowly digested to glucose, and where significant quantities of starch which reach the lower gut ("resistant starch"), can mitigate, and also delay the onset of, these diseases. These digestibility properties depend to some extent on starch molecular structure. The characterization of this structure is therefore significant for understanding and developing healthier slower digestible rice. In this chapter, a series of techniques used for characterizing starch structure are reviewed and the procedure for preparing rice starch samples with minimum degradation for characterizing starch chain length distribution (CLD) and overall molecular structure is given. Some methods for choosing or developing plants showing desirable structural characteristics are briefly summarized.

Key words Rice quality, Starch structure, Size-exclusion chromatography, Chain-length distribution, Digestibility

1 Introduction

Starch digestibility is one of the important qualities for rice. High levels of rapidly digestible starch (RDS), defined here as the amount of starch digested in the first 20 min as measured by the method of Englyst [1], have been associated with increasing incidence of metabolic syndrome, including obesity and diabetes. On the other hand, rice varieties containing larger amounts of slowly digestible starch (SDS) and enzyme-resistant starch (RS), which are starches digested between 20 and 120 min and that not digested after 120 min, respectively, are desirable to prevent and alleviate these complications [2–4].

Starch is a complex branched homopolymer of glucose, having a broad distribution of molecular weights and sizes, and with two major components: amylose and amylopectin. Amylose (weight-average molar mass~10^5–10^6 Da) is a slightly branched polymer with a small number of long branches, while amylopectin (weight-average molar mass ~10^7–10^9 Da) is a highly branched polymer

Nese Sreenivasulu (ed.), *Rice Grain Quality: Methods and Protocols*, Methods in Molecular Biology, vol. 1892,
https://doi.org/10.1007/978-1-4939-8914-0_10,

containing a vast number of short chains. Some starches also contain an intermediate component, which consist of highly branched molecules with smaller molecular weights than amylopectin, but similar to that of amylose [5–7]. All these starch components have polymer backbones that are linked by α-(1⟶4) glycosidic bonds; the branch points are formed by α-(1⟶6) glycosidic bonds. Different plant species have different amylose and amylopectin contents. The normal amylose content of starch is approximately 15–30%, whereas waxy starch is essentially devoid of amylose [8]. Amylose contents as high as 50% have been found in high-amylose starches [6, 9]; note in this context that the value of "amylose content" depends on the method of measurement [10].

Starch structure can be divided into multiple structural levels [11]. Individual starch chains (branches) form the first level. The second level consists of fully branched individual amylose and amylopectin molecules without any chain alignment. Level 3 describes the conformation of starch molecules, encompassing the following features in the native unprocessed grain: starch chains aggregate, entwining into a helical structure; the helices aggregate to form crystallites; and finally the crystallites form alternating amorphous and crystalline lamellae. Higher structural levels of native starch granules comprise growth rings, individual granules, and association with non-starch components [12].

The first two levels of structure can be significant determinants of both the higher-level structures and a number of structure-property relations for starch-containing substances [13, 14]. For instance, the first level structure (the starch chain-length distribution, the CLD) influences starch digestibility: the ratio of short and long chains in amylopectin has been found to be related to the amount of SDS [15]. This has two mechanistic reasons. First, the higher proportion of longer amylopectin chains leads to a greater amount of retrogradation (the partial crystallization of these longer chains after some branches have been removed during the digestion process). Second, the reduction in digestion rate of the higher proportion of shorter chains can be attributed to an increase in the number of branch points which inhibit enzymatic hydrolysis [16]. Both native uncooked and cooked rice starch granules containing a greater amount of chains with degree of polymerization (DP) in the range 6–12 have increased starch digestions rates, whereas the opposite trend is observed for longer amylopectin chains [17, 18]. Furthermore, higher amounts of amylose and extra-long amylopectin chains can greatly decrease the digestibility of raw starch granules and gelatinized starch [19].

Different techniques have been used to characterize starch molecular structure. Size-exclusion chromatography (SEC), fluorophore-assisted carbohydrate electrophoresis (FACE), and high-performance anion-exchange chromatography (HPAEC) are used to characterize level 1 structure, giving the CLD of starch molecules. These techniques use enzymatically debranched starch as the analyte. FACE, which separates molecules based on mass-to-charge ratio, is currently the best technique to characterize short starch branches with a DP up to ~180 [20, 21]. The first part of the experimental procedure for FACE involves labeling the reducing end of the short starch branches with a fluorophore (typically 8-amino-1, 3, 6-pyrenetrisulfonic acid, APTS); separation is then performed using capillary electrophoresis with fluorescence detection. The labeling has been proven to be independent of starch chain size, at least up to DP ~135 [22]. The size distribution derived from FACE is the number distribution (relative number) of chains of DP X, $N_{de}(X)$. HPAEC, usually with pulsed amperometric detection (PAD), also provides the number CLD of starch. Both FACE and HPAEC can only give information on shorter, largely amylopectin, chains; HPAEC also suffers from a mass bias which is very laborious to correct. While SEC is able to characterize the larger amylose chains, it suffers from problems such as band broadening, which is the effect where molecules of a given size "leak" into a range of elution volumes, instead of all eluting at the same elution volume. SEC can also be used to characterize the second level of starch structure, with multiple-detection, the number, weight-average molecular weight, and z-average radius of gyration distributions [20]. Unfortunately, there is unavoidable shear scission of the amylopectin molecules in current SEC systems and therefore it cannot give accurate molecular size characterization of the amylopectin component [23]. Multi-angle laser light scattering (MALLS) without size separation can measure the weight-average molecular weight and average radius of gyration of starch; with careful sample handling, there would be no shear scission (in the absence of size separation). However, neither dynamic nor static light scattering is able to give reliable size distributions without size separation [11].

Field-flow fractionation (FFF) is a range of related techniques that has attracted considerable interest recently in regards to the characterization of high molecular weight macromolecules. It separates macromolecules without a stationary phase, so shear scission is minimal, which gives this technique the potential to reveal the full size distribution of undegraded native starch. However, without major technical advances, it is not possible to obtain the size distribution of fully dissolved and undegraded native starch using FFF techniques, due to problems with obtaining adequate signal to noise ratio for starch in an optimal eluent [11].

It is essential to be aware that (contrary to what is sometimes stated incorrectly) SEC separates by molecular size, not by molecular weight. For complex branched polymers such as starch, there is no relation between molecular weight and molecular size: molecules of the same size can have different branching structures and hence different molecular weights. The size separation parameter with SEC is the so-called hydrodynamic volume V_h or the corresponding hydrodynamic radius R_h. V_h is defined [24] as the volume of a hypothetical impenetrable sphere displaying in a hydrodynamic field the same frictional effect as an actual polymer molecule. For SEC, V_h is a complicated quantity with the dimensions of molecular size, proportional to the product of the weight-average intrinsic viscosity and the number-average molecular weight [25]: $V_h = \frac{2}{5}\left[\bar{\eta}(V_h)\right]_w \overline{M}_n(V_h) / N_A$, where N_A is Avogadro's constant.

To obtain SEC data which can be reproduced, it is essential to calibrate with appropriate standards, and either report the results in terms of molecular size or give the calibration curve; data reported only in terms of elution time or volume are irreproducible, as this varies from day-to-day and with the SEC setup, and it is scientifically inadmissable to publish irreproducible data. Starch CLD methods involve one more step, debranching, which removes the branching points of starch with isoamylase [21, 26].

2 Materials

We now give a typical procedure for obtaining the SEC distribution of debranched starch.

2.1 Rice

Nipponbare rice grains.

2.2 Buffer and Solutions

Tricine buffer (250 mM, pH 7.5)

1. Add 44.80 g tricine to 1 L deionized water.
2. Prepare 1 M NaOH by dissolving 4 g NaOH in 100 mL deionized water.
3. Adjust the pH of 1). To 7.5 with ~20 mL of 2).
4. Store the tricine buffer at 4°C.

Sodium bisulfite solution (0.45% w/w)

1. Add 0.45 g sodium bisulfite to 100 mL deionized water.
2. Store at room temperature (RT).

Dimethyl sulfoxide (DMSO)/LiBr

1. Add 13.75 g LiBr to 2.5 L pure DMSO with dissolved LiBr (0.5% w/w).
2. Store at RT.

Acetate buffer (0.1 M, pH ~3.5)

1. Acetic acid (0.1 M)—diluted 1.15 mL of 17.4 M acetic acid with deionized water to 200 mL.
2. Sodium acetate (0.1 M)—dissolve 1.64 g of anhydrous sodium acetate (or 2.72 g of sodium acetate trihydrate) in deionized water and dilute to 200 mL.
3. Adjust the pH of acetic acid solution (0.1 M) with sodium acetate solution (0.1 M) to pH 3.5 (the ratio is 18.9:1).
4. Store acetate buffer at 4°C. It expires in 6 months.

Sodium azide solution (0.04 g/mL)

1. Add 8 g of sodium azide to 200 mL deionized water.
2. Store at RT.

Protease from Streptomyces griseus *(Type XIV, ≥3.5 units/mg solid, powder,* Sigma-Aldrich *P5147).*

1. Store protease in −20 °C Freezer.

Isoamylase from Pseudomonas sp. *(Lot 130104b, Megazyme).*

1. Store Isoamylase at 4 °C.

Pullulan standards (peak molecular weight ranging from 342 to 2.35 × 10^6, Polymer Standard Services, Mainz, Germany)

1. Dissolve 1 mg of pullulan standards in 1 mL DMSO/LiBr (0.5%, w/w) solution in the thermomixer at 80 °C at 350 rpm for 3 h.
2. Store at RT.

2.3 Equipment

Hydrophilic Teflon membrane filter

Millipore, Billerica, MA, USA.

Cryogrinder

Freezer/Mill 6850 SPEX, Metuchen, NJ, USA.

Thermomixer

Thermomixer Comfort, Hamburg, Germany.

Centrifuge

Avanti® J-E, Beckman Coulter, USA.

Freeze-drier

BenchTop Pro with Omnitronics, VirTis SP Scientific, USA.

SEC

Agilent 1100 Series SEC, Agilent Technologies, Waldbronn, Germany

1. Separation columns (GRAM precolumn, GRAM30, and 1000 analytical columns, Polymer Standard Services, PSS, Mainz, Germany).
2. Refractive index detector (RID, 235RID-10A, Shimadzu, Kyoto, Japan).

PSS WinGPC Unity software

Polymer Standard Services, Mainz, Germany.

3 Methods

3.1 Sample Preparation

An essential preliminary for characterizing starch structure using SEC is that the sample be prepared in a molecularly dispersed (unaggregated) form, without loss (or at least without loss that is biased in any structural property such as size) and without degradation. This is particularly hard for native starch, as distinct from modified or processed starch: the presence of non-starch material (non-starch polysaccharides, cell-wall material, proteins, and lipids) and high crystallinity complicates nondestructive extraction and dissolution. Moreover, the isolation of starch from a natural source (e.g., a cereal grain) requires grinding, which can easily induce molecular degradation [26, 27]. Dissolving previously extracted or degraded/processed starch is a simpler task to implement without further degradation, because extraction (and/or degradation and/or processing) is highly likely to reduce molecular size, and partially or wholly remove non-starch materials. Another difficulty in dispersing starch is its propensity to hydrogen-bond with itself.

Water-based systems (without any other hydrogen-bonding solvent) can be readily implemented for the starch extraction/dissolution. This however normally involves using aggressive conditions, for example, high temperature, microwave heating, pressure, and high or low pH, although other techniques involve initial dissolution in DMSO followed by redispersion in a water-based solvent, and/or dispersion using a microwave oven [28]. However, thus far, there is no firm evidence that any such techniques avoid significant degradation, re-association and/or loss [29, 30].

Methods have been described [21, 26, 31] for ensuring that dissolution is complete and with minimum degradation using a non-aqueous hydrogen-bonding solvent, specifically dimethyl sulfoxide (DMSO) with small amounts of a hydrogen-bond disrupter, for example LiBr. The methodology needs some degree of optimization for each type of grain. A typical procedure uses a combination of protease, sodium bisulfite, DMSO with 0.5% w/w LiBr, and ethanol solutions to completely dissolve starch and remove non-starch components. For this method, different characterization procedures have been applied to test for loss and degradation of starch, and where the final amounts of non-starch components are minimized. For obtaining the first-level starch structure, isoamylase is used to remove the branching points of starch [21, 26].

3.1.1 Cryogenic Milling of Rice Grains

1. Pour enough liquid nitrogen into the cryogrinder (*see* **Note 1**).
2. Put rice grain samples in the plastic vials containing a grinding bead.
3. Lock the plastic vials in position in the cryogrinder.
4. Close the lid of the cryogrinder.
5. Freeze the rice grain samples in liquid nitrogen for 2 min.
6. Pulverize samples for 5 min with grinding speed at 10 s^{-1}.
7. Open the lid and remove the samples.
8. Transfer the rice flour to another tube.
9. Store samples at RT until use.

3.1.2 Starch Extraction (For Pure Starch, Skip **Steps 1** *and* **3–6***)*

1. Transfer 7–9 mg flour (equivalent to ~4 mg dry starch) to a 2 mL microcentrifuge (Eppendorf) tube.
2. Mix the flour with 0.5 mL protease in tricine buffer (2.5 Units/mL) using a vortex mixer at 37 °C for 30 min.
3. Centrifuge at 4000 × *g* for 10 min and discard the supernatant.

   (**steps 4** and **5** are recommended only for grains containing proteins with high disulfide bonds)
4. Mix the precipitate with 0.5 mL sodium bisulfite solution (0.45%) using a vortex mixer, and incubate at 37 °C for 30 min.
5. Centrifuge at 4000× *g* for 10 min and discard the supernatant.
6. Suspend the precipitate in 1.5 mL DMSO/LiBr solution (0.5% w/w) and shake it gently by hand to make a homogenous mixture (*see* **Note 2**).
7. Heat in the thermomixer at 80 °C 350 rpm for 24 h, inverting occasionally by hand to ensure a homogenous mixture.
8. Mix well, centrifuge at 4000 × *g* for 10 min, and collect the supernatant in a 15 mL centrifuge tube (this step is to remove insoluble materials, mostly non-starch polysaccharides and lignin).
9. Precipitate the starch in the supernatant using 10 mL absolute ethanol.
10. Centrifuge at 4000 × *g* for 10 min and discard the supernatant (this step is to remove non-starch soluble materials, mostly lipids and any residual proteins).
11. Add another 10 mL absolute ethanol and shake gently.
12. Centrifuge at 4000 × *g* for 10 min and discard the supernatant (this step is to remove residual DMSO/LiBr).
13. Trap the last drop of ethanol by inverting the tube on a paper towel.

(For molecular analysis of fully branched starch, after the procedure of starch extraction above, dissolve the samples in DMSO/LiBr with a concentration of 2 mg/mL in a thermomixer at 80 °C with shaking for 2 h. Centrifuge the samples at 4000 × *g* for 10 min and transfer the supernatant to SEC vials to store until SEC analysis (*see* **Note 3**)).

3.1.3 Debranching (Also Prepare a Blank Without Starch)

1. Disperse the precipitate in 0.9 mL of warm or hot deionized water and heat in a boiling water bath for 15 min.
2. Cool the dispersion to room temperature and add 5 μL of sodium azide solution (0.04 g/mL) as preservative, 0.1 mL of 0.1 M acetate buffer (pH ~3.5).
3. Transfer 2.5 μL isoamylase solution to the starch dispersion.
4. Vortex the whole mixture and incubate in a water bath at 37 °C for 3 h.
5. Neutralize the mixture to pH ~7 (measured using pH paper) with 0.1 M NaOH (about 0.1 mL).
6. Transfer the mixture to a 2 mL microcentrifuge (Eppendorf) tube.
7. Heat in the thermomixer at 80 °C 350 rpm for 1 h.
8. Lift up the lid of the microcentrifuge tube and cover with a punctured cap (two holes).
9. Freeze the starch dispersion in liquid nitrogen until reduced or no bubbling is observed.
10. Freeze-dry the starch overnight (dried samples may be stored in desiccator until use; best to tightly seal the microcentrifuge tube with parafilm).
11. Dissolve the dry starch in 1.0 mL DMSO solution containing LiBr (0.5% w/w) in the same microcentrifuge tube in the thermomixer at 80 °C 350 rpm for 24 h and invert it by hand occasionally to ensure complete dissolution.
12. Mix well, centrifuge at 4000 × *g* for 10 min (*see* **Note 4**).
13. Transfer the supernatant in a SEC vial.
14. Store it at room temperature until SEC analysis.

3.2 SEC Separation and Detection

Separation. The main component of the SEC setup is the separation column, which contains a stationary phase of porous polymer packing material. Polymer molecules in the mobile phase with a large hydrodynamic volume cannot penetrate into the pores and will therefore elute out of the column first, while smaller polymers take a longer time, as they have access to the extra volume inside the pores. The different detectors give complementary information on the whole distribution of polymer molecules. The analytes must not interact enthalpically with the stationary phase in order to have the proper size-separation regime.

There are technical problems with SEC, particularly band broadening and shear scission. Band broadening is the effect where molecules with the same V_h elute at a finite range of elution volumes. Although this broadening is quite small for good modern columns, it can significantly affect the distribution shape [32] and hide some important features, such as shoulders or subsidiary maxima that could represent different starch components and enzyme sets. Although some methods have been suggested to correct the band-broadening problem [33–37], these are not yet readily available to be implemented for starch systems. Another major problem with SEC is shear scission for large branched molecules like amylopectin. Shear forces inside the column will break large molecules into two roughly equal size parts [20, 23]. Some efforts have been made to reduce the amount of shear scission, such as changing elution conditions, e.g., flow rate [23, 29]. While shear scission is reduced with a slower flow rate, the effect cannot be completely removed satisfactorily (as shown by dimensionless-number analysis [23]), because the lower the flow rate, the greater the band-broadening. There is also the limitation of finding an appropriate solvent system with sufficiently low viscosity. Despite these limitations, the degree of shear scission for different polymer molecules in the same SEC setup and same run would vary only slightly, which allows for some meaningful comparison between the amylopectin components of starches.

Contamination from those non-starch components in the sample which have not been removed in the dissolution procedure could result in elution in a region which overlaps that of the starch in the sample. Further, contaminants such as lipids and phospholipids can cause starch molecules to form helical complexes and entangle with one another, which can also result in artifacts. Further, it is essential to check that there are no artifacts in the apparent distribution because of column exclusion limits: for example, if different samples all show a rapid drop to the baseline at the same size, it is highly likely that this is an artifact arising because all molecules larger than a certain size elute at the same time, irrespective of their true molecular size, because all larger sizes are too big to enter the pores in the column.

Detection. The objective is to detect the amount of starch present in a given elution slice, with the signal dependent only on a chosen property of the starch (e.g., distribution of the number or weight of chains) in that elution slice.

The common detectors for SEC are viscometric, differential refractive index (DRI), and multiple-angle laser light scattering (MALLS), which give, respectively, the number ($N(V_h)$) and weight ($w(V_h)$) distributions, plus weight-average molecular weight $M_w(V_h)$ and z-average radius of gyration $R_{g,z}(V_h)$ [20, 38]. The more types of detectors, the more information about starch structure can be obtained by size separation, although one-dimensional methods are always limited to single moments of an infinite-dimensional distribution [38]. The SEC weight distribution $w(\log V_h)$ is defined by $w(\log V_h)d(\log V_h)$ being the weight of chains in the volume increment $\log V_h$ to $\log V_h + d(\log V_h)$. This definition is because this quantity is the signal given by an ideal SEC with no band broadening and with a linear calibration curve [39]. For linear polymers, DRI detection is able to give both number and weight distributions, as there is a trivial relationship between $N(V_h)$ and $w(\log V_h)$, given by $w(\log M) = M^2\ N(M)$, where M is the molecular weight [39]. However, this relationship is not valid for complex branched polymers.

A MALLS detector used with a DRI detector enables one to obtain $M_w(V_h)$, and thus enables the X axis to be put in terms of M_w rather than V_h or R_h. However, this is the weight-average molecular weight in that elution slice, not the actual molecular weight. No SEC device can give a molecular weight distribution for a branched polymer, because the size separation is by size, not molecular weight. (In principle, mass-spectrometric techniques could give a true molecular weight distribution, but with problems of fragmentation and molar-mass-dependent signal intensity, this is impractical except for special cases such as relatively low molecular weights using MALDI or electrospray MS.)

For debranched starch, as with any linear polymer, there is a unique relationship between molecular size (V_h or R_h) and molecular weight. In principle, the MALLS detector could give absolute molecular weight for this, but the problem is that the light-scattering signal is weak with small molecules such as debranched amylopectin chains. Instead, the principle of universal calibration is invoked, which is the assumption that separation is only a function of R_h and is independent of polymer composition. One then calibrates with essentially monodisperse standards of known R_h, as directly measurable from the intrinsic viscosity of an individual monodisperse sample. Knowing R_h for a given polymer, one can then find the corresponding molecular weight using the Mark-Houwink equation: $V_h = {}^{2}/_{5}\ KM^{1} + {}^{\alpha}/N_A$. Here K and α are the Mark-Houwink parameters, which depend on the polymer, solvent and temperature, and can be obtained from independent viscometry measurements (and are often available in the literature). Note that the Mark-Houwink equation is an empirical one, and may not work well under certain circumstances, such as quite low molecular weights.

Here is a typical procedure.

1. DMSO/LiBr (0.5% w/w) solution after filtration through a 0.45 μm hydrophilic Teflon membrane filter is used as mobile phase (the Mark—Houwink parameters for this eluent at 80 °C are $K = 2.424 \times 10^{-4}$ dL g^{-1} and $\alpha = 0.68$).

2. Separation columns are PSS GRAM precolumn, GRAM30, and 1000 analytical columns.
3. Only a refractive index detector is used.
4. The separation columns are held at 80 °C and the detector is set at 45 °C.
5. The flow rate is set at 0.5 mL min^{-1}.
6. A series of pullulan standards with molecular weights ranging from 342 to 2.35×10^6 is used for calibration.

3.3 Data Processing and Reporting

Data processing. Data processing is needed to convert the raw signals from one or more detectors (these signals depend on the particular setup of the separation and detection system) into one or more distribution functions (e.g., a number or weight distribution). The distribution functions are expressed as V_h or R_h distributions, which depend only on the sample and not on the separation/detection system. Note that $w(\log V_h)$ and $w(\log R_h)$ are the same, to within an arbitrary normalization constant.

Data reporting. The objective for data reporting is to present the distribution data along with sufficient information that the experiment is reproducible. It should contain a complete description of the starting material and the dissolution technique. A full description of the separation setup should also be provided, including for example flow rate, material used for the column, and so on. The calibration information should also be reported, which can be deposited as supplementary information. It is strongly preferable to report results in terms of R_h rather than V_h, as scientists have a "feel" for the size (diameter) or a molecular entity rather than for its volume.

1. The SEC chromatograms are analyzed using appropriate software, such as Astra (Wyatt) or WinGPC (PSS).
2. A calibration curve is made to relate the elution time to V_h and thus R_h, with $V_h = {}^4/_3 \pi R_h^3$.
3. The size distribution $w(\log R_h)$ is calculated from the detector signal (*see* Fig. 1 for an example).

The normalization of any distribution, including those from SEC or FACE, is arbitrary, and can be chosen for convenience to bring out whatever feature is being considered. For example, to look at amylopectin CLDs from different samples, the CLD can conveniently be normalized to the overall maximum (which will almost always be the amylopectin component in a CLD). For looking at amylose features, it is convenient to normalize to the maximum in the amylose component of the CLD. Under other circumstances, it can be convenient to normalize to the amount of starch present in the sample, e.g., when comparing changes in the debranched distributions as enzymatic digestion proceeds [40].

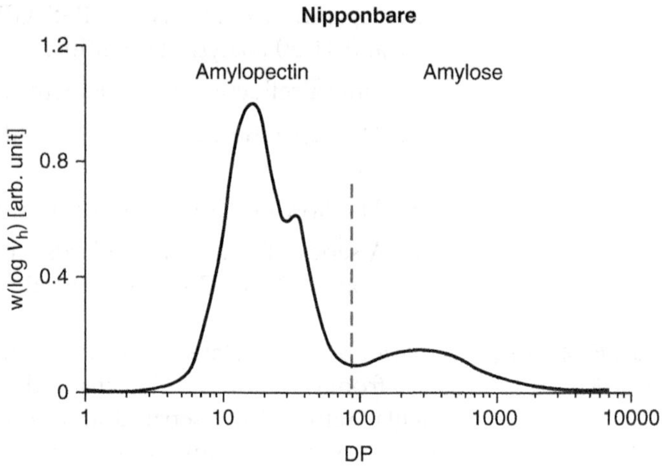

Fig. 1 A typical example of starch CLD from Nipponbare rice grain, obtained with SEC. The amylopectin and amylose chain ranges are shown, with the dividing line set at the minimum in $w(\log V_h)$ (which is close to DP 100)

The CLD can be expressed in terms of the weight or number distributions, related as given above: $w(\log M) = M^2\ N(M)$. Molecular weight for a linear homopolymer can also be expressed in terms of the degree of polymerization X, where for linear starch, one has $M = 162.2X + 18.0$, 162.2 being the molecular weight of the anhydroglucose monomeric unit and 18.0 that of the additional water in the reducing end group.

For reasons given in detail elsewhere [41, 42], it is strongly preferable to plot the number CLD, $N_{de}(X)$, as log $N_{de}(X)$, which voids many artifacts of normalization and brings out features at higher DPs which are essentially invisible in a non-logarithmic plot (as unfortunately is often done in the literature). On the other hand, an amylose CLD is best plotted as $w(\log X)$ against $\log X$.

4 Notes

1. Make sure the cryogrinder is used in a well-ventilated area as a large amount of liquid nitrogen is consumed by the cryogrinder.
2. The grinding step might take a few days, depending on the nature and the amount of grain in the sample. For less than 30 samples, the time for starch extraction will take an afternoon and the debranching procedure a whole day. For more than 30 samples, it is better to divide the samples into different sets and do the starch extraction and debranching accordingly. The freeze-dry procedure for starch samples normally takes overnight followed by another 24 h to dissolve the dried samples in DMSO/LiBr (0.5% w/w) solution. Finally, the separation time for a single starch sample using SEC is ~1 h.

3. This protocol also applies to measuring fully branched starch molecular structure (except of course that the debranching procedure is omitted, and usually the columns will be different).
4. Carrying out the centrifugation step must be done just before transferring the starch solution into the SEC vial. This step is to remove the impurities in the solution which could block the SEC columns from blocking, and this step also reduces the height of contamination peaks in the chromatograms.

FACE

Many of the practical procedures shown above are also used in FACE measurements, especially the starch isolation, dissolution, and debranching steps. The most important difference is the need to label with a charged fluorophore. This technology was developed in elegant work by Morell and coworkers over some years [22, 43]. Recently, some improvements in these techniques have been developed by Wu et al. [21]. Figure 2 gives an example of a FACE CLD. This was obtained with a PA-800 Plus System and monitored with a solid-state laser-induced fluorescence (LIF) detector with an argon-ion laser for excitation (Beckman-Coulter, Brea, CA, USA), a 50 μm diameter N-CHO coated capillary with effective capillary separation length of 40 cm, and the Beckman carbohydrate separation buffer as the separating medium. The separation used an applied voltage of 30 kV (current ~14 mA) at 25 °C over 90 min of total separation time. The areas under the peaks give the number CLD directly, as this device gives baseline separation between each DP. This setup enables CLDs to be measured as high as DP ~170, which is much higher than the range in earlier types of FACE equipment. A reminder that FACE gives the number CLD $N_{de}(X)$ directly, whereas SEC gives the SEC weight distribution with $w(\log X) = X^2\, N_{de}(X)$.

FULLY BRANCHED DISTRIBUTIONS

The techniques used above, but without the debranching step, give the fully branched distributions. As stated above, it is essential to be aware that SEC characterization cannot avoid shear scission of the larger amylopectin molecules. The example given in Fig. 3 shows a typical result (recalculated from data in [44]). It shows the smaller amylose molecules, the larger amylopectin ones, and also the region where shear scission of the amylopectin component is significant. The extraneous features for smaller sizes are non-starch impurities.

USING THIS METHODOLOGY IN THE DEVELOPMENT OF NUTRITIONALLY IMPROVED STARCHES

Although not the main topic of this chapter, we now give a brief overview showing how this methodology can be used to develop improved plant varieties. The simplest example is to choose or breed varieties where the CLD is known to be nutritionally advantageous, e.g., with longer amylopectin branches [45]. Another is

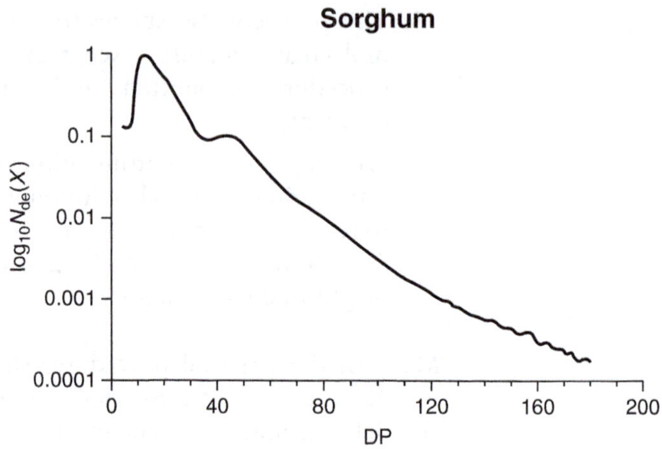

Fig. 2 A typical example of starch CLD from sorghum grain, obtained with FACE. It gives the number CLD; the characterization range is as high as DP ~170. Although FACE yields $N_{de}(X)$ with resolution of individual X values, for visual ease the data are presented as a continuous line

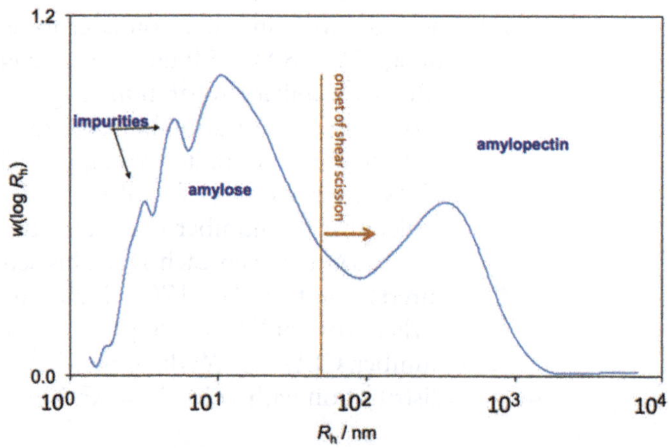

Fig. 3 SEC weight distribution $w(\log R_h)$ (arbitrary units) for a rice starch (data recalculated from V_h distribution given in [44]). Also shown is the size value above which shear scission occurs, and two peaks at low size which are artifacts from non-starch impurities

a transgenic method developed by the present authors [46]. This is based on fundamental theory for the biosynthetic processes controlling the CLD [47, 48]; from this, it is predicted that a slight change in the minimum chain-length requirements for the function of starch branching enzyme would have a significant effect on the CLD. Such a change could be brought about by changing a single residue on the enzyme branching site, and can be implemented by standard transgenic techniques in molecular biology. At each step of the process, it was important to monitor, using the techniques given in this chapter, the CLDs in vitro produced by the modified enzymes. This led to selection of sites which could give reduced activity of the enzyme, as required for this goal. The same technology could be used to develop plants producing starches with slower digestibility and hence improved nutritional performance.

Acknowledgments

Robert G. Gilbert greatly appreciates the support of the Chinese Government's 1000 Talents program of the State Foreign Experts Bureau.

References

1. Englyst HN, Kingman SM, Cummings JH (1992) Classification and measurement of nutritionally important starch fractions. Eur J Clin Nutr 46(Suppl 2):S33–S50
2. Byrnes SE, Miller JCB, Denyer GS (1995) Amylopectin starch promotes the development of insulin-resistance in rats. J Nutr 125:1430–1437
3. Johnston KL, Thomas EL, Bell JD, Frost GS, Robertson MD (2010) Resistant starch improves insulin sensitivity in metabolic syndrome. Diabet Med 27:391–397
4. Lehmann U, Robin F (2007) Slowly digestible starch, its structure and health implications: a review. Trends Food Sci Technol 18:346–355
5. Baba T, Arai Y (1984) Structural features of amylomaize starch. 3. Structural characterization of amylopectin and intermediate material in amylomaize starch granules. Agric Biol Chem 48:1763–1775
6. Li L, Jiang HX, Campbell M, Blanco M, Jane JL (2008) Characterization of maize amylose-extender (ae) mutant starches. Part i: relationship between resistant starch contents and molecular structures. Carbohydr Polym 74:396–404
7. Vilaplana F, Gilbert RG (2010) Two-dimensional size/branch length distributions of a branched polymer. Macromolecules 43:7321–7329
8. Nakamura T, Yamamori M, Hirano H, Hidaka S, Nagamine T (1995) Production of waxy (amylose-free) wheats. Mol Gen Genet 248:253–259
9. Jiang HX, Jane LL, Acevedo D, Green A, Shinn G, Schrenker D, Srichuwong S, Campbell M, Wu YS (2010) Variations in starch physicochemical properties from a generation-means analysis study using amylomaize v and vii parents. J Agric Food Chem 58:5633–5639
10. Vilaplana F, Hasjim J, Gilbert RG (2012) Amylose content in starches: toward optimal definition and validating experimental methods. Carbohydr Polym 88:103–111
11. Gilbert RG (2011) Size-separation characterization of starch and glycogen for biosynthesis-structure-property relationships. Anal Bioanal Chem 399:1425–1438
12. Gilbert RG, Witt T, Hasjim J (2013) What is being learned about starch properties from multiple-level characterization. Cereal Chem 90:312–325
13. Witt T, Gilbert RG (2014) Causal relations between structural features of amylopectin, a semicrystalline hyperbranched polymer. Biomacromolecules 15:2501–2511
14. Witt T, Doutch J, Gilbert EP, Gilbert RG (2012) Relations between molecular, crystalline, and lamellar structures of amylopectin. Biomacromolecules 13:4273–4282
15. Zhang G, Ao Z, Hamaker BR (2008) Nutritional property of endosperm starches from maize mutants: a parabolic relationship between slowly digestible starch and amylopectin fine structure. J Agric Food Chem 56:4686–4694
16. Zhang G, Sofyan M, Hamaker BR (2008) Slowly digestible state of starch: mechanism of slow digestion property of gelatinized maize starch. J Agric Food Chem 56:4695–4702
17. Srichuwong S, Sunarti TC, Mishima T, Isono N, Hisamatsu M (2005) Starches from different botanical sources i: contribution of amylopectin fine structure to thermal properties and

enzyme digestibility. Carbohydr Polym 60: 529–538

18. Benmoussa M, Moldenhauer KA, Hamaker BR (2007) Rice amylopectin fine structure variability affects starch digestion properties. J Agric Food Chem 55:1475–1479
19. Zhu LJ, Liu QQ, Wilson JD, Gu MH, Shi YC (2011) Digestibility and physicochemical properties of rice (oryza sativa l.) flours and starches differing in amylose content. Carbohydr Polym 86:1751–1759
20. Vilaplana F, Gilbert RG (2010) Characterization of branched polysaccharides using multiple-detection size separation techniques. J Sep Sci 33:3537–3554
21. Wu AC, Li E, Gilbert RG (2014) Exploring extraction/dissolution procedures for analysis of starch chain-length distributions. Carbohydr Polym 114:36–42
22. O'Shea MG, Samuel MS, Konik CM, Morell MK (1998) Fluorophore-assisted carbohydrate electrophoresis (face) of oligosaccharides: efficiency of labeling and high-resolution separation. Carbohydr Res 307:1–12
23. Cave RA, Seabrook SA, Gidley MJ, Gilbert RG (2009) Characterization of starch by size-exclusion chromatography: the limitations imposed by shear scission. Biomacromolecules 10:2245–2253
24. Jones RG, Kahovec J, Stepto R, Wilks ES, Hess M, Kitayama T, Metanomski WV (2009) Compendium of polymer terminology and nomenclature. Iupac recommendations 2008. Royal Soc Chem, Cambridge
25. Kostanski LK, Keller DM, Hamielec AE (2004) Size-exclusion chromatography-a review of calibration methodologies. J Biochem Biophys Methods 58:159–186
26. Syahariza ZA, Li E, Hasjim J (2010) Extraction and dissolution of starch from rice and sorghum grains for accurate structural analysis. Carbohydr Polym 82:14–20
27. Hasjim J, Jane JL (2009) Production of resistant starch by extrusion cooking of acid-modified normal-maize starch. J Food Sci 74: C556–C562
28. Bello-Perez LA, Roger P, Baud B, Colonna P (1998) Macromolecular features of starches determined by aqueous high-performance size exclusion chromatography. J Cereal Sci 27: 267–278
29. You SG, Lim ST (2000) Molecular characterization of corn starch using an aqueous hpsec-malls-ri system under various dissolution and analytical conditions. Cereal Chem 77:303–308
30. Kim HS, Huber KC, Higley JS (2006) Alkaline dissolution of starch facilitated by microwave heating for analysis by size-exclusion chromatography. J Agric Food Chem 54:9664–9669
31. Schmitz S, Dona AC, Castignolles P, Gilbert RG, Gaborieau M (2009) Quantification of the extent of starch dissolution in dimethylsulfoxide by 1h nmr spectroscopy. Macromol Biosci 9:506–514
32. van Berkel KY, Russell GT, Gilbert RG (2005) Molecular weight distributions and chain-stopping events in the free-radical polymerization of methyl methacrylate. Macromolecules 38:3214–3224
33. Schnoll-Bitai I, Vega J, Mader C (2007) Estimation of the band broadening parameters in single detection size-exclusion chromatography: a comparative study of various column combinations. Anal Chim Acta 604:9–17
34. Gruendling T, Guilhaus M, Barner-Kowollik C (2008) Quantitative lc-ms of polymers: determining accurate molecular weight distributions by combined size exclusion chromatography and electrospray mass spectrometry with maximum entropy data processing. Anal Chem 80:6915–6927
35. Gruendling T, Guilhaus M, Barner-Kowollik C (2009) Fast and accurate determination of absolute individual molecular weight distributions from mixtures of polymers via size exclusion chromatography-electrospray ionization mass spectrometry. Macromolecules 42:6366–6374
36. Castro JV, van Berkel KY, Russell GT, Gilbert RG (2005) General solution to the band-broadening problem in polymer molecular weight distributions. Aust J Chem 58:178–181
37. Konkolewicz D, Taylor JW II, Castignolles P, Gray-Weale AA, Gilbert RG (2007) Towards a more general solution to the band-broadening problem in size separation of polymers. Macromolecules 40:3477–3487
38. Gray-Weale A, Gilbert RG (2009) General description of the structure of branched polymers. J Polymer Sci Polymer Chem 47:3914–3930
39. Clay PA, Gilbert RG (1995) Molecular-weight distributions in free-radical polymerizations.1. Model development and implications for data interpretation. Macromolecules 28:552–569
40. Witt T, Gidley MJ, Gilbert RG (2010) Starch digestion mechanistic information from the time evolution of molecular size distributions. J Agric Food Chem 58:8444–8452
41. Castro JV, Dumas C, Chiou H, Fitzgerald MA, Gilbert RG (2005) Mechanistic information from analysis of molecular weight distributions of starch. Biomacromolecules 6:2248–2259

42. Wu AC, Witt T, Gilbert RG (2013) Characterization methods for starch-based materials: state of the art and perspectives. Aust J Chem 66:1550–1563
43. O'Shea MG, Morell MK (1996) High resolution slab gel electrophoresis of 8-amino-1,3, 6-pyrenetrisulfonic acid (apts) tagged oligosaccharides using a DNA sequencer. Electrophoresis 17:681–686
44. Gray-Weale A, Cave RA, Gilbert RG (2009) Extracting physically useful information from multiple-detection size-separation data for starch. Biomacromolecules 10:2708–2713
45. Ao ZH, Simsek S, Zhang GY, Venkatachalam M, Reuhs BL, Hamaker BR (2007) Starch with a slow digestion property produced by altering its chain length, branch density, and crystalline structure. J Agric Food Chem 55:4540–4547
46. Li C, Wu AC, Go RM, Malouf J, Turner MS, Malde AK, Mark AE, Gilbert RG (2015) The characterization of modified starch branching enzymes: toward the control of starch chain-length distributions. PLoS One 10:e0125507
47. Wu AC, Morell MK, Gilbert RG (2013) A parameterized model of amylopectin synthesis provides key insights into the synthesis of granular starch. PLoS One 8:e65768
48. Wu AC, Gilbert RG (2010) Molecular weight distributions of starch branches reveal genetic constraints on biosynthesis. Biomacromolecules 11:3539–3547

Chapter 11

Rice Grain Quality Benchmarking Through Profiling of Volatiles and Metabolites in Grains Using Gas Chromatography Mass Spectrometry

Cindy Llorente, Rosario Jimenez, Jackie, Yariv Brotman, Alisdair R. Fernie, and Nese Sreenivasulu

Abstract

Gas chromatograph coupled with mass spectrometer is widely used to profile volatiles and metabolites from the homogenized rice flour obtained from mature grains. Rice grains consist of central endosperm which stores majorly starch and, in addition, accumulate various storage proteins as storage reserves. The outer nutritious aleurone layer stores lipids, sugar alcohols, volatiles, antioxidants, vitamins, and various micronutrients. Once paddy sample is dehulled, milled, and ground cryogenically, the brown rice flour is subjected to extraction of primary metabolites and volatiles using an appropriate extraction method. In metabolite profiling of the liquid extract obtained from the rice sample, mixture is initially subjected to methoxyamination then silylation before being subjected to untargeted metabolite profiling. Peaks obtained are processed for noise reduction and specific signal selection. Volatile compounds are initially extracted using a solid phase adsorbent prior to analysis. All these compounds, metabolites, and volatiles are detected in the mass selective detector by fragmentation at 70 eV ionization energy and the resultant mass spectrum compared with a built-in library of compounds. Data mined from the gas chromatography mass spectrometry analysis are then subjected to post-processing statistical analysis.

Key words Rice, Aroma, Metabolites, Metabolite profile analyses

1 Introduction

Chemical composition of seeds comprises a wide array of storage products including starch, storage proteins, lipids, various primary metabolism intermediates, and secondary metabolites carrying human health benefits [1, 2]. Various metabolite profiling techniques such as gas chromatography mass spectrometry (GC-MS), liquid chromatography mass spectrometry (LC-MS), capillary electrophoresis mass spectrometry (CE-MS), and chromatography coupled nuclear magnetic resonance (NMR) technologies are available with various degrees of accuracy and sensitivity [3]. Combining the information obtained from several of these

Nese Sreenivasulu (ed.), *Rice Grain Quality: Methods and Protocols*, Methods in Molecular Biology, vol. 1892, https://doi.org/10.1007/978-1-4939-8914-0_11,

technologies, it is possible to cover the global metabolome index with thousands of metabolites. Among these platforms, the GC-MS is one of the key technologies applied routinely to profile several hundred metabolites that helped reveal the metabolic signature index of different plant species which possess diverse seed storage products [4–6]. Over the last decade, the GC-MS platform is the most widely used due to superior chromatographic resolution, better reproducibility, repeatability of mass fragmentation through electron impact ionization, as well as its low operational and maintenance cost [7]. These attributes make GC-MS reliable in terms of characterization of numerous compounds in different biological systems, especially at various stages of their development.

Rice grains contain up to 90% starch and 6–8% proteins preferentially in endosperm, while the outer nutritious layer bran and endosperm contain lipids, antioxidants, dietary fiber, and other micronutrients [8]. Thus, rice remains an important cereal crop providing a valuable source of energy for human consumption. GC-MS profiling technique has been instrumental in acquiring biological information for the identification of varietal differences in terms of rice grain quality [9], starch quality [10] and for distinguishing aroma through volatile profiling [11–13]. The measured 759 metabolic signatures investigated from the grains of 85 lines from backcrossed inbred population were investigated for association mapping using metabolome quantitative trait loci analysis [14, 15]. Additional application includes the identification of genotypic and phenotypic differences of germinating rice seeds [16]. Elucidation of seed storage composition of varietal differences [17], drought implications on seed storage composition [18], and application of metabolomic techniques as screening method for predicting rice quality traits were reported in the past [19–21]. Metabolomic studies have also helped in the investigation of the molecular basis of rice quality [22] as well as the identification of metabolic variation between transgenic lines and its wild type, mapping populations created from breeding lines due to higher accumulation of sucrose, mannitol, and glutamic acid [23]. The GC-MS platform was also shown to be useful in elucidating the type of mutation in rice as part of food safety assessments [24].

Metabolites in rice grains are widely studied using the GC-MS platform for separation, identification and quantification [7, 25]. By facilitating extraction procedures specific to compounds of interest, the GC-MS aids not only in the identification of more than 100 primary metabolites, but also to unravel the volatile signatures of the sample being analyzed [11, 26, 27]. In this chapter, we describe the GC-MS protocols for volatiles and metabolites found in ground rice grains as performed at the Grain Quality and Nutrition Center, IRRI. The rice paddy is dehulled, milled, and ground with liquid nitrogen using a cryomill prior to extraction

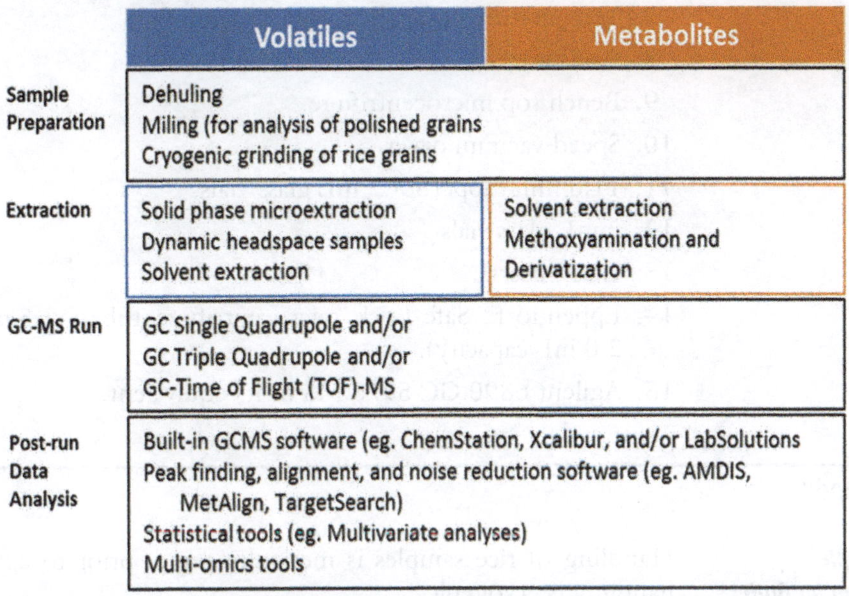

Fig. 1 Workflow of GC-MS in volatile and metabolite profiling in rice grains

and GC-MS analysis. Resulting chromatogram is initially processed for noise reduction and peak alignment before subjecting to statistical analysis (Fig. 1).

2 Materials

2.1 Reagents

1. Methyl tert-butyl ether:Methanol solution (5:1) with (0.3 mg/mL) decanoic acid as internal standard (*see* **Note 1**).
2. Methanol:H_2O (1:3).
3. Methoxyamine hydrochloride.
4. Pyridine.
5. N,O-bis(trimethylsilyl)trifluoroacetamide (BSTFA) with 1% trimethylchlorosilane (TMCS).
6. C4—C24 Even Carbon Saturated FAMEs (1000 μg/mL each component in hexane, analytical standard).

2.2 Equipment

1. Analytical balance (10 mg–220 g ± 0.1 mg capacity).
2. Dehuller (8–10 g capacity).
3. Miller (8–10 g capacity).
4. Ball grinder (equipped with grinding jar with 50 g capacity for handling cryo-cooled samples).
5. Liquid nitrogen container with dispenser.
6. Cold water bath (−10 °C temperature).

7. Temperature-controlled vortex mixer.
8. Sonicator.
9. Bench top microcentrifuge.
10. Speed vacuum dryer.
11. 11.0 mm capper for 2 mL glass vials.
12. 2 mL glass vials.
13. Glass inserts.
14. Eppendorf® Safe-Lock microcentrifuge tubes (1.5 mL and 2.0 mL capacity).
15. Agilent 6890 GC 5975 MS or its equivalent.

3 Methods

3.1 Sample Preparation for Both Volatile and Metabolite Profiling

Handling of rice samples is meticulous and, prior to extraction, handling is cryogenic.

1. Weigh around 2 g of the rice paddy and clean for any impurities.
2. Immediately dehull and place in a grinding jar which is pre-cooled with liquid nitrogen.
3. Pour a generous amount of liquid nitrogen to a grinding jar with the grinding ball to freeze the sample (*see* **Note 2**). Repeat for 2–3 times.
4. Mill the frozen rice sample (*see* **Note 3**).
5. Grind milled samples using the ball grinder for 1 min.
6. Place samples in labeled Ziploc bags or microfuge tubes and immediately store in −80 °C freezer prior to analysis.

3.2 Extraction of Compounds

3.2.1 Volatiles

1. Place 1 g of rice flour sample in 10 mL GC vial.
2. Seal immediately using the suitable crimp cap and septum (*see* **Note 4**).
3. Allow the vials to stand for 30 min to let all the flour settles at the bottom of the vial. This will prevent contamination of the SPME fiber with rice flour.
4. Place the vials in the GC sample tray.
5. From the GC method, heat the rice flour with agitation to allow for the release of the volatiles in the headspace of the sample vial.
6. Insert the SPME fiber to the headspace of the vial for the absorption of the volatiles.
7. Desorb the fiber in the injector and allow for the volatiles pass through the GC column for separation and to the mass spectrometer for fragmentation.

3.2.2 Metabolites

1. Weigh 100 mg of rice flour sample (*see* **Note 5**) per sample in 2.0 mL Eppendorf® Safe-Lock microcentrifuge tubes.
2. Add sample with precooled with MTBE:MeOH containing decanoic acid (0.3 mg/mL) as an internal standard.
3. Place tubes in ice bath while finishing a batch of samples.
4. Vortex mix tubes for 10 min.
5. Sonicate mixture for 15 min making sure that the water bath will not heat up.
6. Add sample mixture with 500 μL MeOH:H_2O solution.
7. Centrifuge samples at 12,000 × *g* for 5 min.
8. Aliquot 1 mL of the polar supernatant into properly labeled 1.5 mL Eppendorf® Safe-Lock microcentrifuge tubes.
9. Dry the extracts overnight in a speed vacuum dryer or its equivalent.
10. Once extracts are dried, remove vials from speed vacuum dryer (*see* **Note 6**). Dried extracts are temporarily stored in dessicator.

3.3 Derivatization of Compounds (See Note 7)

To improve the volatility of some compounds, analytes of particular interest must undergo methoxyamination prior to derivatization where the polar functional groups are modified to decrease their polarity and hence they can be separated in the GC column.

1. Cool the vials to room temperature.
2. Add dried extract with 10 μL methoxyamine hydrochloride (40 mg in 1 mL pyridine), and mix for 90 m in 30 °C dry bath.
3. After the reaction, derivatize the samples using 90 μL BSTFA with 1%TMCS (with 20 μL FAMES in 1 mL BSTFA).
4. Heat the mixture with shaking for 30 min at 37 °C to ensure that derivatization is complete.
5. 60 μL of derivatized mixture is then transferred to 200 μL glass inserts inside 2 mL microfuge tubes.
6. Centrifuge samples at 12,000 × *g* for 30 s to settle down any solid particles into the bottom of the glass insert. This will also avoid clogging the syringe of the GC.
7. Once done, transfer glass inserts into properly labeled 2 mL GC vials and samples are analyzed in the GCMS.

The method described in this procedure employs silylation to alter the functionality of the targeted functional groups with the use of N, O-bis(trimethylsilyl)trifluoroacetamide or BSTFA with 1% trimethylchlorosilane (TMCS). With BSTFA, the active hydrogens of the compounds containing –SH, –OH, –NH, and –COOH are replaced with the trimethylsilyl (TMS) group. This process is catalyzed by trimethylchlorosilane (TMCS) which is already mixed with BSTFA at 1%. Pyridine is used as a solvent which contributes to the reactivity of the solvent by accepting the protons (H^+) during derivatization process. There had been many published methods using different reagents to derivatize compounds of different classes such as MSTFA w/ 1% TMCS [25] in rice and in biological sample extracts [28]; BSTFA w/ 1% TMCS [23] in transgenic rice; MSTFA [24] in rice; hydroxylamine hydrochloride and hexamethylsilylimidazole and trifluoroaceteic acid [29] for sugars in carrots. A comprehensive discussion on the derivatization of sugars [30] provides a selection for the preparation of procedures for the analysis of carbohydrates which may be useful in rice research.

3.4 GC-MS Run (See Note 8)

For every batch of run in the GC a reagent blank and a set of standards in a quality control (QC) mix should be included. Both blank and QC mix should also undergo similar extraction and derivatization procedure together with the sample.

The variability in samples can arise from multiple sources including physiological differences and variability from the analytical method itself. Measuring metabolites using mass spectrometry techniques to explore natural variation from the diversity panel requires appropriate care about homogenous representation of tissue samples, including enough biological and technical replications. In addition, analytical variation caused by suboptimal performance of the chosen apparatus and instrument drift over time are additional major issues in large-scale metabolomics studies, which requires further attention. Batch-to-batch variation is another technical source of variation arising from the sum of both manual and robotic samples handling. The presence of batch-to-batch variation makes it difficult to integrate data from independent batches of samples. This issue is particularly problematic when dealing with a large number of samples such is the case when analyzing structured plant populations. To counter this, several normalization methods have been developed to minimize non-biological variation. For example, normalizations by a single or multiple internal or external standard compounds [7, 31] were considered. Similarly, isotope-labeled internal standard approaches [32] were established to monitor analytical error. While there is no single best way to conduct metabolomic studies, there are a number of pitfalls and known problems that need to be carefully avoided. Detailed guidelines and practice and normalization protocols [33] have been published previously for this purpose. As the number of samples in the data set increases there is a corresponding time-dependent variation in the metabolite data. Removing platform-specific sources of variability such as systematic errors is one of the top priorities in metabolomics data preprocessing. However, metabolite diversity leads to different responses to variations at given experimental conditions, making normalization a very demanding task. The Quality Control (QC) samples are of key importance, and these are best prepared by pooling equal volumes of material from all of the biological samples to be analyzed [34, 35]. Alternatively, a chemically defined mixture of authenticated reference compounds that mimics the metabolic composition of the investigated biological material can be employed. These synthetic mixtures are then subjected to the same sample extraction, subject to instrumental analyses (ideally distributed across the analytical run), and data processing, thus providing quality checks for technical and analytical error, and quantitative calibration to eliminate batch effects for the final processed data. This normalization is a crucial step for minimizing the batch-to-batch data variability across extended periods. As such, this is a crucial requirement for large-scale phenotyping, which facilitate inter-batch data integration.

Depending on the objectives of the study, a single or a triple quadrupole MS can be employed to carry out the profiling for both volatiles and metabolites. Single-quadrupole MS is very useful in giving a full untargeted scan of the compounds while in triple quadrupole the resulting transitions from the precursor ion to the product ion become the fingerprint of the targeted analytes. In a single quadrupole, compounds are first fragmented in the ionization chamber using electron impact ionization set at 70 eV to produce fragment ions. These fragments are then separated by mass in the quadrupole, allowing the ions to be detected at slightly differing time intervals. The resultant mass spectrum can be indexed against the built-in library of compounds commercially available for the identification of analytes.

In the triple quadrupole MS system, there is a second quadrupole or collision cell that further fragments the fragment ions generated from the first quadrupole, also known as the precursor ions. Prior to the collision induced dissociation, the precursor ions are first selected based on their selectivity and sensitivity to produce the product ions. The product ions then pass to the third quadrupole for another mass filtering process. Instead of detecting a specific ion, the transition from a precursor ion to product ion is recorded. This type of detection allows us to properly identify and quantify trace-level compounds in a complex matrix such as that of rice (Fig. 2). Matrix compounds might coincide in having the same precursor ions as the target analytes, but the chances of having the same precursor-to-product transition are rare [36, 37].

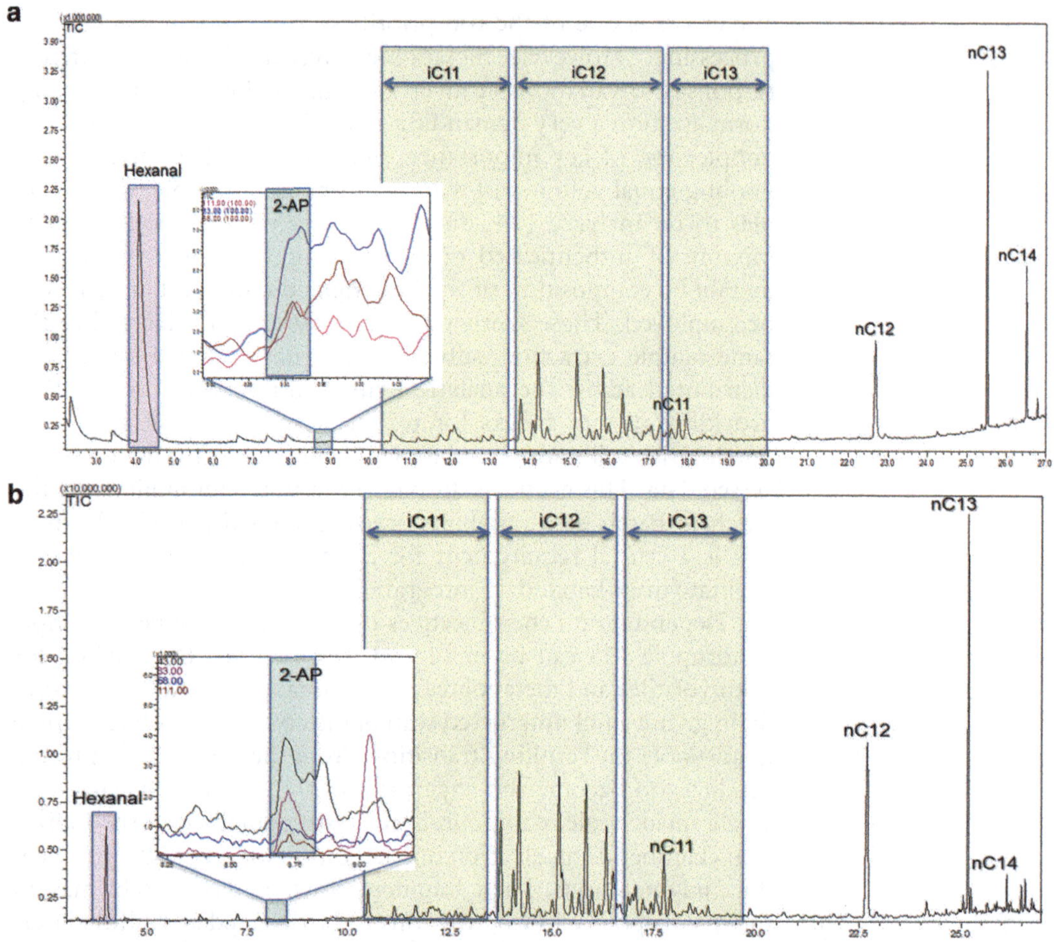

Fig. 2 Total ion chromatogram of a rice sample analyzed for volatiles using (**a**) Dynamic Headspace Sampling and (**b**) SPME for sampling. Both insets show corresponding selected ion monitoring profiles of 2-acetyl-1-pyrroline or 2-AP (chromatogram obtained from Shimadzu GCMS-TQ8040)

Application Box: GCMSMS Analysis of Volatiles and Metabolites in Rice using Triple Quadrupole GC [38]

Targeted metabolites need to be accurately quantified to use them as biomarkers to discriminate differences in traits of diverse rice species. Profiling with a GC-MS/MS (GC triple quadrupole MS) was proven to be an effective tool for complex matrices following 2 phases during profiling. Phase I requires the discovery of the compounds that are present in the sample using the full scan mode. Phase II involves the analysis of selected secondary fragmentation in the multiple reaction monitoring (MRM) mode. Once these MRM transitions are established, they can be used to create targeted screening methods that will scan the whole sample for the identified transitions of each of the specific analytes. This type of workflow allows the

(continued)

researcher to identify and quantify metabolites in a sample without the worry of varying matrix. Metabolomic workflow showing the use of full scan and MRM measurements for discovery and targeted analyses respectively were described earlier [38]. Different systems will have different ways of creating the method for the targeted phase of metabolite profiling using the GC-MS/MS. Generally, compounds of interest, particularly those that may be considered potential biomarkers are chosen for the precursor ions study based on the fragment's intensity and selectivity. From the target compound, product ions are selected and transitions from the precursor ions to the product ions are optimized. That is, the collision energy that will produce the maximum ion intensity of the compound will then be used as the suitable collision energy of the compound for the actual analysis of the samples.

In metabolites analysis, common fragments with m/z ratio such as 147 and 73 may not be a good precursor ion as these fragments of trimethylsilyl (TMS) group are common in all the derivatized compounds. Once the transition parameters are identified for each compound, this information is then used by the GC-MS/MS system to identify and quantify the compounds (*see* Fig. 3).

Aside from volatiles, rice metabolites have already been investigated showing the capacity of the GC triple quadrupole MS system to identify useful mass fragment transitions of compounds. This can be used as a guide in selecting transitions for the analytes of interest.

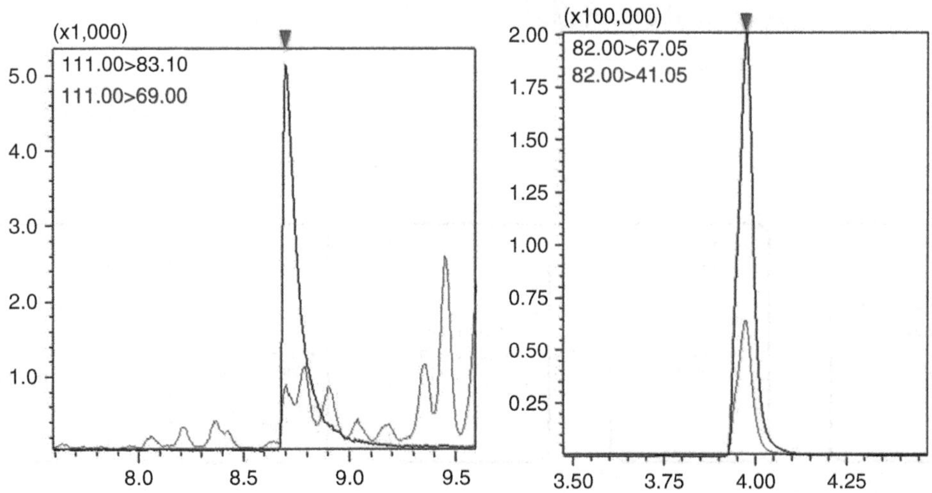

Fig. 3 Multiple reaction monitoring chromatograms showing peaks corresponding to transition ions of (**a**) 2-AP and (**b**) hexanal generated from a GC-MS/MS analysis (chromatogram obtained from Shimadzu GCMS-TQ8040)

4 Post-Run Data Analysis

Raw data obtained after every GC run can be analyzed one-at-a-time or as a batch. Processing data in batches will ensure the analyst that similar data treatment methods areapplied to all the chromatograms as suggested by Fig. 4. Several chromatography data systems are available like LabSolutions for Shimadzu and ChemStation for Agilent. Generally, integration parameters are set up and applied automatically to most of the chromatograms. It is possible that since the samples can be largely varied in terms of sample matrices, they would have slight differences due to retention time shifts in the chromatogram. While the built-in library provides the spectrum of the desired compound, the value of running the actual standard compound and comparing the analyte's mass spectra to that of the standard from pooled libraries around the world [39] is still important for the correct identification of the compound. There are additional softwares available for deconvolution, baseline correction, and peak alignment that could be useful during the data mining process such as AMDIS [40]. The TargetSearch Package is a preprocessing package from Bioconductor for searching and identifying metabolites using corrected retention time indices [41]. The identification process requires the use of retention index markers or standards to obtain aligned peaks and to identify outliers all throughout the sample runs (Fig. 4).

Moreover, several data pretreatment methods down the pipeline such as centering, scaling, and transformations are in place to further improve the biological interpretability of the biological data set [42]. After the preprocessing, multivariate statistical analysis is often used to initially describe and explain the profiles obtained in the analysis.

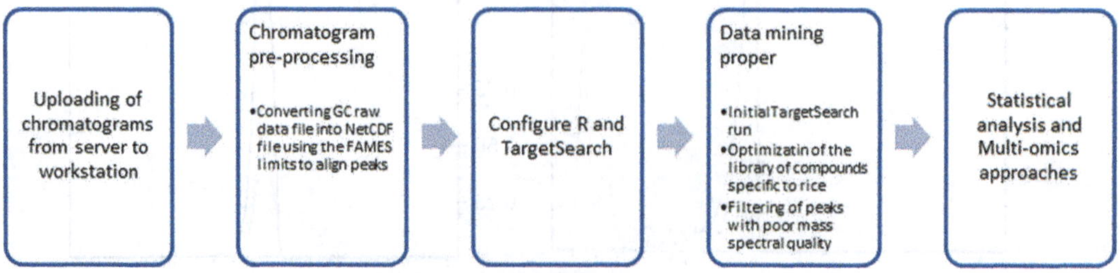

Fig. 4 Workflow for GCMS post run data analysis using TargetSearch

5 Notes

1. Prepare the extraction solution for the whole set of samples for analysis to minimize day effects.
2. Ultra-cooled samples and environment during extraction will ensure that any enzymes present are quenched and any metabolic process will no longer occur.
3. If the interest of the researcher includes the bran, there is no need to mill or polish the rice sample. Cryo-grinding of the sample must be immediately done.
4. Make sure that the septum used for the vials is suitable for SPME analysis to avoid breakage of the SPME fiber.
5. For mature grains, and depending on the sensitivity of the equipment, amount of sample recommended for extraction is 300 mg while for developing grains and germinating seeds, smaller amount (20 mg) is desired.
6. Moisture will hydrolyze the derivatizing reagents rendering the latter to lose its integrity for sample derivatization. Make sure that there is no water or moisture adhering on the surfaces of the vials, glass inserts, and pipette tips and all throughout the process especially right after freeze-drying. If the extracts will not be derivatized immediately, store vials in a dessicator. Store the derivatizing reagents in a cool, dry place when not in use.
7. Derivatize only extracts that will be analyzed in the GC on the same day.
8. Make sure that the GC-MS has no leak or moisture in the system. Perform auto-tune and replace filaments, septa, liners as needed.

Acknowledgments

This work has been supported under the CGIAR thematic area Global Rice Agri-Food System CRP, RICE, Stress-Tolerant Rice for Africa and South Asia (STRASA) Phase III funding.

References

1. Sreenivasulu N, Wobus U (2013) Seed-development programs: a systems biology-based comparison between dicots and monocots. Annu Rev Plant Biol 64:189–217
2. Weckwerth W (2003) Metabolomics in systems biology. Annu Rev Plant Biol 54:669–689
3. Patti GJ, Yanes O, Siuzdak G (2012) Innovation: metabolomics: the apogee of the omics trilogy. Nat Rev Mol Cell Biol 13(4):263–269
4. Zhen S, Dong K, Deng X, Zhou J, Xu X, Han C, Zhang W, Xu Y, Wang Z, Yan Y (2016) Dynamic metabolome profiling reveals significant metabolic changes during grain development of bread wheat (triticum aestivum L.). J Sci Food Agric 96(11):3731–3740

5. Q-q Y, C-q Z, M-l C, D-s Z, J-z C, Wang Q, Li Q-f, H-x Y, Gu M-h, SS-m S, Q-q L (2016) Biofortification of rice with the essential amino acid lysine: molecular characterization, nutritional evaluation, and field performance. J Exp Bot 67:4285–4296
6. Tian JS, Liu CC, Xiang H, Zheng XF, Peng GJ, Zhang X, Du GH, Qin XM (2015) Investigation on the antidepressant effect of sea buckthorn seed oil through the GC-MS-based metabolomics approach coupled with multivariate analysis. Food Funct 6(11):3585–3592
7. Lisec J, Schauer N, Kopka J, Willmitzer L, Fernie AR (2006) Gas chromatography mass spectrometry–based metabolite profiling in plants. Nat Prot 1:387–396
8. Butardo VM Jr, Sreenivasulu N (2016) Tailoring grain storage reserves for a healthier rice diet and its comparative status with other cereals. Int Rev Cell Mol Biol 323:31–70
9. Calingacion MN, Boualaphanh C, Daygon VD, Anacleto R, Sackville-Hamilton NR, Biais B, Deborde C, Maucourt M, Moing A, Mumm R, de Vos RCH, Erban A, Kopka J, Hansen TH, Laursen KH, Schjoerring JK, Hall RD, Fitzgerald MA (2012) A genomics and multiplatform metabolomics approach to identify new traits of rice quality in traditional and improved varieties. Metabolomics 8:771–783
10. Kusano M, Fukushima A, Fujita N, Okazaki Y, Kobayashi M, Oitome NF, Ebana K, Saito K (2012) Deciphering starch quality of rice kernels using metabolite profiling and pedigree network analysis. Mol Plant 5:442–451
11. Bryant RJ, McClung AM (2011) Volatile profiles of aromatic and non-aromatic rice cultivars using SPME/GC-MS. Food Chem 124(2):501–513
12. Zeng Y-X, Hu C-Y, Lu Y-G, Li J-Q, Liu X-D (2009) Abnormalities occurring during female gametophyte development result in the diversity of abnormal embryo sacs and leads to abnormal fertilization in indica/japonica hybrids in rice. J Integr Plant Biol 51(1):3–12
13. Kim JK, Park S-Y, Lim S-H, Yeo Y, Cho HS, Ha S-H (2013) Comparative metabolic profiling of pigmented rice (Oryza sativa L.) cultivars reveals primary metabolites are correlated with secondary metabolites. J Cereal Sci 57:14–20
14. Matsuda F, Okazaki Y, Oikawa A, Kusano M, Nakabayashi R, Kikuchi J, J-i Y, Ebana K, Yano M, Saito K (2012) Dissection of genotype–phenotype associations in rice grains using metabolome quantitative trait loci analysis. The Plant J 70:624–636
15. Zhou XH, Dong Y, Xiao X, Wang Y, Xu Y, Xu B, Shi WD, Zhang Y, Zhu LJ, Liu QQ (2011) A 90-day toxicology study of high-amylose transgenic rice grain in Sprague–Dawley rats. Food Chem Toxic 49(12):3112–3118
16. Vu NT, Shimada H, Kakuta Y, Nakashima T, Ida H, Omori T, Nishi A, Satoh H, Kimura M (2008) Biochemical and crystallographic characterization of the starch branching enzyme I (BEI) from Oryza sativa L. Biosci Biotech Biochem 72(11):2858–2866
17. C-y L, Lai VMF, Shen M-C (2004) Changes in retrogradation properties of rice starches with amylose content and molecular properties. Cereal Chem 81(3):392–398
18. Calingacion M, Fang L, Quiatchon-Baeza L, Mumm R, Riedel A, Hall RD, Fitzgerald M (2015) Delving deeper into technological innovations to understand differences in rice quality. Rice 8:43
19. Annison G, Illman RJ, Topping DL (2003) Acetylated, propionylated or butyrylated starches raise large bowel short-chain fatty acids preferentially when fed to rats. J Nutr 133(11):3523–3528
20. Champagne ET, Bett-Garber KL, Grimm CC, McClung AM, Moldenhauer KA, Linscombe S, McKenzie KS, Barton FE (2001) Near-infrared reflectance analysis for prediction of cooked rice texture. Cereal Chem 78(3):358–362
21. Laguerre M, Mestres C, Davrieux F, Ringuet J, Boulanger R (2007) Rapid discrimination of scented rice by solid-phase microextraction, mass spectrometry, and mutivariate analysis used as a mass sensor. J Agric Food Chem 55:1077–1083
22. Redestig H, Kusano M, Ebana K, Kobayashi M, Oikawa A, Okazaki Y, Matsuda F, Arita M, Fujita N, Saito K (2011) Exploring molecular backgrounds of quality traits in rice by predictive models based on high-coverage metabolomics. BMC Syst Biol 5:176
23. Zhou J, Ma C, Xu H, Yuan K, Lu X, Zhu Z, Wu Y, Xu G (2009) Metabolic profiling of transgenic rice with cryIAc and sck genes: an evaluation of unintended effects at metabolic level by using GC-FID and GC-MS. J Chromatogr B: Analyt Technol Biomed Life Sci 877:725–732
24. Frank T, Meuleye BS, Miller A, Shu Q-Y, Engel K-H (2007) Metabolite profiling of two low phytic acid (lpa) rice mutants. J Agric Food Chem 55:11011–11019
25. Heuberger AL, Lewis MR, Chen M-H, Brick MA, Leach JE, Ryan EP (2010) Metabolomic and functional genomic analyses reveal varietal differences in bioactive compounds of cooked rice. PLoS One 5(9): e12915

26. Yang DS, Shewfelt RL, Lee K-S, Kays SJ (2008) Comparison of odor-active compounds from six distinctly different rice flavor types. J Agric Food Chem 56(8):2780–2787
27. Verhoeven HA, Jonker H, RCH DV, Hall RD (2012) Solid phase micro-extraction GC–MS analysis of natural volatile components in melon and Rice. In: Hardy NW, Hall RD (eds) Plant metabolomics: methods and protocols. Humana Press, Totowa, NJ, pp 85–99
28. Agilent Technologies, Inc (2013) Agilent fiehn GC/MS metabolomics RTL library a complete GC/MS solution for metabolite identification. Agilent Technologies
29. Soria AC, Sanz ML, Villamiel M (2009) Determination of minor carbohydrates in carrot (Daucus carota L.) by GC-MS. Food Chem 114:758–762
30. Ruiz-Matute AI, Hernández-Hernández O, Rodríguez-Sánchez S, Sanz ML, Martínez-Castro I (2011) Derivatization of carbohydrates for GC and GC–MS analyses. J Chromatogr B Analyt Technol Biomed Life Sci 879:1226–1240
31. Lou Q, Ma C, Wen W, Zhou J, Chen L, Feng F, Xu X, Lu X, Luo L, Mei H, Xu G (2011) Profiling and association mapping of grain metabolites in a subset of the core collection of Chinese rice Germplasm (Oryza sativa L.). J Agric Food Chem 59:9257–9264
32. Chang WT, Lin DL, Liu RH (2001) Isotopic analogs as internal standards for quantitative analyses by GC/MS—evaluation of cross-contribution to ions designated for the analyte and the isotopic internal standard. Forensic Sci Int 121(3):174–182
33. Chen M, Rao RSP, Zhang Y, Zhong CX, Thelen JJ (2014) A modified data normalization method for GC-MS-based metabolomics to minimize batch variation. Springerplus 3:439
34. Winnike JH, Wei X, Knagge KJ, Colman SD, Gregory SG, Zhang X (2015) Comparison of GC-MS and GC×GC-MS in the analysis of human serum samples for biomarker discovery. J Proteome Res 14(4):1810–1817
35. Zhao X, Zhang G, Wang Y, Zhang F, Wang W, Zhang W, Fu B, Xu J, Li Z (2016) Metabolic profiling and physiological analysis of a novel Rice introgression line with broad leaf size. PLoS One 10(12):e0145646
36. Li J, Dong F, Xu J, Liu X, Li Y, Shan W, Zheng Y (2011) Enantioselective determination of triazole fungicide simeconazole in vegetables, fruits, and cereals using modified QuEChERS (quick, easy, cheap, effective, rugged and safe) coupled to gas chromatography/tandem mass spectrometry. Anal Chim Acta 702(1):127–135
37. Tranchida PQ, Zoccali M, Schipilliti L, Sciarrone D, Dugo P, Mondello L (2013) Solid-phase microextraction with fast GC combined with a high-speed triple quadrupole mass spectrometer for targeted and untargeted food analysis. J Sep Sci 36(13):2145–2150
38. Matthews J, Gemme S, Huebschmann H-J, Llorente C, Jimenez R, Sreenivasulu N (2015) Metabolomics of rice genotypes using GC-MS/MS. ThermoScientific Application Note 1–9
39. Schauer N, Steinhauser D, Strelkov S, Schomburg D, Allison G, Moritz T, Lundgren K, Roessner-Tunali U, Forbes MG, Willmitzer L, Fernie AR, Kopka J (2005) GC–MS libraries for the rapid identification of metabolites in complex biological samples. FEBS Lett 579(6):1332–1337
40. Agilent Technologies, Inc (2005) Automated mass spectral deconvolution and identification system AMDIS 2.64, in Released December, NIST
41. Cuadros-Inostroza Á, Caldana C, Redestig H, Kusano M, Lisec J, Peña-Cortés H, Willmitzer L, Hannah MA (2009) TargetSearch-a bioconductor package for the efficient preprocessing of GC-MS metabolite profiling data. BMC Bioinformatics 10:428
42. van den Berg RA, Hoefsloot HCJ, Westerhuis JA, Smilde AK, van der Werf MJ (2006) Centering, scaling, and transformations: improving the biological information content of metabolomics data. BMC Genomics 7:142

Chapter 12

Re-sequencing Resources to Improve Starch and Grain Quality in Rice

Gopala Krishnan Subbaiyan, Ardashir K. Masouleh, Agnelo Furtado, Daniel L. E. Waters, and Robert J. Henry

Abstract

Next-generation sequencing can identify differences in the rice genome that explain the genetic basis of grain quality variation. Differences in rice grain quality are mainly associated with differences in the major component of the grain, starch. Association of rice quality variation with rice genome variation can be conducted at the gene or whole-genome level. Re-sequencing of specific genes or whole genomes can be used depending on the extent to which candidate genes for the traits of interest are known. Amplicon sequencing of genes involved in starch metabolism can help in targeted discovery of the molecular genetic basis of differences in starch related quality attributes. Whole-genome re-sequencing can complement this, when the genetic basis of the trait is expected to be outside the coding region of starch metabolism genes. These approaches have been used successfully to understand the rice genome at specific loci and over the whole genome.

Key words Amplicon sequencing, Association analysis, Genotyping, Grain quality, Re-sequencing, Rice, SNPs, Starch

1 Introduction

Sequencing has been utilized to identify genetic variations in genes determining starch quality in rice. Rapid advances in sequencing technologies have enabled the discovery of millions of polymorphisms in the genome such as Single-Nucleotide Polymorphisms (SNPs) and Insertion-Deletion Polymorphisms (InDels) by comparing the whole-genome sequences of genotypes with high-quality reference genome sequences. Re-sequencing can be used to identify differences in the genome that explain the genetic basis of grain quality variation in rice [1]. Differences in the grain and cooking quality of rice are strongly associated with differences in the major component of the grain, starch [2]. Association of variation in rice quality with variation in the genome of the rice genotype can be conducted either at specific gene(s) or at the

Nese Sreenivasulu (ed.), *Rice Grain Quality: Methods and Protocols*, Methods in Molecular Biology, vol. 1892, https://doi.org/10.1007/978-1-4939-8914-0_12,

whole-genome level. Re-sequencing of specific genes [3] or whole genomes [4] can be used depending on the extent to which candidate genes for the traits of interest are known.

Starch chemistry is a key contributor to rice grain quality since starch forms the major component of the rice grain. In rice, as many as 18 genes involved in starch biosynthesis have been identified and their sequence information is also available. Amplicon sequencing of these genes involved in starch metabolism can help in targeted discovery of the molecular genetic basis of differences in starch related quality attributes [5, 6]. Whole-genome re-sequencing can not only be used to study variations in the genes already known to be involved in starch biosynthesis but also in identifying the genetic basis of the trait at the genome level thereby enabling us to discover novel polymorphisms influencing starch quality other than the already known genes involved in starch metabolism genes. These approaches have been used successfully to understand the rice genome at specific loci [7] and over the whole genome [8] to unravel the genetic basis of starch quality in rice. In this chapter, various sequencing approaches used in analyzing the genomic variations determining starch quality in rice are discussed with details of methods for sequencing, data analysis, and association of genotypic data and phenotypic data.

2 Materials

2.1 Re-sequencing of Starch Metabolism Genes

Chemicals—DNeasy Plant Mini Kit (QIAGEN), QIAshredder Mini spin columns, TBE buffer, BioRadiProof™ High-Fidelity DNA polymerase Kit, iProof polymerase, Ethidium bromide/PicoGreendsDNA quantification kit (Invitrogen), iPLEX gold reaction cocktail.

Instruments—Tissue lyser, Thermal cyclers, ABI Sequencer, Illumina Sequencer, Nanodrop/Spectrometer.

Sequencher (Gene Codes Corporation, Ann Arbor, MI) and CLUSTAL W (http://www.ebi.ac.uk/Tools/clustalw2/index.html), Bio Edit (http://www.mbio.ncsu.edu/bioedit/bioedit.html), Primer3 (http://bioinfo.ut.ee/primer3-0.4.0/). However, we suggest the Clone Manager V9.1 (Sci-Ed Software, Cary, NC) (http://www.scied.com/pr_cmbas.htm), CLC bio.

2.2 Whole-Genome Re-sequencing

DNA extraction for whole-genome re-sequencing by CTAB method

1. 1 M, Tris–HCl, pH 8.0: Dissolve 121.1 g of Tris base in 800 mL of distilled water, adjust the pH to 8.0 with concentrated HCl and add distilled water to a final volume of 1 L.
2. 0.5 M Ethylenediaminetetraacetic acid (EDTA). Dissolve 93.05 g of EDTA disodium salt in 400 mL, adjust the pH to 8.0 then adjust the volume to 500 mL.

3. CTAB extraction buffer (2% CTAB, 100 mM Tris–HCl, pH 8.0; 20 mM EDTA; 1.4 M NaCl): Dissolve 20 g of CTAB in 200 mL of 1 M Tris–HCl, 100 mL of 0.5 M EDTA and 500 mL of distilled water with a magnetic stir bar.
4. 3 M sodium acetate, pH 5.0. Dissolve 24.6 g of anhydrous sodium acetate in 80 mL of distilled MilliQwater, adjust the pH to 5.0 with glacial acetic acid and then the volume to 100 mL with distilled MilliQ water.
5. TE buffer (10 mM Tris–HCl, pH 8.0, 1 mM EDTA): Take 10 mL of 1 M Tris–HCl and add 2 mL of 500 mM EDTA pH 8.0. Adjust the final volume to 1 L with distilled water.
6. Mortar and pestle.
7. Liquid nitrogen.
8. Steel spatula.
9. 50 mL Falcon tubes.
10. Fume hood.
11. 65 °C water bath.
12. Thermometer.
13. Chloroform:isoamylalcohol (24:1).
14. Chloroform.
15. RNase A 10 mg/mL, aqueous solution.
16. Ethanol (Ethyl alcohol, ≥99.5% pure).
17. 70% ethanol (v/v) ethanol/distilled water.
18. Centrifuge for 50 mL Falcon tubes and 12,000 × *g*.
19. DNA electrophoresis apparatus and gel documentation system.
20. DNA stain compatible with agarose gel (e.g., ethidium bromide or SYBRsafe).
21. Software—CLC bio Genomics Workbench, CLUSTAL W (http://www.ebi.ac.uk/Tools/clustalw2/index.html), Bio Edit (http://www.mbio.ncsu.edu/bioedit/bioedit.html).

2.3 SNP Genotyping

1. Sequenom®, MassARRAYAnalyzer Compact mass spectrometer.

2.4 Association of SNP with Phenotypes

1. Phenotypic data of the genotypes on starch quality.
2. Genotypic data from amplicon sequencing/whole-genome re-sequencing.
3. Software—TASSEL (http://www.maizegenetics.net/#!tassel/c17q9).

3 Methods

3.1 DNA Extraction and Quantification

The protocol describes how to extract DNA from rice plants for amplification of target genes and DNA for whole-genome re-sequencing. Extracting DNA through a rapid and inexpensive DNA extraction protocol is suitable for amplification of target genes but this method is not advisable for high-throughput sequencing. The DNA for next-generation sequencing must be of high quality; minimal shearing (no observable shearing on the gel) and with an absorbance ratio between 1.8 and 2.0 [9]. Therefore, protocols such as Cetyltrimethylammonium bromide (CTAB) based DNA extraction protocols which extract high-quality DNA from many tissues and most species at high yield for minimal cost must be used for DNA extraction.

3.1.1 DNA Extraction from Rice for Amplification of Target Genes

A fast and easy DNA extraction method can guarantee accurate PCR amplification of the target gene. The DNeasy Plant Mini Kit (or Maxi kit) (QIAGEN) is recommended for high-throughput extraction and purification of genomic DNA from plant tissues. The kit is supplied in several sizes (*see* **Note 1** with Mini Spin Columns which are suitable depending on the number of samples to be extracted). Kits contain several contents which can be found at: https://www.qiagen.com/au/resources/resourcedetail?id=95dec8a9-ec37-4457-8884-5dedd8ba9448&lang=en

A complete and quick disruption of leaf tissues is necessary to achieve high-quality DNA free from degradation.

The manufacturer's DNA extraction protocol is outlined below:

1. Sterilize 5–10 seeds with 10% hypochloride (commercial) solution for 3 min. A drop of Tween 20 or commercial dish washing liquid can assist sterilization.
2. Rinse the seeds with tap water and finally with sterilized water for 5 min.
3. Wash the sterilized seeds in 70% Ethanol for 1 min.
4. Rinse again with sterilized water for 2 min.
5. Place the sterilized seeds (5–10 seeds) on wet Watson paper in the bottom of a petri dish. Ensure the seeds are not placed over each other and there is sufficient space to grow the seeds.
6. Incubate petri dishes in a germinator at 25–28 °C with 12 h day/night photoperiod.
7. Check the petri dishes regularly and discard fungi-infected seeds/peri dishes.
8. Grow the seeds for a week until the seedlings are 10–15 cm in height.

9. Normally, 100 mg, 50 mg, and 1 g of plant leaf tissues (fresh weight) are sufficient for Mini, DNeasy 96, and Maxi kits, respectively.
10. Plant materials can be kept for an extended period of time at −80 °C until DNA extraction.
11. Mechanically disrupt the plant materials to powder using a mortar and pestle and liquid nitrogen.
12. Add liquid nitrogen as required to keep the sample frozen and grind until obtaining a fine powder.
13. For a larger number of samples, the Tissue Ruptor with 3 mm tungsten carbide beads in a 2 mL safe-lock microcentrifuge tube (*see* **Note 2**) containing 100 mg of fresh leaf tissue should be used.
14. Liquid nitrogen is added to the tube (safe-lock microcentrifuge tube), and the sample frozen for 30 s. The sample should be kept submerged in liquid nitrogen and disrupted for approximately 30 s at full speed of Tissue Ruptor/TissueLyser.
15. Allow the liquid nitrogen to evaporate and immediately proceed to the next step.
16. Add 400 μL Buffer AP1 and 4 μL RNaseA stock solution (100 mg/mL) to a maximum of 100 mg (wet weight) or 20 mg (dried) disrupted leaf tissue in tubes, then vortex vigorously (*see* **Note 3**).
17. Incubate the mixture for 10 min at 65 °C and mix the tubes two or three times during incubation by inverting to lyse the cells.
18. Add 130 μL Buffer P3 to the lysate, mix by inverting, and incubate the tubes for 5 min on ice.
19. Centrifugation of the lysate for 5 min at 20,000 × *g* (14,000 rpm) in room temperature is recommended.
20. Place the QIAshredder Mini spin column(s) (lilac colour) in 2 mL collection tubes.
21. Pipet supernatants into the QIAshredder Mini spin column (*see* **Note 4**).
22. Centrifuge for 2 min at 20,000 × *g* (14,000 rpm).
23. Supply new tubes (not included in the kit) and transfer the flow-through fraction (supernatants). Be careful not to disturb the debris pellet. Normally, up to 450 μL of lysate is recovered.
24. Prepare buffer AW1 with addition of ethanol with the appropriate amount of ethanol (96–100%) as indicated on the bottle to obtain a working solution (*see* **Note 5**).
25. Add 1.5 volumes of Buffer AW1 to the cleared lysate, and mix by pipetting. For example, if 450 μL of lysate is recovered, add 675 μL of Buffer AW1. Quickly pipette Buffer AW1 onto the cleared lysate and mix immediately.

26. Place the DNeasy Mini spin column in 2 mL collection tubes (supplied).
27. Pipet 650 μL of the mixture from **step 25**, including any formed precipitates, into DNeasy Mini spin columns.
28. Centrifuge the Mini spin columns for 1 min at ≥6000 × *g* (corresponds to ≥8000 rpm for most microcentrifuges), and discard the flow-through (see next step)*.
29. *Repeat **steps 27** and **28** by reusing the collection tubes until all lysates are processed. Finally, discard the last flow-through along with collection tube.
30. Add Ethanol (96–100%) to Buffer AW2 (96–100%) as indicated on the bottle to obtain a working solution.
31. Place the DNeasy Mini spin column into a new 2 mL collection tube (supplied in the kit).
32. Add 500 μL Buffer AW2.
33. Centrifuge the tube(s) for 1 min at ≥6000 × *g* (≥8000 rpm) and discard the flow-through and reuse the collection tube(s) for the next step.
34. Add 500 μL Buffer AW2 to the DNeasy Mini spin column again.
35. Centrifuge for 2 min at 20,000 × *g* (14,000 rpm). This will dry the DNeasy Mini spin column membrane which contains the DNA.
36. Remove the DNeasy Mini spin column and discard flow-through and collection tube very carefully to ensure no residual ethanol is carried over to the next step.
37. Supply 1.5 or 2 mL microcentrifuge tube(s) (not provided in the kit).
38. Place the dried DNeasy Mini spin column (containing DNA from the previous step) onto the new tube(s).
39. Add 100 μL Buffer AE directly onto the DNeasy membrane.
40. For elution, incubate the tubes for 5 min at room temperature (15–25 °C), and then centrifuge for 1 min at ≥6000 × *g* (≥8000 rpm).
41. A second elution by repeating the **steps 39** and **40** is recommended (*see* **Note 6**).

DNA is extracted from each individual sample separately and pooled equimolarly to create a mega pool for the whole population.

3.1.2 DNA Extraction for Whole-Genome Re-sequencing

Next-generation sequencing requires the use of an appropriate DNA extraction method [10]. There are many rapid and inexpensive DNA extraction protocols that extract poor quality DNA, but they are not suitable for high-throughput sequencing. Commercial kits generally extract a high yield with superior quality DNA, however, they are relatively expensive and this is an issue when surveying plant breeding populations. Cetyltrimethylammonium bromide (CTAB)-based DNA extraction protocols extract high-quality DNA from many tissues and most species at high yield for minimal cost. The DNA for next-generation sequencing has to be of very good quality; minimal shearing (no observable shearing on the gel) and with an absorbance ratio between 1.8 and 2.0.

3.1.3 CTAB Method for Extraction of DNA

1. Heat a water bath to 65 °C (*see* **Notes 7** and **8**). For each sample, add 2 mL of CTAB extraction buffer to a 15 mL Falcon tube and pre-heat to 650 °C by placing in the water bath.
2. Take 2–3 g of young leaf material and place in a mortar. Pour liquid nitrogen over the sample (*see* **Notes 9** and **10**). As the liquid nitrogen evaporates, grind the material with a pestle. Ensure the sample remains frozen for the entire procedure by adding liquid nitrogen within 20 s of the liquid nitrogen evaporating (*see* **Note 11**).
3. Using a small spatula, scrape the ground powdered leaf material into a 50 mL Falcon tube containing 8 mL of CTAB extraction buffer which has been preheated to 65 °C.
4. Incubate the mixture for 1 h at 65 °C, manually mix the contents gently by inverting the tubes at least every 10 min.
5. After the 1 h's incubation, remove the tubes containing the plant/buffer mixture from the 65 °C water bath and centrifuge for 5 min at 12,000 × *g*.
6. Carefully use a pipette to remove the supernatant from the tube to a fresh 50 mL Falcon tube, taking care not to disturb the plant material pellet.
7. Add RNase A to a final concentration of 1 μg/mL, mix well gently by inverting the tubes three or four times, and then incubate at 37 °C for 15 min (*see* **Note 12**).
8. In a fume hood, extract the supernatant with an equal volume of chloroform:isoamylalcohol (24:1) (*see* **Note 13**). The chloroform:isoamylalcohol is added to the supernatant and mixed gently by inverting the tubes about 50 times. This step is to be carried out in a fume hood (*see* **Note 14**).
9. Centrifuge for 1 min at 12,000 × *g* (*see* **Note 15**).
10. In a fume hood, remove the upper aqueous phase using a suitable pipette. Take care not to remove any chloroform:isopropanol. In the fume hood, discard the waste chloroform:isopropanol into an appropriate waste bottle with a lid which is resistant to organic solvents and is gas tight.
11. Repeat **steps 7–9**.

12. Precipitate the nucleic acids by the addition of a 1/10 volume of 3 M sodium acetate (pH 5.0) and three volumes of 95% ethanol (*see* **Notes 16** and **17**). Mix gently by inverting the tubes several times.
13. Centrifuge at 12,000 × *g*. Orient the tubes so the side of the tube which faced the outside of the centrifuge can be identified following centrifugation.
14. The DNA will be near the bottom of the tube on the side of the tube which was oriented toward the outside of the centrifuge. Remove and discard the supernatant, taking care not to disturb the DNA pellet.
15. Add 5 mL of 70% ethanol to remove residual salts from the DNA pellet. Incubate at room temperature for 10 min.
16. Centrifuge at 12,000 × *g*. Orient the tubes so the side of the tube which faced the outside of the centrifuge can be identified following centrifugation.
17. The DNA will be near the bottom of the tube on the side of the tube which was oriented toward the outside of the centrifuge. Remove and discard the supernatant, taking care not to disturb the DNA pellet.
18. Partially dry the DNA pellet at room temperature for a few minutes. Over drying the DNA or the presence of ethanol in the undried DNA can lead to difficulty in dissolving the DNA in TE buffer.
19. Dissolve the DNA pellet in 200 μL of TE buffer (Tris-EDTA, pH 8.0). Leaving the tubes overnight at room temperature will assist in dissolving the DNA. This can be accelerated by a 65 °C incubation which has the added benefit of removing any nuclease activity.
20. Determine the quality of the DNA by measuring the absorbance ratio (should be between 1.8 and 2.0) using a spectrophotometer. Resolve the extracted DNA using electrophoresis on a 0.7% or 1% agarose gel, to determine the quantity (by comparing the intensity to a known amount high-molecular weight DNA e.g., lambda DNA) and the level of shearing of the extracted DNA.
21. Store the DNA at −20 °C.

3.1.4 Measurement of Genomic DNA Concentrations

There are a number of ways of measuring DNA concentration which differ by convenience, accuracy, and precision. These methods include simple UV (260 nm) absorbance, UV fluorescence of dyes bound to double stranded DNA, and real-time PCR probes. Although simple UV absorbance at 260 nm using devices such as Nanodrop or conventional cuvette-based spectrophotometer is convenient, accuracy is hampered by other UV absorbance of contaminating molecules and the pH of the solution. Probe-based real-time PCR quantification methods are extremely accurate. However, they require significant time and effort to optimize. UV fluorescence of dyes bound to double stranded DNA strikes an appropriate balance between accuracy and convenience and is described here. We have used both ethidium bromide stained DNA in agarose gels and Pico-Green (PicoGreendsDNA Quantification kit, Invitrogen) to measure the concentration of PCR products.

The protocol for quantification of the amplified DNA with Ethidium bromide stained DNA in agarose gel electrophoresis is detailed below:

1. The combs must create wells which have a volume which is greater than the volume of the sample to be analyzed. A well volume of around 20 μL will suffice for most analytical purposes. Determine the volume of gel required to fill the gel tray to a depth that satisfies this requirement.
2. Take a gel tray and seal it with masking tape at each end. It is important to ensure agarose gel does not leak from the tray (*see* **Note 18**).
3. Mix 1 g of agarose per 100 mL of Tris-Borate EDTA (TBE) buffer in a Schott bottle. Mix by shaking (*see* **Note 19**).
4. Place the bottle in a microwave oven with the lid slightly open. Heat the mixture on high setting, mixing occasionally, until the agarose has melted. Check the agarose has melted by swirling the liquid, holding it up to the light and looking for small lenticular, clear pieces of unmelted agarose.
5. When the agarose has melted, carefully pour it into the prepared gel tray. Take care to avoid bubbles and make sure the combs are covered to sufficient depth.
6. Let the agarose set by leaving it at room temperature for at least 30 min.
7. Remove the masking tape and place the gel tray in an electrophoresis tank of the appropriate size, making sure the wells are covered by TBE buffer.
8. Mix the samples with loading buffer on a piece of parafilm or in Eppendorf or PCR tubes. Loading buffer contains dense material such as glycerol or ficoll which ensures the sample falls to the bottom of the well, and dyes such as bromophenol blue and xylene cyanol so the sample can be seen. Depending on the efficiency of the PCR, 2 μL of sample should be sufficient.
9. Carefully load the samples into the wells of the gel. This is best done by holding the pipette at approximately 45 °C, using both hands, one to steady the pipette tip, and letting the sample drop into the well.

10. Load at least one of the wells with a DNA sample of known size and concentration. More accurate concentration estimation is achieved by using a dilution series of the sample of known concentration.
11. Close the lid on the gel tank. Ensure the electrodes are correctly orientated (*see* **Note 20**).
12. Set the power pack to 5 V/cm. For example, if the distance is between the electrodes is 20 cm, set the voltage to 100 V.
13. Run the gel until the bromophenol blue dye reaches the end of the gel.
14. Remove the gel from the gel tank and place in a TBE solution containing 2–5 μg/mL ethidium bromide (*see* **Note 21**). Use TBE as the staining solution because this allows the gel to be re-run if the results are not satisfactory.
15. Allow the gel to stain for 15 min.
16. Rinse the gel with tap water. This removes excess ethidium bromide from the surface of the gel. Excess ethidium bromide can be removed by placing the gel in TBE which does not contain ethidium bromide. The DNA will retain the dye while the surrounding gel destains, improving the contrast between the DNA and background.
17. Take the gel and use a trans-illuminator to irradiate the gel with UV light. Record the results of the experiment by capturing an image of the gel digitally or on film.
18. Estimate the concentration of each amplicon by comparison with the DNA standards.

There are a number of fluorescent dsDNA binding dyes which can be used to measure the concentration of DNA, many of which are proprietary. The Invitrogen Quant-iT™ PicoGreen® (Picogreen) dye is a popular option because of its high sensitivity and specificity for dsDNA and its linear response over several orders of magnitude, even in the presence of contaminates such as salts, detergents, proteins, and ssDNA (oligonucleotides) used in PCR.

In common with all other DNA binding dyes, Picogreen is a potential mutagen and carcinogen. Picogreen is supplied in a DMSO solution which should be handled with care as DMSO facilitates entry of organic molecules into human tissues. Suitable safety glasses, laboratory coat, and gloves should be worn at all times. The Picogreen/DMSO undiluted stock solution should be opened and diluted in a fume hood. Picogreen/DMSO must be disposed of in accordance with the relevant regulations which govern the laboratory in which they are used.

The protocol for quantification of the amplified DNA with PicoGreendsDNA Quantification kit, Invitrogen is explained below:

Table 1
Standard curve dilution series

TE volume (μL)	(2 μg/mL DNA) PicoGreen® (μL)	Stock DNA (μL)	Quant-iT™ concentration (ng/mL)
0	100	100	1000
90	10	100	100
99	1	100	10
999	1	1000	1
100	0	100	0 (Blank)

1. Warm the Picogreen to room temperature before opening the vial.
2. If using a Invitrogen Quant-iT™ PicoGreen® kit, prepare a 1× TE buffer (10 mM Tris–HCl, 1 mM EDTA, pH 7.5) by diluting the concentrated 20× buffer which comes with the kit 20-fold with sterile, distilled, DNase-free water. If using Picogreen which is not sourced from a kit, take care to ensure the TE buffer used to dilute the Picogreen is free of nucleic acids since they bind with Picogreen, even at very low (picogram) concentrations and interfere with the assay (*see* **Note 22**).
3. Prepare the Picogreen by diluting the concentrated solution 1 in 200 in 1× TE (e.g., 100 μL Picogreen in 19.9 mL TE or 10 μL Picogreen in 1.9 mL) (*see* **Note 23**).
4. First dilute the DNA standard to a final concentration of 2 μg/mL (or 2 ng/μL). This 2 μg/mL is the Stock DNA.
5. Make a series of duplicate dilutions in the first two rows of each 96-well microplate, or the first row of a 384-well microplate, which correspond to the amounts of TE, 2 μg/mL Stock DNA, and Picogreen as described in Table 1.
6. If using a 100 μL/well 96-well microplate, add 4 μL of DNA to each well followed by 96 μL of TE and 100 μL of the Picogreen. If using a 25 μL/well 384 well microplate, add 1 μL of DNA to each well followed by 24 μL of TE and 25 μL of the Picogreen (*see* **Note 24**).
7. Mix standards and amplicon samples by pipetting up and down several times, ensuring no air bubbles are formed in the wells.
8. Incubate at room temperature for 5 min protected from light to ensure the DNA binding sites for Picogreen are fully saturated with Picogreen.

9. Measure DNA standard and fluorescence with a microplate reader at the fluorescein wavelengths of ~480 nm (excitation) and ~520 nm (emission). To ensure readings remain in the detection range of the fluorometer, ensure the instrument's gain is set so the DNA standard solution of 1 μg/mL has fluorescence intensity just below the plate reader maximum.
10. Subtract fluorescence of the reagent blank from DNA standards. Either create a standard curve by plotting fluorescence versus DNA concentration or, depending on the plate reader, enter the data into an Excel spreadsheet which has a macro that carries out the appropriate calculations.
11. Determine DNA concentration with reference to the standard curve or by entering the data into an Excel spreadsheet which has a macro that carries out the appropriate calculations.

3.2 Re-sequencing of Starch Biosynthesis Genes

This protocol describes how to bulk the DNA, primer design, and amplification of target genes by long-range PCR, amplicon pooling, sequencing, and polymorphism (SNP) identification.

If the ultimate objective of measuring starch biosynthesis gene allele frequencies is to be realized, it is important to ensure bias toward any one sample, either random or systematic, is minimized at each step of the process. The risk of bias arising during amplicon sequencing is minimized when all amplicons are present in pools in equimolar amounts.

3.2.1 Designing of PCR Primers

It is essential to choose the target genes which have the greatest effects on starch synthesis, composition, and quality. For this purpose, a complete literature review was done and specific starch quality genes chosen for targeted amplification and deep sequencing. The correct gene sequence (gDNA and cDNA) is needed to design primers for amplification (*see* **Note 25**). The genes involved in starch biosynthesis and the sequence of the primers of each of the rice starch genes have been reported by Kharabian-Masouleh et al. [11] and are publicly available here: http://onlinelibrary.wiley.com/doi/10.1111/j.1467-7652.2011.00629.x/suppinfo (Table S3).

The following major databases can be used to find the sequence of target genes (*see* **Note 26**):

1. National Centre of Biotechnology Information (NCBI): (http://www.ncbi.nlm.nih.gov/).
2. Rice Genome Annotation Project: (http://rice.plantbiology.msu.edu/cgi-bin/putative_function_search.pl).
3. The sequence data can be saved in separate MS Word files or in FASTA format in Notepads.
4. For alignment purposes, it is preferable to save the data in FASTA format in Notepad files with recognizable naming.

3.2.2 Primer Design for Long-Range PCR (LR-PCR) Capture and Amplification of Target Gene Sequence

1. This protocol can be used for any gene varying in length from 1 to 20 kb or even beyond.
2. Each gene sequence can be split into half or more if the length is greater than >5 kb. However, this can be changed depending on the performance and efficiency of polymerase used for amplification. High-fidelity polymerases enzymes will produce better results because they can amplify genes with greater speed and accuracy. This will be explained later in this chapter.
3. The following criteria should be met during the primer design: Each sequence should include ~500–1000 bp extra from up- and downstream of the coding regions to ensure that no variant will be missed from 5′ and 3′ UTRs. Approximately, 150 bp overlap at the end of neighboring fragments is required to minimize the bias which normally occurs at the both end of sequencing sides. This will enhance accuracy of contig construction.
4. Specific primers can be designed for each gene (of gene fragment) using software available for free online such as Primer3 (http://bioinfo.ut.ee/primer3-0.4.0/). However, we recommend Clone Manager V9.1 (Sci-Ed Software, Cary, NC) (http://www.scied.com/pr_cmbas.htm), which produces high quality and efficient primers and also encompasses a comprehensive set of tools for sequence manipulation and analysis, sequence alignment, and primer design.
5. To begin primer design, divide the whole sequence (in length) into half. For example, if gene X is 10,000 bp in length, it can be divided into two fragments of approximately 5000 bp, with a 150 bp overlap in the middle (Fig. 1).
6. Save the halved sequence in the same or different files on your computer.
7. Run Clone manager software, click New (New Molecule Wizard) on top menu bar (left corner) to create a new molecule.
8. Choose "Enter Complete Information."
9. Insert a molecule name; choose DNA from drop down list. Molecule description and notes are optional.
10. Click next; then paste the DNA molecule sequence and length.
11. Click next; then choose linear DNA.
12. In the next step, the molecule features are optional. Click next and now your molecule has been created.
13. Go to "Primer" on top bar menu and then click on "Design."

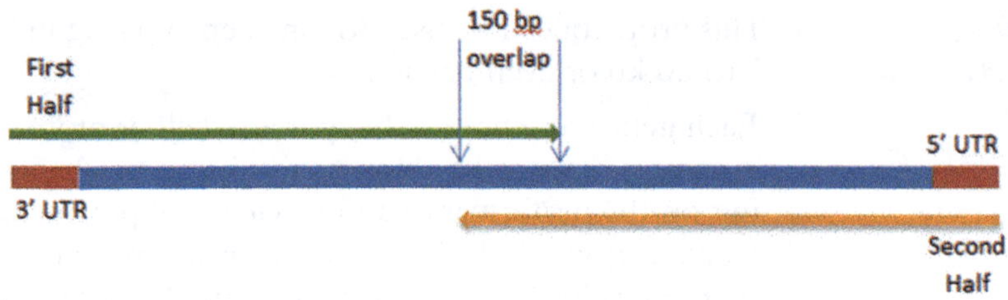

Fig. 1 The gene of interest (blue color) may be divided into two sections, depending on the length of gene. This is recommended for genes with length more than 5 kb. The red segments are 3′ and 5′ UTRs. A 150 bp (or less) overlap might be in the middle

14. Give your primer design details such as primer type (PCR Primer Pair), choose primer length for forward and reverse (usually ~20–25 bp) and then click next.
15. In "Source Molecule," browse and choose your molecule that you already have created.
16. For primer target position type number 1 and then 100.
17. Click finish.
18. The program will match the created DNA sequence and position a primer of the length you specify at the initial position you specify. You can then move the primer along the sequence (by tapping on left and right arrows of keyboard) or adjust its length until the primer evaluation meets the criteria set.
19. Order primers.

3.2.3 Long-Range PCR to Amplify Target Genes

For amplification, a semi- to long-range PCR (LR-PCR) needs to be employed in which each gene is amplified with 1–2 pairs of primers (*see* **Note 27**).

The following procedure should be used for accurate amplification of starch genes:

1. Concentration of extracted DNA should be quantified, using the automated flurometric protocol of PicoGreen (PicoGreendsDNA Quantification kit; Invitrogen, San Diego, CA).
2. The concentration of each individual DNA sample must be diluted to 10–20 ng/μL. DNA of 20 individuals should be mixed for each reaction.
3. Each fragment gene (halved) can be amplified separately to find the optimum PCR condition. However, as the number of gene fragments and genotype individuals are often quite large, it is better to use a unified LR-PCR approach for all genes and fragments.

4. It is preferable to apply the LR-PCR protocol to one standardized DNA sample (e.g., Nipponbare) (*see* **Note 28**). This will help determine whether the PCR conditions are optimum and the primers are functional or not.
5. After finding the suitable protocol and efficient primers for amplification of fragments, the protocols and primers can be applied to different DNA samples.
6. The BioRadiProof™ High-Fidelity DNA polymerase kit (BioRad Laboratories, Hercules, CA) is strongly recommended for LR-PCR (*see* **Note 29**).
7. TheBioRadiProof™ High-Fidelity DNA polymerase kit contains two amplification buffers (HF and GC). The HF (High Fidelity) is used initially and if the fragment does not amplify the GC buffer can be used. The GC buffer is normally used for GC-rich, difficult to amplify fragments.

The details of reaction mixture preparation for LR-PCR amplification is presented below:

1. DNA of 20 individuals (15 ng/μL) or less should be mixed for each reaction.
2. All primers (Forward and Reverse separately) should be diluted to 2.5 μM and stored in −20 °C until use.
3. Prepare a master mix then dispense into separate 1.5 mL tube and add different primers (Forward and Reverse for each target gene amplicon), as required.
4. Two sets of reactions can be prepared; first with HF. If some primers did not amplify then a master mix should be prepared with GC buffers as follows:

Components	Concentration	Volume for 20 μL reaction (μL)
Buffer HF and GC	1× (exactly as provided in kit)	4
dNTPs (mix)	10 mM	0.4
Primer 1 (Forward)	2.5 μM	4
Primer 2 (Reverse)	2.5 μM	4
DNA template	15 ng/μL	2
iProof DNA polymerase	0.02 U/μL	0.2
Sterile water		5.4
Total volume		20

5. HF buffer can amplify most starch synthesis genes. However, some such as SSIIa may need GC buffer for amplification.

LR-PCR amplification thermo-cycling parameters are presented below:

1. Elevated denaturation and annealing temperatures might be required for iProof as its reaction buffers contain more salts.
2. Each gene/primer has its specific annealing temperature. Therefore, too many different cycling temperatures might be "optimum" for each primer. To avoid this problem and shorten the time to find the optimum thermo-cycling condition, we recommend a hot touch-down protocol to amplify all genes in one cycling condition (*see* **Note 30**).

Cycle step	Temp (°C)	Time	Number of cycles
Initial denaturation	98	1 min	1
Touch down denaturation	98	10 s	
Touch down annealing	72–62	20 s (in each cycle temp will reduce 1 °C)	10 Cycles
Touch down extension	72	4 min	
Denaturation	98	10 s	
Annealing	62	20 s	28 Cycles
Extension	72	4 min	
Final extension	72	10 min	

3. Prior to pooling the DNA for Illumina sequencing, the PCR products should be subjected to Sanger sequencing using BigDye Terminator version 3.1 (Applied Biosystems, Foster City, CA) to ensure that the target gene is captured.
4. The sequences generated through Sanger sequencing should be aligned with the reference sequence. This will help ensure the target gene has been captured and avoid the cost of high-throughput sequencing of non-target sequences.

3.2.4 Quantification of Amplicons for Creation of Equimolar Pool

Precise quantification of amplicons is a prerequisite for creation of equimolar pool as any error in this step may lead to bias in the representation of the samples in the sequences. Although there are several methods for measurement of DNA concentration, PicoGreen (PicoGreendsDNA Quantification kit, Invitrogen) is one of the preferred methods.

3.2.5 Amplicon Pooling (Pooling of Amplified DNA)

A uniform DNA pooling approach should be applied for all samples, before and after amplification. This is called equimolar pooling, in which normalized DNA samples (15–20 or 15 ng/μL) from each sample should be pooled in equimolar concentration to make a megapool. For amplification (capture stage), DNA of 20 individuals (or less) should be mixed for each reaction tube (gene segment). The following procedure should be followed after amplification:

1. The concentration of PCR products should be measured, preferably using Pico-Green (PicoGreendsDNA Quantification kit, Invitrogen).
2. Prepare a second pool for each amplicon (amplified fragment). This means that PCR products of all tubes (each tube contains DNA of 20 individuals) related to one amplified fragment should be pooled. This will facilitate the final equimolar pooling of PCR products to prepare the megapool.
3. After mixing PCR products related to each fragment (second pool), concentration of second pools must be measured again using Pico-Green (PicoGreendsDNA Quantification kit, Invitrogen) or Nanodrop, using the protocol here: http://nanodrop.com/Library/CPMB-1st.pdf.
4. The concentration of each second pool should be normalized to 25 ng/μL, using distilled-deionized water.
5. For preparation of mega pools, the length of each fragment (bp) must be taken into consideration. Longer amplicons require more DNA to be added than smaller fragments to ensure the numbers of fragment copies are equal. The relative amount added is directly proportional to length, twice as much of a 10,000 bp a fragment is added relative to a 5000 bp fragment.
6. As 2.5 ng of amplicons in 2500 μL (Megapool including all amplicons of all fragments) are needed for massive parallel sequencing, the following formula must be used for calculations:

   Amount of amplification product required (μL) = (Length of each fragment (bp))/(Total length of all genes (bp)) × 2500 (μL).
7. Example: The length of first fragment (half of gene) of GBSSIH1 is 2164 bp. If we consider the total length including all genes that needs to be sequenced (e.g., 17 genes with totally

34 halved fragments) as 125,582 bp (~125 kb) then the amount required will be calculated as follows:

Amount required for GBSSIH1 (μL) = (2164 (bp))/(125,582 (bp)) × 2500 = 43.079 μL.

8. All other fragments must be calculated as above and added to the final Megapool for Illumina sequencing.

3.2.6 Amplicon Sequencing

Sequence the pooled amplicons using a MiSeq (Illumina) or similar sequencing platform. Data volumes need to be large enough to obtain sufficient coverage of the total length of amplicons in the bulk to allow distinction of SNP represented by a single individual (in the bulk) from a sequencing error.

3.2.7 Data Analysis to Discovery Polymorphisms in Population

After massive parallel sequencing (Illumina), the output data will be generated, which the data is analyzed with the software package CLC Bio. The following parameters will be given for each gene separately:

1. Total length of assembled sequence for each gene, including coordinates and chromosome number.
2. Total number of variants for each gene, including nonsynonymous SNPs (nsSNPs), synonymous SNPs (Functional SNPs), and Indels (insertion deletions) in coding or noncoding regions. All the information about SNP/Indels comes with frequency of SNPs/Indels, their positions, coordinates, and type of amino acid change that they may cause. Figure 2 has been generated by CLC Bio and shows distribution of SNP across the gene of interest.

The following parameters can be calculated manually, using the data output supplied by software:

1. The total rate of polymorphism is calculated as: TSI/TL × 100, where TSI = total number of SNPs and Indels, and TL is the total length of each candidate gene.
2. The rate of functional (nonsynonymous SNPs) can be calculated as: NS/TL × 100, where NS = number of nonsynonymous SNPs in each locus and TL is the total length of each candidate gene.
3. Number of SNPs per kb length of gene, which indicates the variability of point mutations in specific locus of the population.
4. Number of InDels per kb length of gene, which indicates the variability of larger mutations in specific locus of the population.
5. Ka/Ks ratio, which is the proportion of nonsynonymous (Ka) relative to synonymous (Ks) SNPs. The Ka/Ks ratio indicates

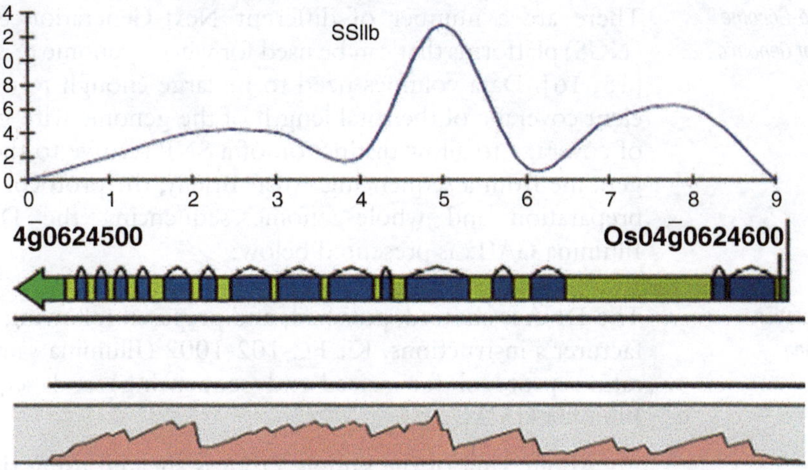

Fig. 2 SNP/Indel distribution and short read coverage pattern across SSIIb gene generated by CLC Bio after massively parallel sequencing (Illumina). The *X*-axis indicates the length of sequenced area (genes) in kb and *Y*-axis shows the number of detected SNPs/InDels. The graphs show the distribution of SNPs across the gene (The values under zero must be regarded as zero). Graphics in the middle side show the relevant gene (Blue = introns and Yellow = exons). Graphics in the down side (pink color) show the coverage pattern of each gene (Adapted from Kharabian-Masouleh et al. [11])

nature of selection pressure exerted on the gene during evolution. The Ka/Ks ratio = 1 means that selective condition of evolution (perhaps by environment or human manipulation) has been neutral. The Ka/Ks ratio < 1 shows the negative or purifying selection pressure, where Ka/Ks ratio > 1 indicates positive or diversifying (adaptive) effects on gene during selection [12].

3.3 Whole-Genome Re-sequencing

Whole-genome polymorphism detection by sequencing [10, 12] is a powerful tool for association genetics. High-throughput and cost effective sequencing technologies have aided discovery of millions of genome-wide DNA polymorphisms such as Single-Nucleotide Polymorphisms (SNPs) and Insertion-Deletion (InDels) which are an invaluable resource for marker-assisted breeding [13]. SNPs and InDels are becoming the preferred markers in molecular breeding due to multiple advantages such as high frequency, stability, high-throughput capability, and cost effectiveness over other DNA markers [13]. Next-generation sequencing technologies make possible the discovery of a massive number of DNA polymorphisms by comparing the whole-genome sequences of individuals with high-quality reference genome sequences [13]. SNPs have been employed in breeding programs for marker-assisted and genomic selection, association and QTL mapping, positional cloning, haplotype and pedigree analysis, seed purity analysis, and variety identification [14].

3.3.1 Whole-Genome Sequencing of Genomic DNA

There are a number of different Next-Generation Sequencing (NGS) platforms that can be used for whole-genome re-sequencing [15, 16]. Data volumes need to be large enough to obtain sufficient coverage of the total length of the genome with good depth of coverage to allow distinction of a SNP relative to the reference genome from a sequencing error. Briefly, the protocol for sample preparation and whole-genome sequencing the DNA using Illumina GAIIx is presented below:

Sample Preparation and Sequencing

The DNA is sheared, polished, and prepared following the manufacturer's instructions, Kit FC-102-1002 (Illumina sample preparation protocol for paired-end sequencing) and sequenced in Illumina GAIIx.

1. About 1 μg of the sample DNA is sheared using the adaptive focused acoustics method on a Covaris S2 device with the following settings: duty cycle 10%; intensity 5; cycles per burst 200 for 180 s at 6 °C.
2. Ligation products are purified by electrophoresis on Invitrogen E-gel Size Select 2%.
3. The fragments in the range of predominantly 450-bp size are selected from the gel.
4. PCR products are further purified with a QIAquick PCR Purification Kit and quantified using a DNA 1000 chip on an Agilent BioAnalyzer 2100.
5. Approximately 4 pmol per individual 76 × 2 cycles on an Illumina Genome Analyzer (GAIIx) using sequencing kit v4.
6. Base calling was performed with Illumina software Pipeline 1.4 (Illumina, San Diego, CA, US).

3.3.2 Analysis of Whole-Genome Re-sequencing Data

The bioinformatics tools presently available help discovery of genome-wide polymorphisms by comparing the whole-genome sequence of individual genotypes with high-quality reference genome sequences [13]. CLC Genomics workbench is used to assemble the trimmed, high-quality reads from each of the genotypes individually to the reference genome. A schematic overview of the steps involved in whole-genome sequence data analysis and discovery of variants (SNPs and InDels) in comparison with the high-quality reference genome sequence with whole-genome re-sequencing data is presented in Fig. 3.

3.3.3 Quality Control of the Whole-Genome Sequence Data

The Paired-end sequence reads needs to be checked for their quality and the low-quality sequence reads, as well as the adapter sequences should be removed before further analysis. In order to achieve this, the short reads are subjected to the process of trimming in CLC genomics workbench 5.0 (http://www.clcbio.com), which involves the following steps:

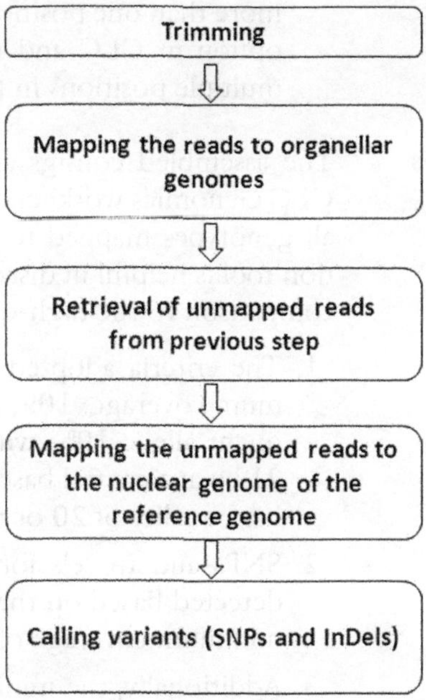

Fig. 3 An overview of whole-genome sequence data analysis and discovery of variants

1. The paired-end sequences are trimmed using trim function with a quality score limit of 0.01.
2. The adaptor sequences are also removed using the trim option in CLC Genomics Workbench 11.0.1 (http://www.clcbio.com).
3. Additionally, the reads of less than 30 base pairs (bp) in length are also discarded from the short reads, after which the short read sequences are ready to mapping.

Mapping of Reads

1. The trimmed short-read sequences are first aligned to the organellar genomes of the rice reference genome (Chloroplast genome, mitochondrial genome) (*see* **Note 31**).
2. The unmapped reads are then taken up for further assembly against the nuclear genome of the reference genome. In case of rice, the reads were assembled to the Nipponbare reference pseudomolecule 4.0 with CLC Genomics workbench.
3. The mapping parameters generally adopted in CLC genomics work bench involve mismatch cost—2, insertion cost—3, deletion cost—3, length fraction—0.9, and similarity—0.9 (*see* **Note 32**).
4. Then, the reads that align to unique genomic positions need to be identified through the removal of the reads that aligned to

more than one position of the reference genome using filtering option in CLC and used for determining reads mapping to multiple positions in the reference and unmapped reads.

Calling of Variants (SNPs and InDels)

The assembled contigs are screened for SNPs and InDels using CLC Genomics workbench 11.0.1 for detecting the variants across all genotypes mapped to the reference genome. The SNP detection tool is helpful in discovering SNPs in the assembled contigs in comparison to the high-quality reference genome.

1. The criteria adopted for detection of variants include: minimum coverage: 10×, and minimum frequency of the least frequent allele: 10%, with additional parameters of a window of 21 bp at a central base quality score of 30 with the surrounding base quality of 20 or more for SNP detection.
2. SNPs and InDels for each of the DNA samples can also be detected based on the contigs from individual assemblies with minimum coverage of 4×.
3. Additionally, the minimum variant frequency is very stringent with >90% for SNP and >30% for InDel detection in individual DNA samples.
4. The SNPs from repetitive regions can be eliminated through selection of "No repeats" option in the overlapping annotation tab in CLC Genome workbench.
5. The SNPs and InDels belonging to specific starch related genes can be filtered out based on the LOC-Ids and the genomic coordinates (Table 2) using filtering option in CLC bio.
6. In order to estimate the gene diversity and the other parameters as described in the above section, the complete sequence of the gene for each genotype can be extracted from the assembled reads as consensus assembly.
7. The SNPs/InDels and their association with the starch phenotype can be assessed using TASSEL software as described in the association analysis section (Subheading 3.5).

3.4 Genotyping of SNP in Rice Genotypes

Usually, a large number of samples and many single-nucleotide polymorphisms (SNPs) involve in starch bio-synthesis analyses; therefore a rapid, inexpensive, and highly automated multiplex method is required to genotype the sequence variants. For example, Kharabian-Masouleh et al. [3] discovered 66 functional (coding) SNPs in 233 Australian rice samples, encompassing 18 starch bio-synthesis genes. Using conventional Sanger sequencing method, a total of 66 × 233 = 15,378 separate sequencing reactions are required to genotype SNP-individuals. In this section, we explain an optimized high-throughput multiplexed SNP assay (Sequenom®) to detect SNP polymorphisms in a large number of samples. Then, we explain how the genotyping data can be matched with phenotype to distinguish association between SNPs and traits.

Table 2
List of genes involved in starch biosynthesis and their chromosomal location

S. No.	Name of the gene/enzyme	Gene ID (NCBI)	Gene ID (MSU)	Chromosomal location	Genome coordinates	
					Start	End
1	AGPS2b, [ADP-glucose pyrophosphorylase (Small Unit)]	Os08g0345800	LOC_Os08g25734	8	15760599	15754206
2	SPHOL (alpha 1,4-glucan phosphorylase)	Os03g0758100	LOC_Os03g55090	3	32183093	32190581
3	GPT1 (Glucose-6-phosphate/phosphate-translocator)	Os08g0187800	LOC_Os08g08840	8	5138640	5142712
4	Granule Bound Starch Synthase I (Waxy gene)	Os06g0133000	LOC_Os06g04200	6	1764623	1769657
5	Granule Bound Starch Synthase II	Os07g0412100	LOC_Os07g22930	7	13584483	13576435
6	Starch Synthase I	Os06g0160700	LOC_Os06g06560	6	3078060	3078060
7	Starch Synthase IIa	Os06g0229800	LOC_Os06g12450	6	6747562	6751981
8	Starch Synthase IIb	Os02g0744700	LOC_Os02g51070	2	32125071	32119749
9	Starch Synthase IIIa	Os08g0191500	LOC_Os08g09230	8	5351108	5362370
10	Starch Synthase IIIb	Os04g0624600	LOC_Os04g53310	4	32149493	32158120
11	Starch Synthase Iva	Os01g0720500	LOC_Os01g52250	1	31786842	31797321
12	Branching enzyme I	Os06g0726400	LOC_Os06g51084	6	31775431	31782688
13	Branching enzyme IIa	Os04g0409200	LOC_Os04g33460	4	20260837	20265349
14	Branching enzyme IIb	Os02g0528200	LOC_Os02g32660	2	20213965	20224864
15	Debranching enzyme-isoamylase 1	Os08g0520900	LOC_Os08g40930	8	25981756	25988347

3.4.1 DNA Extraction

The DNA should be extracted from leaves of seedlings and normalized for PCR reactions as described in previous sections. The Qiagen (Valencia, CA, USA) DNeasy Plant Kit might be used according to the instructions provided earlier.

3.4.2 Primer Design/Generation of SNP Markers

Primer design should be done using MassARRAY® Assay design v3.1 software. The user's manual can be accessed here: http://cancerseqbase.uchicago.edu/documents/AssayDesign3.1Guide.pdf

In this system, there are two primers. First, amplification primers (Forward and reverse), which amplify the target gene fragment, including SNP(s); and second, the extension primer. Generally, the extension primer is designed to hybridize directly adjacent to a SNP on a single (amplified) fragment of DNA and then to extend by one of four mass-modified nucleotides. The mass-modified nucleotides are terminator molecules specific to the DNA nucleotide 3′ of the primer (Fig. 4).

The steps must be taken for capture and extension design:

1. Install Sequenom® MassARRAY® Assay design 3.1 software on your computer. The package usually comes with the purchased platform.
2. Open the software by double-click and choose your desired MassEXTEND Assay Type.
3. Click on Browse and select the file (.txt) that you have already prepared in .txt format (NotePad). The text should include the sequence of all genes and involving SNPs placed in brackets (Fig. 5).
4. There should be a heading as SNP_ID SEQUENCE in the first line. Each gene should have a separate name (Fig. 6).
5. Usually two assay types are offered by software: MBE Mass Extend (Multiple Base Extension) and SBE Mass Extend (Single-Base Extension). These terms refer to how the extend primers are designed.
6. For starch genes, use SBE (Single-Base Extension), which is chemically coupled with choices for a specified stop mix (Fig. 4c). As shown in Fig. 4c, the SBE assays analyze SNPs by a single-nucleotide extension after the extend primer bound into the region of DNA sequence variation.
7. Choose the multiplex level (1–20). We do not recommend the multiplex beyond 20. This means that the maximum combined assays will not be more than 20 in one PCR (5 μL).
8. Choose your SNP Capture settings. Set the amplicon size as 80, 100, and 120 bp, for Minimum, Optimum, and Maximum, respectively.
9. Sequence annotation should be set to Scan and Restrict.
10. Two (10-mer) tags will be added to the 5′ end of each amplification primer by default to avoid confusion in the mass spectrum and to improve PCR performance (5-ACGTTGGATG-3).

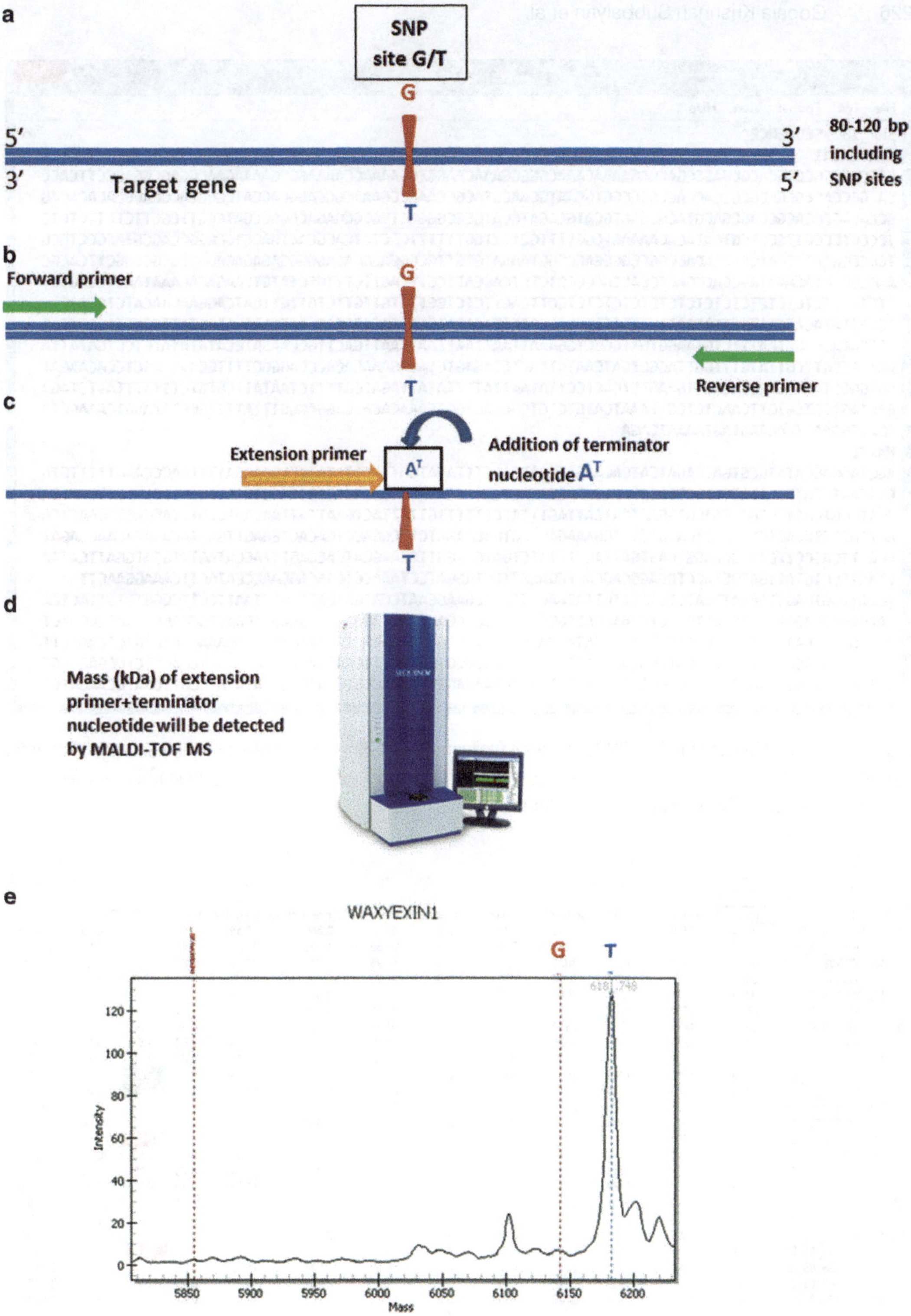

Fig. 4 Application of multiplexed-MALDI-TOF MS assays to detect SNP (genotyping). (**a**) Double strand DNA including SNP site (G/T). A targeted length of 80–120 bp (spanning SNP site) should be amplified. (**b**) This will be done using specific forward and reverse primers (including 10-mer tag). (**c**) Amplification with extension primer (single-strand DNA) and termination using mass-modified terminator nucleotide. (**d**) The mass of extended molecule plus mass-modified terminator nucleotide will be detected using MALDI-TOF MS. (**e**) The output data will pass through a computer software (Typer) and peak show the SNP genotype. The results will also be available in EXCEL spreadsheet

N_163SNPs2 - Notepad

File Edit Format View Help

```
SNP_ID  SEQUENCE
WAXYEXIN1
GAGCGCGCGCGCACGCGAGCCGAGGAGAAAACAAACGGGGGAAACAAGCTGGAAAAGCAAAAGGGGAAAAGAACGGAGCGGAGGCTTCACC
CACGGCCACCGCGACGCGCCACCAGCGTGCGGTGCAATGCAACGTACGCCAAGCCGAAACGGCAGGCAGCATCGCGCACGCACGCACACACAG
GCCACAGCACACGCGAGCGACGTACGCGAGTGCATGCAGATGCATGCGCGGGGCTCGCGCGAGACCGGCCGATGGGTTCGCTTCTCTTCTCTC
TCCCGTCCCGTTGCGTCGTCATAGACAAAAGTCGGTTTTGCTTTTGGTTTTTTGGCTCTGAGGCACTGACGTGCGGGCCAGCGTACGCCTGCG
TGCCCCGCATGTCATCGTCGACACCGGCCGGGGACCGGGTAAAATGTGTTGCGGGAGGGAGAGGGGGAGAGAGAGATCGCGCGGGCTTCACGC
AACGGCGCTACAAATAGCCACCCACACCACCACCCCCTCTCTCACCATTCCTTCAGTTCTTTGTCTATCTCAAGACACAAATAACTGCAGTCT
CTCTCTCTCTCTCTCTCTCTCTCTCTCTCTCTCTCTGCTTCACTTCTCTGCTTGTGTTGTTCTGTTGTTCATCAGGAAGAACATCTGCAAG
[G/T]TATACATATATGTTTATAATTCTTTGTTTCCCCTCTTATTCAGATCGATCACATGCATCTTTCATTGCTCGTTTTTCCTTACAAGTAG
TCTCATACATGCTAATTTCTGTAAGGTGTTGGGCTGGAAATTAATTAATTAATTAATTGACTTGCCAAGATCCATATATATGTCCTGATATTA
AATCTTCGTTCGTTATGTTTGGTTAGGCTGATCAATGTTATTCTAGAGTCTAGAGAAACACACCCAGGGGTTTTCCAACTAGCTCCACAAGAT
GGTGGGCTAGCTGACCTAGATTTGAAGTCTCACTCCTTATAATTATTTTATATTAGATCATTTTCTAATATTCGTGTCTTTTTTTATTCTAGA
GTCTAGATCTTGTGTTCAACTCTCGTTAAATCATGTCTCTCGCCACTGGAGAAACAGATCAGGAGGGTTTATTTTGGGTATAGGTCAAAGCTA
AGATTGAAATTCACAAATAGTAAAATCAGA
WAXYEX6
AGGTAGGAGCATATGCGTGATCAGATCATCACAAGATCGATTAGCTTTAGATGATTTGTTACATTTCGCAAGATTTTAACCCAAGTTTTTGTG
GTGCAATTCATTGCAGATCAAGGTTGCAGACAGGTACGAGAGGGTGAGGTTTTTCCATTGCTACAAGCGTGGAGTCGACCGTGTGTTCATCGA
CCATCCGTCATTCCTGGAGAAGGTGGAGTCATCATTAGTTTACCTTTTTTGTTTTTACTGAATTATTAACAGTGCATTTAGCAGTTGGACTGA
GCTTAGCTTCCACTGGTGATTTCAGGTTTGGGGAAAGACCGGTGAGAAGATCTACGGACCTGACACTGGAGTTGATTACAAAGACAACCAGAT
GCGTTTCAGCCTTCTTTGCCAGGTCAGTGATTACTTCTATCTGATGATGGTTGGAAGCATCACGAGTTTACCATAGTATGTATGGATTCATAA
CTAATTCGTGTATTGATGCTACCTGCAGGCAGCACTCGAGGCTCCTAGGATCCTAAACCTCAACAACAACCCATACTTCAAAGGAACTT
[C/A]TGGTGAGTTACAATTGATCTCAAGATCTTATAACTTTCTTCGAAGGAATCCATGATGATCAGACTAATTCCTTCCGGTTTGTTACTGA
CAACAGGTGAGGATGTTGTGTTCGTCTGCAACGACTGGCACACTGGCCCACTGGCGAGCTACCTGAAGAACAACTACCAGCCCAATGGCATCT
ACAGGAATGCAAAGGTCTATGCTTGTTCTTGCCATACCAACTCAAATCTGCATGCACACTGCATTCTGTTCAGAAACTGACTGTCTGAATCTT
TTTCACTGCAGGTTGCTTTCTGCATCCACAACATCTCCTACCAGGGCCGTTTCGCTTTCGAGGATTACCCTGAGCTGAACCTCTCCGAGAGGT
TCAGGTCATCCTTCGATTTCATCGACGGGTATGAGTAAGATTCTAAGAGTAACTTACTGTCAATTCGCCATATATCGATTCAATCCAAGATCC
```

Fig. 5 Input texts file showing two SNPs in starch waxy gene (GBSSI). Please note how the headings should be named and SNP sites must be placed in brackets [G/T]. All SNP-sequences can be placed in one file to be imported into ASSAY Design software for primer design

	A	B	C	D	E	F	G	H	I	J	K	L	M	N
1	233	13	1											
2	Peak1	Trough1	Breakdown	FinalVisc	Setback	PeakTime	Past_temp	Martin_N	AC_percent	predicted_N	diff	GT	Chalk%	
3	1_YRR08_01_03	3044	1688	1356	2895	-149	6.0	66.5	0.865	17.49	0.75	0.12	70.12	39.6675
4	10_YRR08_02_16	2987	2005	982	3202	215	6.3	66.55	1.110	19.52	1.01	0.10	71.07	36.9825
5	100_YRA08_02_12	3050	1659	1391	3021	-29	5.9	67.25	0.776	18.35	0.86	-0.08	69.89	3.5779
6	101_YRA08_03_01	3199	1777	1422	3102	-97	5.9	68.05	0.695	18.18	0.84	-0.14	66.60	2.8713
7	102_YRA08_03_02	3195	1981	1214	3094	-101	6.2	68.9	0.607	17.23	0.72	-0.11	70.15	7.1152
8	103_YRA08_03_03	2976	1774	1202	3129	153	6.1	67.3	0.775	18.23	0.84	-0.07	65.75	4.3151
9	104_YRA08_03_04	2872	1480	1392	2560	-312	5.9	68.9	0.599	16.92	0.67	-0.08	70.10	6.9856
10	105_YRA08_03_05	3339	1819	1520	3025	-314	5.9	66.4	0.603	16.46	0.62	-0.01	69.22	3.7925
11	106_YRA08_03_06	3029	1751	1278	2909	-120	6.0	67.3	0.571	16.57	0.63	-0.06	71.91	4.8824
12	107_YRA08_03_08	2918	1658	1260	2958	40	6.0	66.6	0.675	18.59	0.89	-0.21	67.35	3.7207
13	108_YRA08_03_10	3136	1693	1443	3048	-88	5.9	68.15	0.756	18.82	0.92	-0.16	69.02	1.6734
14	109_YRA08_03_11	3551	1748	1803	2921	-630	5.8	67.85	0.669	17.15	0.71	-0.04	-999	2.4485
15	11_YRR08_02_19	3030	1626	1404	2895	-135	5.9	66.5	0.831	17.54	0.75	0.08	69.56	11.2425
16	110_YRA08_04_02	3179	2064	1115	3232	53	6.3	67.25	0.679	15.70	0.52	0.16	70.91	3.6835
17	111_YRA08_04_04	3241	1733	1508	3001	-240	5.9	67.25	0.788	17.93	0.81	-0.02	70.61	3.1544
18	112_YRA08_04_05	2969	1676	1293	3086	117	5.9	66.5	0.815	17.54	0.75	0.06	69.15	3.3158
19	113_YRA08_04_07	3113	1735	1378	3132	19	5.9	67.45	0.673	18.63	0.90	-0.22	68.81	3.7188
20	114_YRA08_04_10	3184	1750	1434	2903	-281	5.9	67.25	0.573	16.12	0.57	0.00	69.75	3.1782
21	115_YRA08_04_11	3156	1883	1273	3095	-61	6.1	67.35	0.649	17.68	0.77	-0.12	69.19	3.324
22	116_YRA08_04_12	3042	1691	1351	2805	-237	5.9	67.3	0.591	16.74	0.65	-0.06	67.35	4.6994
23	117_YRA08_05_02	3109	1724	1385	2875	-234	6.1	66.45	0.717	16.02	0.56	0.16	70.18	4.5795
24	118_YRA08_05_03	3089	1769	1320	2998	-91	6.0	68.95	0.644	17.02	0.69	-0.04	66.55	5.1133
25	119_YRA08_05_04	3268	1635	1633	2712	-556	5.8	69	0.562	16.32	0.60	-0.04	69.00	8.8188
26	12_YRR08_02_20	3263	1630	1633	2746	-517	5.8	67.25	0.607	16.62	0.64	-0.03	69.00	7.5925

Fig. 6 Preparation of trait file (Phenotypic data input)

Table 3
Preparation of PCR reactions to capture starch genes for multiplex SNP assays

Components	Concentration	Volume for 20 μL reaction (μL)	Volume for 5 μL reaction (μL)
10× PCR buffer (InviTrogen)	10× (provided in kit)	2	0.5
$MgCl_2$	50 mM	1.4	0.35
dNTPs (mix)	2.5 mM	4	1
Primer 1 (Forward)	1 μM	2	0.5
Primer 2 (Reverse)	1 μM	2	0.5
DNA template	3–5 ng/μL	1	0.25
Platinum® Taq DNA Polymerase	2 U/rxn	0.3	0.075
Sterile water	–	7.3	1.825
Total volume		20	5

11. For extend primer design (extension primer), choose the following: Tm (°C) 45–100, by "NN", Length 17–28 and leave the others as default.
12. Run the program and you will get all primers designed in an Excel file ready to be ordered.

3.4.3 Gene Capture PCR

Using the aid of amplification primers, we will be able to capture each starch gene in uniplexed or multiplexed assay. Each -plex is referred to a single reaction including a forward and reverse primer able to amplify a gene fragment spanning the SNP site. Before starting the multiplexed genotyping, if there is sufficient time and labor, it is recommended to set up a series of uniplex reactions (assays) to test the performance of primers. Otherwise, the multiplexed experiments can be conducted. The detailed instruction for iplex genotyping can be found here: http://cancerseqbase.uchicago.edu/documents/iPLEXGoldApplicationGuide.pdf

The PCR reaction mixture is recommended for an 8-plex experiment (to capture eight SNP sites):

1. The final volume can be 5 μL but volumes more than this up to 20 μL can be used.
2. The Platinum® Taq DNA Polymerase (Invitrogen, Carlsbad, CA, USA) is recommended for amplifications but any other high-quality Taq DNA Polymerases might be successfully applied.
3. Table 3 shows how to prepare a 20 μL (only for one assay); therefore, for a 5 μL reaction the volume must be divided by 4.

Table 4
Thermocycling to capture starch genes

Cycle step	Temp (°C)	Time	Number of cycles
Initial denaturation	94	15 min	1
Denaturation	94	20 s	
Annealing	56	30 s	45
Extension	72	1 min	
Final extension	72	3 min	1
Storage	4	Forever	–

4. Primers 1 and 2, and template DNA of each genotype should be added separately (should not be added to the master mix).

After preparing the master mix (PCR cocktail), the following steps need to be carried out:

1. Dispense 15 μL of PCR cocktail (without primers and DNA template) into each well of the 384-well plate.
2. Centrifuge the microtiter plate at 110 × *g* (1000 rpm) for 1 min.
3. Add relevant primers (F + R) and DNA template of genotype to be assayed.
4. Now, the total volume will be 20 μL (or 5 μL).
5. Gently mix the plate over vortex and spin down before thermocycling.
6. The first PCR cycling to amplify the gene is as follows (Table 4).

3.4.4 Shrimp Alkaline Phosphatase (SAP) Incubation

Before applying extension primer, remaining unincorporated dNTPs from PCR products should be neutralized by SAP incubation. This will allow the mass modified terminators to bind adjacent to extension primer to recognize SNP variations by MALDI-TOF MS.

Preparation of SAP Enzyme Master Mix

1. Use a 1.5 mL tube to prepare the SAP enzyme solution (master mix) (Table 5).
2. Gently vortex the 1.5 mL tube for 5 s.
3. Centrifuge the 1.5 mL tube for 10 s at 2793 × *g* (5000 rpm).
4. Take a 96-well plate (preferably Vee bottom Sarstedt, Inc. #82.1583).
5. Choose the last bottom row (H row) (*see* **Note 33**) and dispense 85 μL of SAP enzyme into each well of this row (*see* **Note 34**).
6. Use a 12-channel pipettor and draw the SAP enzyme from the row H and distribute into each well in rows G to A (10 μL per well, each time) (*see* **Note 35**).

Table 5
Preparation of SAP solution

Components	Volume (1rxn) (μL)	Volume (384rxn)[a] (μL)
Water (HPLC grade)	1.53	810.9
SAP Buffer (10×)	0.17	90.1
SAP enzyme (1.7 U/μL)	0.30	159.0
Total volume	2.00	1060.0

[a]This includes a 38% "overhang"

SAP Enzyme Addition

The SAP enzyme dispensed in 96-well plate microtiter should be used to add the enzyme into the PCR reactions. For this reason, a liquid handler machine should be used (supplied along with Sequenom platform) as follows:

1. Place the 96-well microtiter plate containing SAP solution on position 1 (left-up corner) of the liquid-handler deck.
2. Spin down the 384-well microtiter sample plate containing amplification products at 110 × *g* (1000 rpm) for 1 min.
3. Put the 384-well microtiter sample plate on a plate flattener (provided by Sequenom manufacturer) and then place it on liquid-handler deck position 3 (left-down).
4. Run the "SAP Addition (96 to 384)" method on the liquid-handler controller PC.
5. Enter number 1 (default) for the number of plates.
6. By default 2 μL of SAP solution will be added to each well of 384-well, containing PCR products (*see* **Note 36**).
7. Remove the 384-well sample microtiter plate and flattener from position 3, and then seal the plate with plate film.
8. Remove the microtiter plate from its plate flattener and centrifuge for 1 min at 110 × *g* (1000 rpm).
9. Incubate the 384-well sample microtiter plate in thermocycler, as follows: 37 °C for 40 min, 85 °C for 5 min, and 4 °C forever.
10. When the SAP incubation is being done, try to prepare the iPLEX Gold reaction cocktail.

3.4.5 Preparation of iPLEX Gold Reaction for Different DNAs of 96–384 Samples (Well-Plate)

1. The iPLEX Gold reaction should be prepared in a 1.5 mL tube as described in Table 6 (*see* **Note 37**).
2. Gently vortex the 1.5 mL tube for 5 s.
3. Centrifuge the 1.5 mL tube for 10 s at 2793 × *g* (5000 rpm).
4. Take a 96-well plate (preferably Vee bottom Sarstedt, Inc. #82.1583).

Table 6
Preparation of multiplexed iPLEX Gold reaction cocktail

Components	Concentration in 9 μL	Volume (1rxn) (μL)	Volume (384rxns)[a] (μL)
Water (HPLC grade)	N/A	0.7395	391.97
iPLEX Buffer Plus (10×)	0.222×	0.200	106.00
iPLEX Termination mix	0.5×	0.100	53.00
Primer mix (14 μM)	1.25 μM	0.94	498.12
iPLEX enzyme	0.5×	0.0205	10.87
Total volume		2.000	1060.0

[a]This includes 38% "overhang"

5. Choose the last bottom row (H row) and dispense 85 μL of iPLEX Gold reaction cocktail into each well of this row.
6. Use a 12-channel pipettor and draw the iPLEX Gold reaction cocktail from the row H and distribute into each well in rows G to A (10 μL per well, each time) (*see* **Notes 38** and **39**).
7. Centrifuge the microtiter plate for 1 min at 280 × *g* (1600 rpm).

Addition of iPLEX Gold Reaction Cocktail into PCR Amplification Products from Gene Capture Stage Using the Liquid-Handler Deck

1. Place the 96-well microtiter plate containing iPLEX Gold reaction solution on position 1 (left-up corner) of the liquid-handler deck.
2. Spin down the 384-well microtiter sample plate containing amplification products at 110 × *g* (1000 rpm) for 1 min.
3. Put the 384-well microtiter sample plate on a plate flattener (provided by Sequenom manufacturer) and then place it on liquid-handler deck position 3 (left-down).
4. Run the "Cocktail Addition (96 to 384)" method on the liquid-handler controller PC.
5. Enter number 1 (default) for the number of plates.
6. By default 2 μL of iPLEX Gold cocktail solution will be added to each well of 384-well, already SAP-treated PCR products from gene capture stage (*see* **Note 36**).
7. Remove the 384-well sample microtiter plate and flattener from position 3, and then seal the plate with plate film.
8. Remove the microtiter plate from its plate flattener and centrifuge for 1 min at 110 × *g* (1000 rpm).

Addition of iPLEX Gold Reaction Cocktail into PCR Amplification Products from Gene Capture Stage Using 12-Channel Pipettor (Manual)

The same procedure as above can be manually repeated using a 12-channel pipettor. 2 μL of iPLEX Gold reaction cocktail should be added to each well of the sample microtiter plate.

3.4.6 Thermocycling the Mixture of iPLEX Gold Reaction and PCR Amplification Products (Gene Captured)

In this stage, the extension primer(s) binds adjacent to SNP site and the mass-modified terminator terminates the extension reaction. The mass of amplified molecule will be detected by MALDI-TOF, which will be converted by computer program to reveal the SNP genotype of individual.

The microtiter plate should be placed in a thermocycler programmed with the parameters described in Table 7.

Cleaning Up the Reaction Products

The next step is to clean up the iPLEX Gold reaction products using Clean Resin (resin). This step optimizes the mass spectrometry analysis.

1. Take a 6 mg resin 384-well dimple plate (reservoirs).
2. Transfer the resin from its container onto the 6 mg resin 384-well dimple plate (provided by Sequenom Company).
3. Spread the resin into the wells of the dimple plate using a scraper and make sure that resin sits inside each well.
4. This can be done by scraping the resin back and forth.
5. Remove excessive resin of the dimple plate using the same scraper.
6. The excessive resin should be returned to the container for further uses.
7. Leave the resin in the dimple plate for at least 20 min.

Adding Water to the 384-Well Sample Microtiter Plate (iPLEX Gold Reaction Products)

1. Place a reservoir of 80 mL including nanopure water on position 1 of the liquid-handler deck (top-left corner).
2. In the same time, centrifuge the 384-well sample microtiter plate for 1 min at 110 × *g* (1000 rpm).

Table 7
Thermocycling of iPLEX Gold reaction and PCR amplification products (*see* Note 40)

Cycle step	Temp (°C)	Time	Number of cycles
Initial denaturation	94	30 s	1
Denaturation	94	5 s	40 Cycles
Annealing	52	5 s	5 Cycles
Extension (for extend primer)	80	5 s	
Final extension	72	3 min	1
Storage	4	Forever	–

3. Remove the plate sealing film and place the 384-well sample microtiter plate on a plate flattener. Then, place sample microtiter and flattener on position 3 of the liquid-handler deck (Down-left side).
4. Run the "16 μL Water Addition (384)" method from the liquid handler controller PC.
5. Leave the default setting unchanged, for 1 destination plate.
6. The liquid handler will add 16 μL of water into each well of sample microtiter plate.
7. Remove the 384-well sample microtiter plate and flattener from position 3, then seal the 384-well sample plate with sealing film.
8. Remove the plate flattener.
9. Centrifuge the microtiter plate for 30 s at 280 × *g* (1600 rpm) to remove air bubbles.

Addition of Resin to the Sample Microtiter Plate

1. Remove the sealing film from the sample microtiter.
2. Carefully place the sample microtiter plate, upside-down, onto the dimple plate (including resin) (*see* **Note 41**).
3. After flipping over, gently tap the dimple plate. This makes the resin fall out from the dimple plate into the wells of the microtiter plate.

Rotate and Centrifuge the iPLEX Gold Reaction Products (Resin Added)

1. Rotate the sample microtiter plate (360° about its long axis) on a rotator at room temperature for 5 min.
2. Remove the plate from rotator and centrifuge the sample microtiter plate at 3200 × *g* for 5 min.
3. Now, the iPLEX Gold reaction products should be transferred to a SpectroCHIP, using the MassARRAYNanodispenser. For instructions, see the MassARRAYNanodispenser User's Guide (Samsung).

3.4.7 Defining Assays and Plates

This should be done using Typer 4 software after nanodispensing. Please see the "Defining Assays" and "Defining Plates" chapters in the MassARRAYTyper Software User's Guide.

3.4.8 Acquiring Spectra and Genotyping

After defining assays and plates, spectra (genotyping) should be acquired using the MassARRAYAnalyzer Compact mass spectrometer. For instructions, see the "Acquiring Spectra" chapter in the MassARRAYAnalyzer Compact User's Guide.

All the above-mentioned information can be accessed here: http://cancer-seqbase.uchicago.edu/documents/iPLEXSoftwareGuide.pdf

The results of genotyping will come out as Excel sheet tables.

3.5 Association of SNP with Phenotype

With the advent of new genotyping technologies such as next-generation sequencing (NGS), the genetic information has become widely available to scientists. Many markers have been used to assess the important agronomic and quality traits. SNP markers are one of the most important and precise molecular markers which have been widely used. However, there is not enough information about association mapping and linkage between these markers and phenotypic data. Relating molecular markers and phenotypic traits is called association analysis or mapping. For this analysis, we need precise phenotypic and sequencing data about the plant's population and its structure. TASSEL is one of the most robust software packages that can be used for association analysis. It is available as stand-alone free software which must be downloaded and installed to your computer.

The process is relatively simple. We have phenotypic and genotypic data (phenotype and SNP variants, respectively) in Excel spread sheets and combine them in TASSEL to find the association among SNPs and traits as follows:

1. Download and install TASSEL http://www.maizegenetics.net/#!tassel/c17q9 (*see* **Note 42**).
2. TASSEL Version 2.1 is suitable for SNP association study and also supports SSR analysis. The newer versions (v3 and higher) do not support SSR marker analysis (*see* **Note 43**).
3. Prepare the trait data in a simple EXCEL sheet using the following format (*see* **Notes 44–47**) (Fig. 6).
4. Click "Save as" and save this phenotypic file in Text (with Tab delimited format). Use a suitable name and save it in your desired folder (e.g., folder A).
5. Prepare the genotypic data in the following format (*see* **Note 48–51**) (Fig. 7).
6. Click "Save as" and save this genotypic input file in Text (with Tab delimited format). Use a suitable name and save it in your desired folder (e.g., folder A).
7. Now, run TASSEL by double clicking on file sTASSEL.jar in the TASSEL 2-1 directory (wherever you already downloaded and installed).
8. Click on data icon, then a File Loader window will be opened.
9. Choose the option button "I will make my best guess and try" the click OK.
10. When the "Open file" window prompted, choose your phenotype input file (Text Tab delimited).
11. Repeat the previous step to load the genotypic file (Text Tab delimited).

	A	B	C	D	E	F	G	H	I	J	K	L	M	N
1	288	28:02:00												
2	AGPL2a_TBGU388647	BEI_TBGU:	BEIIb_GA_	GBSSII_GA	GPT1_GA_	Iso1_TBGU	Iso2_GA_F	Iso2_GA-F	Iso2_TBGU	Pullu_TBG	Pullu_TBG	SPHOL_TB	SPHOL_TB	SSIIa_GA_
3	1_YRR08_01_03	G:G	G:G	C:C	A:A	C:C	G:G	?:?	?:?	?:?	G:G	A:A	G:G	C:C
4	10_YRR08_02_16	G:G	G:G	C:C	A:A	C:C	?:?	?:?	?:?	?:?	G:G	A:A	G:G	C:C
5	100_YRA08_02_12	G:G	G:G	C:C	A:A	C:C	?:?	?:?	?:?	?:?	G:G	?:?	G:G	C:C
6	101_YRA08_03_01	G:G	G:G	C:C	G:A	C:C	?:?	?:?	?:?	C:C	G:G	A:A	G:G	C:C
7	102_YRA08_03_02	G:G	G:G	C:C	A:A	C:C	G:G	?:?	?:?	C:C	G:G	A:A	G:G	C:C
8	103_YRA08_03_03	G:G	G:G	C:C	A:A	C:C	G:G	?:?	?:?	?:?	G:G	?:?	G:G	C:C
9	104_YRA08_03_04	G:G	?:?	C:C	G:G	C:C	G:G	?:?	?:?	C:C	G:G	A:A	G:G	C:C
10	105_YRA08_03_05	G:G	G:G	C:C	A:A	C:C	G:G	?:?	?:?	?:?	G:G	?:?	G:G	C:C
11	106_YRA08_03_06	G:G	G:G	C:C	A:A	C:C	?:?	?:?	?:?	C:C	G:G	A:A	?:?	C:C
12	107_YRA08_03_08	G:G	G:G	C:C	A:A	C:C	?:?	?:?	?:?	?:?	G:G	A:A	G:G	C:C
13	108_YRA08_03_10	G:G	G:G	C:C	G:G	C:C	?:?	?:?	?:?	?:?	G:G	A:A	G:G	C:C
14	109_YRA08_03_11	G:G	G:G	C:C	A:A	C:C	G:G	?:?	?:?	?:?	G:G	A:A	G:G	C:C
15	11_YRR08_02_19	G:G	G:G	C:C	A:A	C:C	G:G	?:?	?:?	?:?	G:G	A:A	G:G	C:C
16	110_YRA08_04_02	G:G	G:G	C:C	?:?	C:C	?:?	?:?	?:?	?:?	G:G	A:A	G:G	C:C
17	111_YRA08_04_04	G:G	G:G	C:C	A:A	C:C	?:?	?:?	?:?	?:?	G:G	A:A	G:G	C:C
18	112_YRA08_04_05	G:G	G:G	C:C	A:A	C:C	?:?	?:?	?:?	C:C	G:G	?:?	G:G	C:C
19	113_YRA08_04_07	G:G	G:G	C:C	A:A	C:C	G:G	?:?	?:?	?:?	G:G	A:A	G:G	C:C
20	114_YRA08_04_10	G:G	G:G	C:C	G:G	C:C	?:?	?:?	?:?	C:C	G:G	A:A	G:G	C:C
21	115_YRA08_04_11	G:G	G:G	C:C	G:G	C:C	?:?	?:?	?:?	C:C	G:G	A:A	G:G	C:C
22	116_YRA08_04_12	G:G	G:G	C:C	G:G	C:C	?:?	?:?	?:?	C:C	G:G	A:A	G:G	C:C
23	117_YRA08_05_02	G:G	G:G	C:C	G:G	C:C	?:?	?:?	?:?	?:?	G:G	?:?	G:G	C:C
24	118_YRA08_05_03	G:G	G:G	A:A	G:G	C:C	G:G	?:?	?:?	C:C	G:G	A:A	G:G	C:C
25	119_YRA08_05_04	G:G	G:G	C:C	G:G	C:C	G:G	?:?	?:?	C:C	G:G	A:A	G:G	C:C

Fig. 7 Preparation of genotypic input file using SNP variant calls

12. Once the data were loaded, they will be appeared in the "Data Tree Panel" (*see* **Note 52**).
13. Click on "A:a Genotype" button. This will run the "Genotype Converter" to convert the raw format into genotype state.
14. On the "Genotype Converter," choose "Create alignment based on genotypic state (e.g., A:a>Aa)" and click OK.
15. This will create a new dataset file, automatically named "GenoStates" in Data Tree Panel.
16. Ctrl + click (highlight to choose) the phenotypic, genotypic, and GenoStates files together in Data Tree Panel (Left-top side).
17. Click on the "U Join" in the "Options Panel."
18. A new dataset named "GenoStates+Phenotypic+Genotypic" will appear in Data Tree Panel (Left-top side).
19. Select/highlight all files already created in Data Tree Panel holding Ctrl + click.
20. Click on "Analysis" button and choose "GLM" (General Linear Model) and then tick "Analyze Each Data Column Separately."
21. Leave the other option as default and click OK.
22. On "Define Output" window change the "#permutations" to 1000 then click Run.
23. A new dataset will be appeared in Results Panel, named GLM+GenoStates+Traits.
24. Click on "Results" button from Option panel and view the results in Excel sheet (*see* **Note 53**).

GBSSI	Section 1														
Trait	Locus	Site	Chr	Chr_pos	df_Marker	F-test	p-value	#perm_Ma	p-perm_Marke	p-adjusted value	df_Model	df_Error	MS_Error	R2_Model	R2_Marker
Peak1	WAXYEXIN1	0	0	0	2	34.346	8.87E-14	1000	9.99E-04	9.99E-04	2	230	38700.42	0.23	0.23
Peak1	WAXYEX10	0	0	0	1	0.1324	0.7163	1000	0.6773	1	1	230	49818.5	5.75E-04	5.75E-04
Peak1	WAXYEX6	0	0	0	1	6.6892	0.0103	1000	0.017	0.1578	1	230	45097.01	0.0283	0.0283
Trough1	WAXYEXIN1	0	0	0	2	3.0984	0.047	1000	0.045	0.6703	2	230	18738.28	0.0262	0.0262
Trough1	WAXYEX10	0	0	0	1	36.9498	4.99E-09	1000	9.99E-04	9.99E-04	1	230	16578.65	0.1384	0.1384
Trough1	WAXYEX6	0	0	0	1	0.4746	0.4916	1000	0.4236	1	1	230	19013.37	0.0021	0.0021
Breakdown	WAXYEXIN1	0	0	0	2	35.1893	4.64E-14	1000	9.99E-04	9.99E-04	2	230	33032.83	0.2343	0.2343
Breakdown	WAXYEX10	0	0	0	1	18.9223	2.05E-05	1000	0.002	9.99E-04	1	230	39440.62	0.076	0.076
Breakdown	WAXYEX6	0	0	0	1	5.1693	0.0239	1000	0.05	0.4246	1	230	39915.87	0.022	0.022
Final Viscosity	WAXYEXIN1	0	0	0	2	15.0534	7.18E-07	1000	9.99E-04	9.99E-04	2	230	36055.07	0.1157	0.1157
Final Viscosity	WAXYEX10	0	0	0	1	106.0684	1.06E-20	1000	9.99E-04	9.99E-04	1	230	27888.24	0.3156	0.3156
Final Viscosity	WAXYEX6	0	0	0	1	0.6859	0.4084	1000	0.2987	1	1	230	40551.6	0.003	0.003
Setback	WAXYEXIN1	0	0	0	2	76.2739	3.89E-26	1000	9.99E-04	9.99E-04	2	230	41231.96	0.3988	0.3988
Setback	WAXYEX10	0	0	0	1	59.8068	3.26E-13	1000	9.99E-04	9.99E-04	1	230	54241.79	0.2064	0.2064
Setback	WAXYEX6	0	0	0	1	8.0543	0.0049	1000	0.028	0.0829	1	230	63651.93	0.0338	0.0338
PeakTime	WAXYEXIN1	0	0	0	2	0.2847	0.7525	1000	0.7283	1	2	230	0.0199	0.0025	0.0025
PeakTime	WAXYEX10	0	0	0	1	0.6186	0.4324	1000	0.3946	1	1	230	0.0199	0.0027	0.0027
PeakTime	WAXYEX6	0	0	0	1	0.021	0.8849	1000	0.7273	1	1	230	0.0196	9.13E-05	9.13E-05
Past_temp	WAXYEXIN1	0	0	0	2	1.51	0.2231	1000	0.2428	0.999	2	230	19.1	0.013	0.013
Past_temp	WAXYEX10	0	0	0	1	0.6619	0.4167	1000	0.4855	1	1	230	19.1607	0.0029	0.0029
Past_temp	WAXYEX6	0	0	0	1	0.3101	0.5781	1000	0.8881	1	1	230	19.1859	0.0013	0.0013
Martin_N	WAXYEXIN1	0	0	0	2	223.2942	1.29E-54	1000	9.99E-04	9.99E-04	2	230	0.041	0.6601	0.6601
Martin_N	WAXYEX10	0	0	0	1	147.7825	1.37E-26	1000	9.99E-04	9.99E-04	1	230	0.0733	0.3912	0.3912
Martin_N	WAXYEX6	0	0	0	1	16.8014	5.76E-05	1000	0.028	9.99E-04	1	230	0.0974	0.0681	0.0681
AC_percent	WAXYEXIN1	0	0	0	2	121.5295	9.63E-37	1000	9.99E-04	9.99E-04	2	230	1.2099	0.5138	0.5138
AC_percent	WAXYEX10	0	0	0	1	44.0661	2.26E-10	1000	9.99E-04	9.99E-04	1	230	2.0869	0.1608	0.1608

Fig. 8 An output file including results of association study

3.5.1 Understanding the Results of the Output File

The result file in Excel format will look like the following (Fig. 8). The first column shows the traits and second column name of SNP variant or marker. The most important columns to be interpreted are *F*-test and R^2_Marker. Generally, in both columns, whatever the values are bigger the association is higher. For example, the last column (R^2 Marker) shows the portion of total variation explained by marker (*see* **Note 54**).

4 Notes

1. Please carefully check the kit storage instruction upon arrival. Most parts of kits can be kept for 1 year at room temperature. However, some parts such as RNaseA stock solution should be stored at 2–8 °C (for storages more than 1 year). The DNeasy Maxi spin columns should be stored dry at 2–8 °C upon arrival.
2. Using the terms "tube" or "tubes" interchangeably, depends on the number of samples.
3. Buffer AP1 and RNase A should not be mixed before use.
4. The QIAshredder Mini spin column will remove precipitates and cell debris. However, some debris may pass through the spin column and form pellet at the bottom of collection tube. Please do not disturb this pellet.
5. Warm up Buffer AW1 to 65 °C using a water bath or heating block to dissolve, if you observed any precipitates created upon storage.
6. Second elution may be done using a new 1.5 or 2 mL microcentrifuge tube(s) to prevent dilution of the first elute and final concentration of extracted DNA.

7. A laboratory coat, gloves, and safety glasses must be worn and correctly fitted at all times.
8. Take care to ensure the water bath has sufficient water to cover the heating element. A laboratory oven can be used in place of a water bath.
9. Liquid nitrogen can cause severe burns. Take great care to ensure the liquid nitrogen is not spilt or splashed and that it does not come in contact with any exposed skin or eyes.
10. The mortar will be very cold and capable of damaging skin so ensure skin does not come in contact with it.
11. Freezing in liquid nitrogen ruptures subcellular organelles and exposes DNA to endogenous nuclease activity. To prevent DNA degradation by these enzymes, the samples must remain frozen at all times until they enter the buffer solution which denatures the nucleases.
12. This step destroys RNA that would otherwise co-precipitate with the DNA at **step 11** (Subheading 3.1.3).
13. Chloroform is a volatile organic solvent which causes brain damage after long-term exposure. Use of a fume hood is therefore a mandatory requirement.
14. This step removes proteins from the sample by precipitation.
15. Chloroform is very volatile. The tubes must be tightly closed to ensure the chloroform does not escape when the samples are outside of the fume hood during centrifugation.
16. The volume of DNA/buffer solution remaining at this step will vary depending on how much of the aqueous phase was removed at **steps 5** and **8** (Subheading 3.1.3).
17. This step separates the DNA from other dissolved molecules in the sample.
18. It is important to disperse the agarose well, otherwise it will form clumps that are very difficult to melt.
19. Gels of 1% or 2% are suitable for most purposes. A 1% gel is suitable for genomic analysis of long amplicons.
20. Correct orientation is when the positive electrode is at the opposite end of the get tank to the samples and the electrodes are plugged into the correct positions in the power pack.
21. Ethidium bromide is a mutagen. Ensure it does not come in contact with the skin.
22. Glass may bind the Picogreen, so Picogreen should be used in plastic labware only. Picogreen is subject to photo degradation; therefore, it needs to be protected from light, either by storing in the dark or by covering it with foil, when not in use. Only prepare sufficient Picogreen which can be used on the day it is prepared.

(a) DNA quantification by Picogreen utilizes a standard curve and when using a plate reader standard curve construction takes place in parallel with PCR production concentration data collection. A standard curve requires a DNA sample of known concentration which can be procured from a range of suppliers such as Sigma, Roche, or Invitrogen.

PCR products are linear and it is preferable to use linear DNA standards such as those found in DNA ladders, which correspond or are similar to the amplicon sizes of ~5 kbp in length. Amplicon concentration will not exceed the concentration of DNA standards supplied by commercial organizations (Sigma, Roche, or Invitrogen and so on) which are in the order of 250 μg/μL because of PCR product inhibition which limits the final concentration of PCR product (amplicons) (Kainz Biochimicaet Biophysica Acta 2000).

23. Although residual salts, oligonucleotides, and gDNA from the PCR mixture may theoretically interfere with the Picogreen assay, all amplicons will have the same concentration of contaminants and will be similarly affected. Given the purpose of quantifying amplicon concentration is to create equimolar pools, it is more important to establish the relative amount of amplicons, not the absolute amount, and so the (minimal) effect of contaminants can be ignored.

24. Failure to identify the correct sequence (gDNA) will result in amplification of the incorrect gene.

25. If there is more than one sequence available in different databases, then alignment of gDNAs may be required to design the best primer based on consensus alignment. Alignment of genes can be by using software packages such as Sequencher (Gene Codes Corporation, Ann Arbor, MI) and CLUSTAL W (http://www.ebi.ac.uk/Tools/clustalw2/index.html), Bio Edit (Free) http://www.mbio.ncsu.edu/bioedit/bioedit.html, etc.

26. If the number of genes and genotypes are large in number, a touch-down protocol coupled with LR-PCR can help reduce time, labor, and costs of the study. This helps to find the optimum PCR condition to capture the target genes in minimum time.

27. It is preferable to use the same genotype, with gDNA/cDNA sequences for standardizing the LR-PCR protocol.

28. There are benefits for using High-fidelity DNA polymerase such as BioRadiProof™ High-Fidelity DNA polymerase. It has a significantly low error rate of amplification (4.4×10^{-7}). The error rate is approximately 50-fold lower than normal

polymerases. It also has high extension efficiency of 5–30 s/kb. The kit content and specifications can be found here: http://www.bio-rad.com/webroot/web/pdf/lsr/literature/10002300B.pdf.

29. Some genes may not amplify with the touch down cycling protocol; then they should be amplified separately with their own optimum conditions.
30. Whole-genome re-sequencing also captures the organellar (chloroplast and mitochondrial) genomes in the sequencing process. There are some similarities in the organellar and nuclear genomes which may lead to erroneous call of variant. Therefore, it is very important to remove the reads mapping to organellar genomes before subjecting the short sequence reads to further analyses.
31. The default length fraction and similarity is 0.5 in CLC genomics workbench. Since the coverage of rice genome sequence, which is comparatively small, in an Illumina sequence run is very good, stringent parameters are affordable. If the coverage is relatively less (as in the case of multiplex barcode sequencing of multiple genomes in a single run), then these parameters have to be adjusted accordingly.
32. The H row will be used as reservoirs to distribute SAP enzyme solution to the rest of the microtiter wells.
33. Care should be taken when pipetting the SAP to minimize loss of solution due to viscosity.
34. After distribution of the SAP enzyme solution, each well in rows G–A should have 10 μL of the SAP enzyme solution, while the wells in row H (initial reservoirs) should have 15 μL.
35. The wash station must be prepared and placed on position 2 of liquid-handler (up-right corner), according to the manufacturer's instruction.
36. Add the reagents as they appeared in the table.
37. During pipetting, ensure that no droplet is placed or adhered to the well-walls. The pipetting must be carefully placed into the centers of microtiter plate wells.
38. After distributing iPLEX Gold reaction cocktail, each well in rows G to A should have 10 μL of the low plexiPLEX Gold reaction cocktail, while the wells in row H have 15 μL.
39. Please be aware that there are two cycling loops, one of five cycles that sits inside a loop of 40 cycles. These two loops result in a 200-cycle program.
40. There is a small-rounded post on dimple plate (including resin). This post should be used to align the wells in microtiter plate with the wells of dimple plate.

41. TASSEL can be run only under Java environment. This means that Java should be installed on your computer. All new Windows operating systems install the latest versions of Java automatically; therefore, this might already be existed on your system. Alternatively, it is freely available online here: http://www.oracle.com/technetwork/java/index.html
42. Preparation of phenotypic and genotypic input files is different in version 5. Please read the instruction of each version before preparing the input files here: https://bitbucket.org/tasseladmin/tassel-5-source/wiki/UserManual
43. The number in the first column (233) is the number of individuals (genotypes). The second number tells TASSEL the number of traits (13 starch quality traits). The last number (1) indicates the number of hear rows.
44. Row number 2 includes the traits names (starch measured characters).
45. Although the first column includes the genotype/line names (e.g., YRR038_01-03), the column header includes the first measured traits name (Peak1). This is due to the left-shift of data. In fact, the column B includes the data of Peak1 measurement and so on to the end of table (column N).
46. There should not be any gap or space among words in header names. Characters, letter, and an underscore might be used, if required.
47. The number in the first row (column A) indicates the number of individuals/genotypes (288). The number in column B (28:02:00) shows the number of markers and diploid nature (2×) of species, respectively.
48. A question mark is commonly used for missing data.
49. Insert the homozygous calls (e.g., G:G) and heterozygous calls (C:A) etc.
50. The second row includes the name of SNP markers.
51. If there are any problems in uploading input files, go back to the Excel file and find the formatting problem (mainly in headers and formatting), then save the corrected file as Text Tab delimited and repeat the steps.
52. If the file was not opened by double click, then right click and "save" the file. Then open by right click using "Open as" Excel spreadsheet.
53. TASSEL Version 5 can generate statistics on association between genes and traits such as contribution of each SNP variant to trait, etc.

References

1. Anacleto R, Cuevas RP, Jimenez R, Llorente C, Nissila E, Henry R, Sreenivasulu N (2015) Prospects of breeding high-quality rice in the post-genomic era. Theor Appl Genet 128(8): 1449–1466
2. Wang K, Henry RJ, Gilbert RG (2014) Causal relations between starch biosynthesis, structure and properties. Springer Sci Rev 2(1):15–33
3. Kharabian-Masouleh A, Waters DLE, Reinke RF, Ward R, Henry RJ (2012) SNP in starch biosynthesis genes associated with nutritional and functional properties of rice. Sci Rep 2:557
4. Gopala Krishnan S, Waters DLE, Katiyar SK, Sadananda AR, Satyadev V, Henry RJ (2012) Genome-wide DNA polymorphisms in elite indica rice inbreds discovered by whole-genome sequencing. Plant Biotechnol J 10: 623–634
5. Kharabian-Masouleh A, Waters DLE, Reinke RF, Henry RJ (2009) A high-throughput assay for rapid and simultaneous analysis of perfect markers for important quality and agronomic traits in rice using multiplexed MALDI-TOF Mass Spectrometry. Plant Biotechnol J 7: 355–363
6. Wang K, Hasjim J, Wu AC, Li E, Henry RJ, Gilbert RG (2015) Roles of GBSSI and SSIIa in determining amylose fine structure. Carbohydr Polym 127:264–274
7. Wang K, Wambugu PW, Zhang B, Wu AC, Henry RJ, Gilbert RG (2015) The biosynthesis, structure and gelatinization properties of starches from wild and cultivated African rice species (*Oryza barthii* and *Oryza glaberrima*). Carbohydr Polym 129(20):92–100
8. Krishnan SG, Waters DLE, Henry RJ (2014) Australian wild rice reveals pre-domestication origin of polymorphism deserts in rice genome. PLoS One 9:e98843
9. Healey A, Furtado A, Cooper T, Henry RJ (2014) Protocol: A simple method for extracting Next-Generation Sequencing quality genomic DNA from recalcitrant plant species. Plant Methods 10:21
10. Henry RJ, Edwards K (2009) New tools for single nucleotide polymorphism (SNP) discovery and analysis accelerating plant biotechnology. Plant Biotechnol J 7:311
11. Kharabian-Masouleh A, Waters DLE, Reinke RF, Henry RJ (2011) Discovery of polymorphisms in starch-related genes in rice germplasm by amplification of pooled DNA and deeply parallel sequencing. Plant Biotechnol J 9(9):1074–1085
12. Gopala Krishnan S, Waters DLE, Henry RJ (2012) Genome-wide variations between elite lines of *indica* rice discovered through whole genome re-sequencing. In: Rangasamy SRS et al (eds) 100 years of rice science and looking beyond. Proceedings of the international symposium held at Tamil Nadu Agricultural University, Coimbatore, Tamil Nadu, India, 9–12 January 2012, pp 118–119
13. Krishnan SG, Daniel LE, Waters DLE, Henry RJ (2014) A method for discovery of pair wise genome-wide SNP variations from whole genome re-sequencing data. In: Furtado A, Henry RJ (eds) Cereal genomics: methods and protocols. Springer, New York, pp 287–294
14. McCouch SR, Zhao K, Wright M, Tung CW (2010) Development of genome-wide SNP assays for rice. Breed Sci 60:524–535
15. Edwards M, Henry RJ (2011) DNA sequencing methods contributing to new directions in cereal research. J Cereal Sci 54:395–400
16. Henry RJ, Edwards M, Waters DLE, Gopala Krishnan S (2012) Molecular markers for plants derived from large scale sequencing. J Biosci 37(5):829–841

Chapter 13

Quantifying Grain Digestibility of Starch Fractions in Milled Rice

Crisline Mae Alhambra, Sushil Dhital, Nese Sreenivasulu, and Vito M. Butardo Jr.

Abstract

Rice is one of the staple foods which serves as the major source of carbohydrate in the human diet. A typical milled rice grain is mainly composed of starch of up to 80–90%, with an average of 6–8% proteins and some trace amounts of dietary fiber. Although cooked white rice can elicit variable glycemic response, a portion of rice starch may evade digestion in the human small intestine. The digested portion of rice can be estimated and characterized in vitro based on starch digestion extent and rate (kinetics). The indigestible portion of starch can also be quantified. This chapter will present micro-scale methods to quantify rice starch digestion rate and extent based on the sugar fractions released after treating the samples with digestive enzymes.

Key words Rice, Grain, Digestibility, Resistant Starch, Nutrition

1 Introduction

Rice is one of the major food crops in the world, feeding almost half of the world's population. The major component of rice grain is carbohydrates in the form of starch and some free sugars [1, 2]. Up to 70% of the human energy intake comes from carbohydrates. Starch can be classified into three fractions based on extent of hydrolysis after treatment with digestive enzymes: (1) Rapidly Digestible Starch (RDS), (2) Slowly Digestible Starch (SDS), and (3) Resistant Starch (RS). RDS and SDS are the fractions of starch which can be digested in the small intestine and subsequently absorbed by the body within 20 min and between 20 and 120 min, respectively. RDS represents rapidly released glucose after consumption of starchy food, while SDS has been associated with reduced sensation of hunger after ingestion due to the slowing down of carbohydrate hydrolysis rate [3]. The undigested fraction of starch is termed RS, which is resistant to small intestinal digestion but mostly fermentable in the colon by the gut microflora

Nese Sreenivasulu (ed.), *Rice Grain Quality: Methods and Protocols*, Methods in Molecular Biology, vol. 1892, https://doi.org/10.1007/978-1-4939-8914-0_13,

[4, 5]. Since RS does not contribute to raise in blood glucose level, it is known to help in the prevention of hyperglycemia, hyperinsulinemia, diabetes, and obesity [6–8].

Both underweight and obesity are major risk factors for the global burden of disease [9]. In both cases, there is a need to investigate the digestibility of foods, mainly the starch portion in common foods such as rice. In this connection, the identification of different fractions of starch in rice is important in determining its effect on human body upon consumption considering the prevalence of diet-induced nutritional concerns which ranges from caloric hunger on one end, to diabetes, overweight, and obesity on the other end of the spectrum.

This method focuses on identifying different starch fractions in polished rice grains to quantify and characterize rice grain digestibility. Microscale methods based on Megazyme kits are presented to determine the amount of total resistant starch (RS) as well as total digestible carbohydrates (DC) in one assay as well as total starch (TS) and total sugars (TSU) in a separate assay. These methods are optimum for large-scale screening of breeding and diversity populations of rice because they utilize a micro method wherein a tenfold decrease in the amount of sample and reagents are used to facilitate assays using microfuge tubes. Screening for digestibility properties in rice breeding and diversity panels are prohibitively expensive due to cost, time, and manpower involved. The methods outlined here greatly streamlined the process. Once superior germplasm is identified, the values obtained from the screening methods can be verified using the conventional AOAC methods for total starch (996-11) and resistant starch (2002-02). In addition, the cooked grain amylolysis method to determine the kinetics of starch digestibility simulating gastrointestinal tract digestion is presented in the last section as proxy measure for glycemic index (GI) measurements. Such an in vitro method helps to characterize diversity rice lines to estimate the values of cooked grain amylolysis. The presented cooked grain amylolysis method here can be used as a pre-screening method prior to actual clinical studies of GI estimation.

2 Materials

2.1 Kits

1. Resistant Starch Assay Kit (Megazyme K-RSTAR 08/11), 1000 assay per kit.
2. Total Starch Assay Kit (Megazyme K-TSTA 07/11), 1000 assay per kit.
3. D-Glucose Assay Kit (Megazyme K-GLUC 07/11), 660 assays per kit.

2.2 Reagents

1. 100 mM sodium maleate buffer pH 6.0 with 5 mM $CaCl_2 \cdot 2H_2O$ (500 mL).
2. 99% Ethanol (500 mL).
3. 50% ethanol (1 L).
4. 2 M KOH (250 mL).
5. 1.2 M sodium acetate buffer pH 3.8 (1 L).
6. 100 mM sodium acetate buffer pH 4.5 (5 L).
7. 80% ethanol (2 L).
8. 50 mM maleic acid buffer pH 5.5 (500 mL).
9. 0.01 M HCl (500 mL).
10. 0.02 M NaOH (500 mL).
11. 200 mM sodium acetate buffer pH 6 with 4 mM $CaCl_2$ and 0.49 mM $MgCl_2$ (1 L).
12. GOPOD reagent (1 L).

2.3 Instruments

1. Analytical balance (10 mg to 220 g capacity ±0.1 mg).
2. Refrigerated Microcentrifuge (4 °C, 24-microfuge tubes capacity).
3. Recirculating water bath (37 °C ± 0.03 °C temperature stability).
4. UV-Vis spectrophotometer or ELISA plate reader (510 nm).
5. Magnetic stirring plate (60–134 × *g*).
6. Speed vacuum dryer (36-microfuge holder).
7. Dehuller (100 g per minute).
8. Miller (8–10 g capacity).
9. Ball Grinder (grinding jar with 50 g capacity).
10. Sieve (30 and 40 mesh).

3 Methods

3.1 Preparation of Rice Samples

Dehull and mill the rice paddy using standard methods to obtain milled rice. Sort the milled rice and discard the broken and discolored grains. Grind polished grain samples (5 g) using a ball grinder set in 25 rpm per second for 1 min. In IRRI, we routinely use Retsch MM400 with 50 mL stainless steel grinder (01.462.0216) with a 25 mm diameter grinding ball (05.368.0105). Use this sample for resistant starch and total starch analysis. For cooked milled rice flour amylolysis, grind milled rice grains for 5 s using a ball grinder set in 20 rpm. Pass the ground sample between 30 and 40 mesh to collect particle size of 400 to 600 μm.

3.2 Total Resistant Starch (RS)

In this step, the total resistant starch (RS) is measured using the RS assay procedure from Megazyme (K-RSTAR 08/11) with modifications. This micro method had been extensively validated in our laboratory to accurately quantify RS comparable with the values obtained using the standard method. The supernatant is also collected to measure total digestible carbohydrates (DC) which is described in Subheading 3.3.

1. Weigh 10 ± 0.1 mg of rice flour directly in a 2 mL microfuge tube.
2. Add 400 μL of enzyme solution (10 mg/mL pancreatin with 3 U/mL amyloglucosidase (AMG) from the Megazyme kit dissolved in 100 mM sodium maleate buffer pH 6.0 with 5 mM $CaCl_2 \cdot 2H_2O$).
3. Incubate horizontally in a shaking water bath (100 strokes/min) at 37 °C for exactly 16 h. We commonly use microfuge rack to hold all the microfuge tubes in place using rubber bands.
4. Remove the tubes and dry each with a paper 12 min before the end of 16 h incubation.
5. Stop the reaction by the addition of 400 μL ethanol (99%).
6. Mix vigorously using a vortex mixer.
7. Centrifuge the tubes at 18,231 × *g* for 30 min.
8. After centrifugation, carefully transfer (*see* **Note 1**) the supernatant using a pipettor and collect in a 15 mL conical tube. Set aside for total digestible carbohydrates (DC) assay (Subheading 3.3).
9. Resuspend the pellets to 200 μL of 50% ethanol and vigorously stir using a vortex mixer.
10. Add a further 600 μL of 50% ethanol and vortex mix.
11. Centrifuge at 18,231 × *g* for 30 min.
12. Carefully remove the supernatant using a pipettor and repeat the suspension and centrifugation of the pellet (**steps 9–11**). Collect and pool all the supernatant obtained in the previous steps using the conical tube in **step 8**.
13. Prepare an ice bath over a stirring plate.
14. Place a stir bar in the tubes containing the pellet and add 200 μL of 2 M KOH while stirring in an ice bath. Stir for approximately 20 min or until complete solubilization of the pellet.
15. Add 800 μL of 1.2 M sodium acetate buffer pH 3.8 while stirring.
16. Immediately add 10 μL of 3300 U/mL of AMG and mix.

17. Place the tubes in water bath at 50 °C for 30 min with intermittent mixing (*see* **Note 2**).
18. After incubation, remove the stir bar and centrifuge at 18,231 × *g* for 10 min and get 100 μL of the supernatant (*see* **Note 3**).
19. Prepare a blank (100 μL of 100 mM sodium acetate buffer pH 4.5) and quadruplicate standards (100 μL of 1 mg/mL glucose, prepared and read four times).
20. Add 3 mL of GOPOD reagent supplied from the Megazyme kit to the 100 μL supernatant, blank and standards. Incubate at 50 °C for 20 min.
21. Read absorbance of the sample and standards against blank at 510 nm.
22. Calculate % RS (dry basis) by using the formula:

$$\%\text{RS} = \Delta A \times F \times \frac{1.03}{0.1} \times \frac{1}{1000} \times \frac{100}{W} \times \frac{162}{180}$$

where ΔA = absorbance of the sample against blank

$$F = \frac{1}{\text{average absorbance of glucose standard}}$$

$\frac{1.03}{0.1}$ = volume correction (0.1 mL aliquot was taken from 1.03 mL final volume)

$\frac{1}{1000}$ = conversion of μg to mg

$\frac{100}{W}$ = factor to present RS as % of sample weight

W = dry weigh of the sample = $\left(\text{flour weight} \times \frac{(100 \quad \text{moisture content})}{100}\right)$

$\frac{162}{180}$ = adjustment from free D-glucose to anhydrous D-glucose (as occurs in starch).

3.3 Total Digestible Carbohydrates (DC)

In this step, total digestible carbohydrates (DC) refer to total digestible starch and glucose present in the flour sample. The supernatant collected from Subheading 3.2 are used as starting materials.

1. Dilute the collected supernatant from RS assay (from Subheading 3.2, **step 12**) to 10 mL with 100 mM sodium acetate buffer pH 4.5.

2. Get 100 μL aliquot, and add 10 μL of AMG (300 U/mL in 100 mM sodium maleate buffer pH 6.0 with 5 mM $CaCl_2 \cdot 2H_2O$).
3. Incubate for 20 min at 50 °C.
4. Prepare a blank (110 μL of 100 mM sodium acetate buffer pH 4.5) and quadruplicate standards (110 μL of 1 mg/mL glucose, prepared and read four times).
5. Add 3 mL of GOPOD reagent the sample (**step 2**), blank and standards. Incubate at 50 °C for 20 min.
6. Read absorbance of the sample and standards against blank at 510 nm.
7. Calculate % DC (dry basis) by using the formula:

$$\%\ \text{DC} = \Delta A \times F \times \frac{10}{0.1} \times \frac{1}{1000} \times \frac{100}{W} \times \frac{162}{180}$$

where

ΔA = absorbance of the sample against blank

$$F = \frac{100}{\text{average absorbance of glucose standard}}$$

$\frac{10}{0.1}$ = volume correction (0.1 mL aliquot was taken from 10 mL final volume)

$\frac{1}{1000}$ = conversion of μg to mg

$\frac{100}{W}$ = factor to present DC as % of sample weight

W = dry weigh of the sample = $\left(\text{flour weight} \times \frac{(100 - \text{moisture content})}{100}\right)$

$\frac{162}{180}$ = adjustment from free D-glucose to anhydrous D-glucose (as occurs in starch).

3.4 Total Starch (TS)

In this section, TS is measured using the Total Starch assay procedure (Megazyme K-TSTA 07/11) with modifications. This method had been extensively validated in our lab to accurately reflect the values obtained using the standard method. Total sugars (TSu) are assayed in Subheading 3.5 using the supernatant obtained in this section.

1. Weigh 10 ± 0.1 mg of starch directly in a 2 mL microfuge tube.
2. Add 500 μL of 80% ethanol and incubate at 85 °C for 5 min.
3. Stir using a vortex mixer after incubation and add 500 μL of 80% ethanol.
4. Mix and centrifuge at 18,231 × *g* for 20 min.
5. Aspirate (*see* **Note 4**) the supernatant and resuspend the pellet by adding 1 mL of 80% ethanol.
6. Mix and centrifuge at 18,231 × *g* for 20 min. Collect all the supernatant obtained for total free sugar quantification in a 2 mL microfuge tube (Subheading 3.5).
7. Prepare an ice bath over a stirring plate.
8. Place a stir bar to the obtained pellet and add 200 μL of 2 M KOH while stirring (700 rpm) in an ice bath.
9. Stir for approximately 20 min or until the complete solubilization of the pellet.
10. Add 800 μL of 1.2 M sodium acetate buffer pH 3.8 while stirring.
11. Immediately add 10 μL of thermo-stable α-amylase (3000 U/mL) and 10 μL of AMG (3300 U/mL).
12. Mix and incubate the tubes at 50 °C for 30 min (*see* **Note 5**).
13. Dilute the liquid contents with water to 10 mL (*see* **Note 6**).
14. Get 1 mL of aliquot, place in a 2 mL microfuge tube.
15. Centrifuge the aliquot at 18,231 × *g* for 10 min.
16. Get 100 μL of the supernatant, place in a separate tube.
17. Prepare a blank (100 μL H_2O) and quadruplicate standards (100 μL of 1 mg/mL glucose, prepared and read four times).
18. Add 3 mL GOPOD to the sample (**step 16**), blank and standards).
19. Incubate at 50 °C for 20 min.
20. Read absorbance of the sample and standards against blank at 510 nm.
21. Calculate % TS (dry basis) by using the formula:

$$\%\ \mathrm{TS} = \Delta A \times F \times \frac{10}{0.1} \times \frac{1}{1000} \times \frac{100}{W} \times \frac{162}{180}$$

where

ΔA = absorbance of the sample against blank

$$F = \frac{100}{\text{average absorbance of glucose standard}}$$

$$\frac{1}{1000} = \text{conversion of μg to mg}$$

$\frac{100}{W}$ = factor to present TS as % of sample weight

W = dry weigh of the sample =

$$\left(\text{flour weight} \times \frac{(100 - \text{moisture content})}{100}\right)$$

$\frac{10}{0.1}$ = volume correction (0.1 mL was taken from 10 mL)

$\frac{162}{180}$ = adjustment from free D-glucose to anhydrous D-glucose (as occurs in starch).

3.5 Total Free Sugar (TSu)

1. Dry the obtained supernatant from TS assay (Subheading 3.4) using a speed vacuum concentrator for approximately 4 h or until completely dry.
2. Reconstitute using 100 μL of 50 mM maleic acid buffer pH 5.5.
3. Add 50 μL of invertase (150 U/mL in 50 mM maleic acid buffer pH 5.5) and incubate for 30 min at 37 °C.
4. Boil in water bath for 4 min and cool immediately in ice bath.
5. Centrifuge at 13,394 × *g* for 10 min, get 50 μL of the supernatant and place in a 2 mL microfuge tube.
6. Prepare a blank (50 μL of 50 mM maleic acid buffer pH 5.5) and quadruplicate standards (50 μL of 1 mg/mL glucose, prepared and read four times).
7. Add 1.5 mL of GOPOD reagent to the sample (**step 5**), blank and standards.
8. Incubate at 50 °C for 20 min.
9. Read absorbance of the sample and standards against blank at 510 nm.
10. Calculate mg of glucose by using the formula:

$$\text{mg of glucose} = \frac{\Delta A}{F} \times \text{volume}$$

where

A = absorbance of the sample against blank

F = average absorbance of glucose standard

V = 0.150 mL

Calculate % sugar (glucose and sucrose) using the formula

$$\%\ \text{sugar} = \frac{\text{mg glucose}}{\text{wt of flour}} \times 100.$$

3.6 Cooked Rice Amylolysis

In this section, the starch digestion rate of cooked rice is determined by the use of enzymes and conditions that mimics the biochemical conditions of the human gastrointestinal digestion. Rice samples used have particle sizes between 400 and 600 μm to minimize the deviation arising from grain fragments, particle size and to mimic the size of chewed rice during mastication. Methods and calculations were developed based on Dhital et al. [10] and Butardo et al. [11].

1. Weigh 500 ± 0.1 mg milled rice (400–600 μm) in a 50 mL glass tube.
2. Add water (*see* **Note 7**) and cover with foil to minimize loss of water.
3. Let it stand for 30 min and cook in a boiling water bath for 30 min.
4. Immediately place vertically at 37 °C water bath over a stirring plate. If not used immediately, store at 50 °C waterbath (*see* **Note 8**). Ensure amylolysis will be done within the day.
5. Add 2 mL of water and dislodge the clumped cooked rice using a metal spatula. Add further 4 mL water to wash the spatula.
6. Add a stir bar (1/2 × 5/16 in) and start stirring at 350 rpm.
7. Immediately start the digestion process (*see* **Note 8**) by adding 5 mL of pepsin (*see* **Note 9**) (1 mg/mL in 0.01 M HCl) and incubate for exactly 30 min.
8. Add 5 mL of 2 M NaOH and incubate for exactly 1 min.
9. Add 20 mL of 200 mM sodium acetate buffer pH 6.0 with 4 mM $CaCl_2$ and 0.49 mM $MgCl_2$ and incubate for exactly 10 min. Get duplicate 200 μL aliquot 2 min before the end of the incubation.
10. Add 5 mL of pancreatin/AMG mixture (2 mg/mL in pancreatin and 28 U/mL AMG in 200 mM sodium acetate buffer pH 6.0 with 4 mM $CaCl_2$ and 0.49 mM $MgCl_2$) and get duplicate 200 μL aliquot at desired sampling points (5, 10, 20, 30, 45, 60, 90, 120, 180 min) and place in a microfuge tube.
11. Immediately place the microfuge tube containing the aliquot in an ice bath or liquid nitrogen to stop the reaction. Stopping the reaction can also be done using ethanol but slight modification in calculation has to be employed.
12. Centrifuge the aliquot at 15,720 × *g* at 4 °C for 10 min.
13. Aspirate 50 μL of the supernatant (*see* **Note 10**) and place in a 2 mL microfuge tube (*see* **Note 11**).
14. Prepare a blank (50 μL of 200 mM sodium acetate buffer pH 6.0 with 4 mM $CaCl_2$ and 0.49 mM $MgCl_2$) and qua-

druplicate standards (55 μL of 1 mg/mL glucose, prepared and read four times).

15. Add 5 μL of AMG (300 U/mL in 200 mM sodium acetate buffer pH 6.0 with 4 mM $CaCl_2$ and 0.49 mM $MgCl_2$) to both blank (**step 14**) and samples (**step 13**).
16. Incubate the samples, blank and standards at 50 °C for 20 min.
17. Add 3 mL GOPOD reagent and incubate at 50 °C for 20 min.
18. Read absorbance of the sample and standards against blank at 510 nm.
19. Calculate % starch hydrolysis (Sh) per sampling time by using the formula:

$$\% \text{ Sh} = \left\{ \frac{\left[\left(\frac{A\,\text{Sx}}{A\,\text{std}}\right) \times \text{d}f\right] \times V \times \frac{162}{180}}{\text{wt Sx} \times \frac{\%\text{ TS}}{100}} \right\} \times 100$$

where:
AS_x = absorbance of the sample at time t
Astd = average absorbance of standard
df = dilution factor at time t
V = volume at time t (in mL)

$\frac{162}{180}$ = adjustment from free D-glucose to anhydrous D-glucose

(as occurs in starch)
wt S_x = weight of sample (mg)
% TS = % total starch (dry basis).

20. Plot the % Sh vs time to get digestion curve.
21. Get the k-value by plotting $\ln[(\%\ \text{Sh}_{x+1} - \%\ \text{Sh}_x)/(t_{x+1} - t_x)]$, $\ln[(\%\ \text{Sh}_{x+2} - \%\ \text{Sh}_{x+1})/(t_{x+2} - t_{x+1})]$...etc. vs $(t_{x+1} - t_x)/2$, $(t_{x+2} - t_{x+1})/2$...etc.

    where:

    $x = 0$ min

    $x + 1 = 5$ min

    $x + 2 = 10$ min

    t = time (in the lab, we usually sample at 5, 10, 15, 20, 30, 45, 60, 90, 120, 180 min).
22. Get the equation of the line of the graph in **step 19**. The absolute value of the slope of the line (m) is the k value or digestion rate and is expressed as min^{-1}.

3.7 Integrating and Interpreting Results

RS and DC assay results would show the proportion of the starch in rice which is resistant to digestion and digestible, respectively. TS and TSu would show the total starch and soluble sugar (glucose and sucrose) contents of rice, respectively. Starch hydrolyzed within certain time points can be identified by using in vitro digestion using the cooked grain amylolysis method. In addition, the k value would determine the rate of digestion. Higher *k* value would indicate faster rate of digestion.

4 Notes

1. Use a 1 mL pipet to aspirate.
2. Mix during the first 15 min of incubation.
3. If RS is >10%, dilute the contents to 10 mL before centrifugation.
4. Similar to RS assay, use 1 mL pipet to aspirate.
5. Similar to RS assay, mix during the first 15 min of incubation.
6. Quantitatively transfer the contents by washing the stir bar with water.
7. Add water depending on amylose content (AC). The water:rice ratio is 1.8:1, 1.6:1 and 1.4:1 for samples with AC of 25+%, 20–24%, and 0–19%, respectively. Alternatively, one can just cook using excess water.
8. Do not allow the cooked rice to cool. Ideally, start the assay within 30 min to avoid starch retrogradation.
9. All enzyme solution to be used should be freshly prepared.
10. Store the supernatant at −20 °C if not used within the day.
11. Dilute the supernatant using 20 mM sodium maleate buffer pH 6.0 and dilution should be higher with increasing time. Get only 50 μL of the diluted supernatant.

References

1. Butardo VM Jr, Sreenivasulu N (2016) Tailoring grain storage reserves for a healthier rice diet and its comparative status with other cereals. Int Rev Cell Mol Biol 323:31–70
2. FAO (1998) Dietary carbohydrate composition. Food and Agriculture Organization of the United Nations, Rome
3. Sparti A, Milon H, Di Vetta V, Schneiter P, Tappy L, Jequier E, Schutz Y (2000) Effect of diets high or low in unavailable and slowly digestible carbohydrates on the pattern of 24-h substrate oxidation and feelings of hunger in humans. Am J Clinical Nut 72(6):1461–1468
4. Englyst HN, Cummings JH (1987) Digestion of polysaccharides of potato in the small-intestine of man. Am J Clin Nutr 45(2):423–431
5. Englyst HN, Macfarlane GT (1986) Breakdown of resistant and readily digestible starch by human gut bacteria. J Sci Food Agric 37(7):699–706
6. Ranganathan S, Champ M, Pechard C, Blanchard P, Nguyen M, Colonna P, Krempf

M (1994) Comparative-study of the acute effects of resistant starch and dietary-fibers on metabolic indexes in men. Am J Clin Nutr 59(4):879–883

7. Topping DL, Clifton PM (2001) Short-chain fatty acids and human colonic function: roles of resistant starch and nonstarch polysaccharides. Physiol Rev 81(3):1031–1064
8. Behall KM, Scholfield DJ, Hallfrisch JG, Liljeberg-Elmstahl HGM (2006) Consumption of both resistant starch and beta-glucan improves postprandial plasma glucose and insulin in women. Diabetes Care 29(5):976–981
9. World Health Organization (2002) The World Health Report: Reducing risks, promoting healthy life
10. Dhital S, Dabit L, Zhang B, Flanagan B, Shrestha AK (2015) In vitro digestibility and physicochemical properties of milled rice. Food Chem 172:757–765
11. Butardo VM Jr, Anacleto R, Parween S, Samson I, de Guzman K, Alhambra CM, Misra G, Sreenivasulu N (2017) Systems genetics identifies a novel regulatory domain of amylose synthesis. Plant Physiol 173(1): 887–906

Chapter 14

Determination of Macronutrient and Micronutrient Content in Rice Grains Using Inductively Coupled Plasma-Optical Emission Spectrometry (ICP-OES)

Lilia Molina, Jennine Rose Lapis, Nese Sreenivasulu, and Rosa Paula O. Cuevas

Abstract

The rice grain endosperm is mostly composed of starch, which serves as the major source of calories for more than half of the world's population. Macro and micronutrients make a minor proportion of the rice grain, which particularly gets accumulated in outer aleurone layer, which are in general eliminated upon milling. Because rice is the major staple, it is seen as an efficient mechanism for delivering both macro- and micronutrients, particularly for the poor who do not have ample access to diversified diets. Enriching micronutrient and macronutrient concentrations in milled rice of endosperm and/or in brown rice, is an important dietary intervention to create health benefits of rice consumers. Efforts are underway to increase the nutritional content of rice through bio/fortification approaches. The plant takes up these same elements from the soil, redirect the transport of these elements into the grain. Thus besides biofortification strategies, scientists can also use the knowledge to design proper soil nutrient management to enrich micronutrients in the grains. Therefore, it is important to be able to determine the macro- and the micronutrient composition of the vegetative parts of the rice plant and of the rice grain. In this chapter, nitric-perchloric acid digestion and inductively coupled plasma-optical emission spectrometry (ICP-OES) methods routinely used in IRRI's Grain Quality and Nutrition Services Laboratory (GQNSL) to determine the concentrations of various macro- and micronutrients found in the rice grain and the rice plant, are described.

Key words Nitric-perchloric digestion, Inductively coupled plasma-optical emission spectrometry (ICP-OES), Macronutrients, Micronutrients

1 Introduction

Food is a means through which nutrition is delivered. Staple crops are of particular importance in improving consumer health because these are consumed routinely and in large amounts. Therefore, biofortification of these crops is a potential means to alleviate macro- and micronutrient malnutrition, particularly for consumers

Nese Sreenivasulu (ed.), *Rice Grain Quality: Methods and Protocols*, Methods in Molecular Biology, vol. 1892, https://doi.org/10.1007/978-1-4939-8914-0_14, © Springer Science+Business Media, LLC, part of Springer Nature 2019

who cannot afford diversified diets [1]. Two global micronutrient deficiency concerns majorly involve iron and zinc. The serious health implications caused by these deficiencies are well-known (e.g., [2–6]). Rice, for instance, is seen as a potential means to address iron and zinc deficiencies because it is a major staple crop. Brown rice contains high levels of proteins, essential vitamins, and minerals [7] and it has high concentrations of iron, zinc, and manganese in its aleurone layer [8]; therefore, brown rice is an option for micronutrient supplementation. However, brown rice is not popular with majority of consumers, who prefer milled rice for various reasons ([7], reviewed in [9]). Milled rice is a poor source of micronutrients, and thus needed to address the large prevalence of micronutrient deficiency, compared to brown rice, because it lost most of its micronutrients during the polishing process. Based on profiling of 95 varieties, the average iron content was reported to average at 2.28 mg/100 g (dry basis); while the average zinc content of 57 varieties was 3.34 g/100 g (dry basis) [10]. Biofortification of iron, zinc, and pro-Vitamin A in staple crops, including rice, are being spearheaded by HarvestPlus and IRRI research programs [11, 12]. On the other hand, transgenic biofortification is reportedly aimed to increase iron content six fold higher and to double zinc content in the milled rice grain [13].

Quantifying the concentrations of various micro- and macronutrients in the milled rice grain, therefore, is increasingly crucial in the context of rice biofortification efforts. Various methods such as microwave digestion, open- and closed-vessel digestion, dry ashing, and wet ashing are used to digest and extract the macro and micronutrients from plant tissues or the grain [14–17]. Profiling of nutrients is then conducted by atomic absorption spectrophotometry (AAS), laser-induced breakdown spectroscopy (LIBS), inductively coupled plasma-optical emission spectrometry (ICP-OES), and inductively coupled plasma-mass spectrometry (ICP-MS) for quantitative nutrient analysis [15, 18–20]. Among the techniques available, X-ray fluorescence (XRF) spectroscopy is employed for quick scan of screening diverse germplasm and ICP-OES techniques to get precise estimations of Fe and Zn concentrations in the narrow enriched germplasm by IRRI and the HarvestPlus group. ICP-OES is considered the gold standard because of its high sensitivity and accuracy; however, it requires tedious sample preparation and digestion prior to ICP-OES analysis [20]. An alternative rapid screening, high sample throughput method was developed by the HarvestPlus Team; it uses the bench-top Energy Dispersive XRF for iron and zinc analysis in whole rice grain or flour samples without the need for the sample digestion step. The XRF method is calibrated using reference values from the ICP-OES method and is adopted for screening a large number of samples where high iron and zinc samples can be further verified by the

ICP-OES technique since XRF is less accurate than ICP-OES analysis [20, 21].

In this chapter, we describe the nitric acid-perchloric acid digestion procedure and the multi-elemental determinations by ICP-OES using simultaneous optical systems and dual viewing of the plasma in rice grains.

2 Materials

2.1 Glasswares and Apparatus

1. Glass culture tubes, 25 × 200 mm.
2. Vortex mixer (Model Vortex Genie 2, Scientific Industries, Inc., USA).
3. Polypropylene tubes, 25 × 100 mm.
4. Narrow-mouth storage bottles (low-density polypropylene) with screw closure, 15 mL to 1000 mL capacities.
5. 2 L volumetric flask.
6. 6 L Erlenmeyer flask.
7. 2.5 L glass bottle.
8. 3.5 L container.
9. Teflon bottle.
10. Analytical balance (Model AW120, Shimadzu Corporation, Japan).
11. Top loading balance (Model BW4200H, Shimadzu Corporation, Japan).
12. Prototype digestor with flexi glass baffle.
13. Inductively Coupled Argon Plasma—Optical Emission Spectrometer with WinLab32 software (Model Optima 5300DV, Perkin Elmer, USA).
14. Autosampler (Model ASX-520, CETAC Technologies, USA) (Fig. 1).
15. Transfer pipette (Model Pipet Lite, Rainin Instruments, LLC, USA).
16. Bottle dispenser (Model Dispensette organic easy calibration, BrandTech Scientific, Inc., USA. Model Optifix Basic 10, EMD Millipore, USA).

2.2 Reagent and Standard Preparation

1. Nitric acid (conc.): 65%, HNO_3 (CAS No. 7697-37-2).
2. Perchloric acid (conc.): 69–72%, $HClO_4$ (CAS No. 7601-90-3).
3. Hydrochloric acid (conc): 36.5–38.0% (CAS No. 7647-01-0).
4. Liquid argon gas supply: high-purity grade (99.99%).

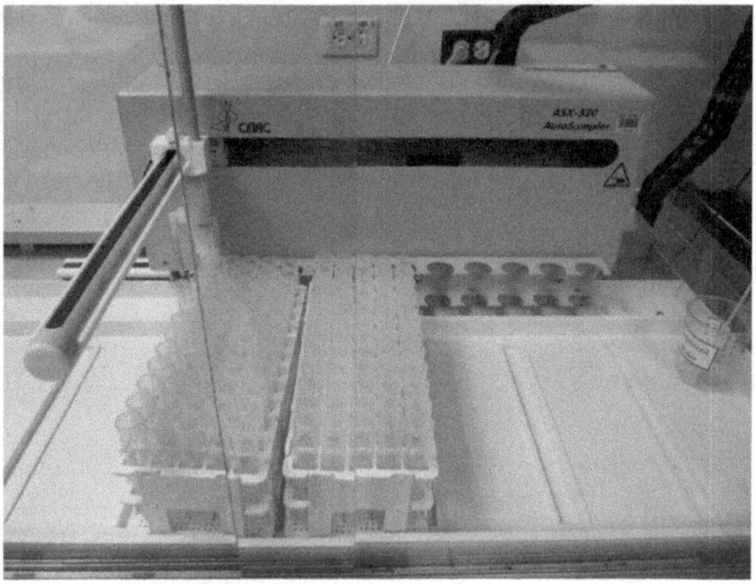

Fig. 1 The ASX-520 autosampler

5. Deionized water: prepared from a deionizing system with cation and anion exchangers to produce deionized water with 18 MΩ resistivity.
6. 1000 mg/L Sc stock standard in 2% HNO_3 (100048-1, High-Purity S,andards, USA).
7. 1000 mg/L Mn stock standard in 2% HNO_3 (100032-1, High-Purity Standards, USA).
8. *1:10 of $HClO_4$:HNO_3 mix acid.* Dispense 100 mL of $HClO_4$ (conc.) in a 2.5 L glass bottle. Add 1000 mL of HNO_3 (conc.). Mix well and transfer to the bottle dispenser labeled "1:10 of $HClO_4$:HNO_3" (*see* **Note 1**).
9. *1% HNO_3 acid.* Dispense 30.76 mL of 65% HNO_3 in a 2 L volumetric flask, and dilute to the mark with deionized water. Mix well and transfer the solution to the Teflon bottle labeled "1% HNO_3" (*see* **Note 2**).
10. *5 N HCl.* Dispense 2.5 L of concentrated HCl in a 6 L container containing 3.5 L of deionized water. Mix well (*see* **Note 3**).
11. *Calibration blank (1% HNO_3 + 2.78% $HClO_4$).* In a 6 L flask, add 92.31 mL of HNO_3 (conc.) and 240 mL of $HClO_4$ (conc.) to 400 mL deionized water. Dilute the solution to the 6 L mark with deionized water and transfer to a designated container. This solution must be used within 1 year.
12. *Mixed Calibration Standards.* For a three-point calibration curve, three stock solutions (lowest standard, middle standard, and highest standard) of mixed elements (Inorganic Ventures, USA) are used (Table 1, *see* **Note 4**).

Table 1
Three-point mixed calibration standard

Element	Lowest standard (mg/L)	Middle standard (mg/L)	Highest standard (mg/L)
Al	1	30	250
Ca	5	60	500
Cu	1	3	25
Fe	1	6	50
K	20	60	500
Mg	5	30	250
Mn	1	6	50
Mo	1	1.5	10
Na	2	12	100
P	10	30	250
S	5	30	250
Zn	1	3	25

13. *Initial Verification Standards.* Another set of three-point standards with similar concentrations as that of the three mixed calibration standard stock solutions but from a different brand (SM-1907-032, SM-1907-033, SM-1907-034, High Purity Standards, USA) or source of the mixed calibration standards (Table 1, *see* **Note 4**).
14. *Continuing Calibration Verification Low standard (CCVL).* Weigh 250 g of the lowest calibration standard stock solution in a 500 mL LDPE bottle, add prepared 1% HNO_3 + 2.78% $HClO_4$ to make up to 500 g.
15. *Continuing Calibration Verification High standard (CCVH).* Weigh 250 g of the highest calibration standard stock solution in a 500 mL LDPE bottle. Then add the prepared calibration blank to make up the weight to 500 g.
16. *Internal Standard (1 mg/kg scandium).* Weigh 1 g of 1000 mg/L Sc stock standard in a 1000 mL LDPE bottle, add the calibration blank to make up the weight to 1000 g. Mix well (*see* **Note 5**).
17. *Instrument Performance Check Solution (10 mg/kg Mn).* Weigh 10 g of 1000 mg/L Mn stock standard in a 1000 mL LDPE bottle and add the calibration blank to make up the weight to 1000 g. Mix well. This solution must be used within a year.

1 mg/kg Mn. Weigh 1 g of 1000 mg/L Mn stock standard in a 1000 mL LDPE bottle and add the calibration blank to make up the weight to 1000 g. Mix well. This solution must be used within a year.

3 Methods

3.1 Sample Preparation

1. Weigh 0.600–0.625 g of oven-dried rice grain sample in a culture tube (25 × 200 mm) (*see* **Note 6**).
2. Add 12 mL of 1:10 of $HClO_4$:HNO_3 mixed acid solution from the dispenser kept in the perchloric fumehood (*see* **Notes 7** and **8**). Wash as much of the sample as possible from the sides of the tube with the acid.

3.2 Digestion

1. A sample is pre-digested from room temperature to 60 °C for 25 min in the digestor with the baffle system in place (*see* **Note 9**). Refer to steps 1–4 of the Digestion temperature scheme in Table 2.
2. Continue digestion following step 5–21 indicated in Table 2. Do not move the tubes and allow the digestion process to complete until about 1 mL of colorless to slight yellow clear digest is left (*see* **Note 10**).
3. Cool the digest and add 20 mL of 1% HNO_3. Warm tubes using the digestor for 35 s and immediately mix using a vortex mixer. Make the final dilution to the 25 mL mark with 1% HNO_3. Mix the tubes using the vortex mixer and transfer the sample immediately into appropriately labeled polypropylene tubes. Analyze digests in the ICP-OES for Na, K, Ca, Mg, Fe, Mn, Zn, Cu, Al, P, S, and Mo (*see* **Note 11**).

3.3 Analyte Measurement via ICP-OES

Detailed instructions on how to operate the ICP-OES can be found in the instrument's manual [22].

1. Turn on the computer and the WinLab software. Turn on the toggle switch for the air supply. Replace tubings in peristaltic pump and replenish water reservoir. Turn on auto sampler, and load software.
2. Perform Hg alignment, instrument performance check, view alignment, background equivalent concentration test (using the instrument performance check solution, 10 mg/kg Mn) and precision test (using calibration blank and 10 mg/kg Mn), according to the manufacturer's recommended procedures for the specific instrument model.
3. Place the calibration samples, blanks, quality control samples, standards, internal standard, and sample tubes in their designated position in the auto sampler.

Table 2
Digestion temperature scheme

Controller step No.	Set point type	Temp setting on controller (°C)	Time (min)	Actual Temp (°C)
1	RAMP	63	25	60
2	HOLD	63	25	60
3	RAMP	89	15	80
4	HOLD	89	20	80
5	RAMP	112	10	100
6	HOLD	112	20	100
7	RAMP	135	10	120
8	HOLD	135	20	120
9	RAMP	147	7	130
10	HOLD	147	60	130
11	RAMP	158	7	140
12	HOLD	158	30	140
13	RAMP	170	7	150
14	HOLD	170	50	150
15	RAMP	193	10	170
16	HOLD	193	40	170
17	RAMP	205	7	180
18	HOLD	205	15	180
19	RAMP	257	15	225
20	HOLD	257	20	225
21	END	OFF		OFF

4. Perform calibration using calibration blank solution and mixed calibration standards.
5. Subject the sample digests for Na, K, Ca, Mg, Fe, Mn, Zn, Cu, Al, P, S, and Mo to ICP-OES. Table 3 summarizes the wavelengths used for each of these elements (*see* **Note 12**).
6. Once analysis is completed transfer all pump tubes into deionized water to flush the system for around 5 min before shutting down the system.
7. Data processing

   The quantitative values must be reported in appropriate units, such as milligrams per kilogram (mg/kg) for micronutrients and % for macro-nutrients. If dilutions were performed,

Table 3
Wavelength used per individual element

Analyte	λ (nm)
Al	394.401
Ca	315.887
Cu	327.393
Fe	259.939
K	766.490
Mg	285.213
Mn	259.372
Mo	202.031
P	213.617
S	181.975
Zn	206.200
Na[a]	589.592, 588.873

[a]There are two wavelengths used for Na. This is for the axial and radial viewing position of the plasma, covering low and high concentration ranges of Na found from different samples in a batch run

the appropriate corrections must be applied to the sample values.

If appropriate, or required, calculate results for solids on a dry-weight basis as follows:

$$\text{Concentration}\left(\text{dry weight}\right)\left(\text{mg}/\text{kg}\right)=\left(C-B\right)\times\left(\frac{V}{W}\right),$$

this is the concentration in ppm.

$$\text{Concentration}\left(\text{dry weight}\right)\left(\%\right)=\left(C-B\right)\times\left(\frac{V\times 0.0001}{W}\right),$$

this is how to convert ppm to %, by multiplying it with 0.0001.

where

C = Digest concentration (mg/L)

B = Average method blank (mg/L)

V = Final volume in mL after sample preparation

W = Weight in g of sample

8. Quality Control.

The following steps are performed as part of the laboratory's quality control program for every ICP-OES run:

(a) Instrument performance check using reagent blank and 10 ppm Mn.

(b) Verify the recovery of the resulting calibration curve with the calibration standard verification (CSV) when run as unknown, the analyzed concentration of the CSV must be within the acceptance criteria of ±3% of the manufacturer's value, otherwise recalibration must be done.

(c) Analyze the initial calibration verification (ICV) standard after calibration. The analyte value should be ±10% of the expected value. If the analyzed value is outside the ±10% limit, the new calibration standards must be used and the instrument is recalibrated prior to sample analysis.

(d) Dilute and rerun samples with analyte concentrations beyond the concentration of the highest calibrant.

(e) Include at least one method blank per sample batch, where analyte value must not exceed the method detection limit; otherwise, the source of contamination should be identified and resolved before continuing analyses.

(f) Analyze at least one reference material with each sample batch. The calculated recovery for each analyte should be within ±2 s.d. of the laboratory's long-term value. If the analyzed value is outside this limit, the source of the problem should be identified and resolved before continuing analyses.

(g) Estimate precision by means of duplicate/replicate analyses for every ten samples. Percent RPD must be ≤20%. However, owing to the heterogeneity of some samples submitted (e.g., whole milled or unpolished rice grain samples), it is not always the case. For such cases, the client is informed prior to conduct of analysis and the % RPD obtained in replicated samples is reflected on the final report.

$$\text{RPD} = \frac{|D1 - D2|}{\left(\frac{D1 + D2}{2}\right)} \times 1000$$

where:
D1 = First sample value
D2 = Second sample value

(h) Unusual situations occurring during sample preparation and analysis must be documented on the work sheet, and non-conformity/corrective action reports must be issued if necessary.

4 Notes

1. Mix and transfer acid under the perchloric hood. Solution must be used within 6 months.
2. This is used for dilution of sample digest. Solution must be used within 6 months.
3. This is used to acid-wash glasswares. Solution must be changed every 6 months or as necessary.
4. These may be manually prepared from individual element stock solutions or can be purchased as mixed standards.
5. Internal standard of 1 mg/kg Sc is used for rice grain samples. This solution must be used within 1 year.
6. The rice grain sample can be milled, unpolished, or paddy (whole grains or ground grains), depending on the customer's requirement. Rice grain samples are oven-dried for 2 h at 70 °C.
7. The mixed acid is needed for digestion since normal concentrated nitric acid only removes the easily oxidizable material in the sample. The more resistant organic matter is acted upon later by the stronger oxidizer, such as perchloric acid [23].
8. If the samples are ground material, mix immediately using a Vortex mixer at its high-speed setting to avoid hardening of sample into balls of dry material that may "spit" when digesting.
9. Pre-digestion with nitric acid and perchloric acid will give protection against possible explosion during the digestion process. An adequate amount of nitric acid must be present and overnight pre-digestion is recommended. Pre-digestion removes easily oxidized organic materials, including hydroxyl compounds that give rise to unstable perchlorate esters [16]. Wear goggles, insulated gloves, and lab coats when doing acid digestion. The improper use of $HClO_4$ has caused numerous explosions during digestion of organic materials. Many explosions were caused by not first predigesting the sample with an adequate amount of HNO_3 [16].
10. Do not let digests boil dry. Samples high in potassium and SiO_2 may have precipitated in the final digest.
11. Most sample residue will dissolve fully if allowed to stand for several hours. Final digest will be colorless. If there is a white precipitate, it may be silica or $KClO_4$. Some samples with very high K may need to be warmed to dissolve. $KClO_4$ may dissolve if it is re-warmed. After tubes are removed, rinse the hood and exhaust stack with the wash-down system for about 30 min with the blower running.

12. During the method development, wavelength or line selection for each analyte is based upon avoidance of spectral interferences, detection limit requirement, and working range requirement. Key instrument parameters such as gas flow rate and applied radiofrequency power must also be optimized [24].

Acknowledgment

The authors are grateful to Mariel Ong, Reah Gonzales, Marnol Santos, Edgar Amoloza, Jose Rosales, and Karen Serdeña, for the assistance in the SOP documentation, sample preparation, validation, and optimization of the digestion and ICP methodologies in GQNSL. This work has been supported under the CGIAR thematic area Global Rice Agri-Food System CRP, RICE, The Indian Council of Agricultural Research, Taiwan support from Council of Agriculture (COA), Ministry of Foreign Affairs (MOFA) and BBSRC Newton grant funding Reference number BB/N013808/1.

References

1. Sharma P, Aggarwal P, Kaur A (2017) Biofortification: a new approach to eradicate hidden hunger. Food Rev Inter 33(1):1–21
2. Walsh CT, Sandstead HH, Prasad AS, Newberne PM, Fraker PJ (1994) Zinc: health effects and research priorities for the 1990s. Envi Health Perspectives 102(Suppl 2):5–46
3. Prasad AS (2001) Recognition of zinc-deficiency syndrome. Nutrition 17(1):61–69
4. Wallwork JC, Milne DB, Sims RL, Sandstead HH (1983) Severe zinc deficiency: Effects on the distribution of nine elements (potassium, phosphorus, sodium, magnesium, calcium, iron, zinc, copper and manganese) in regions of the rat brain. J Nutr 113(10):1895–1905
5. Yip R (1994) Iron deficiency: contemporary scientific issues and international programmatic approaches. J Nutr 124:1479S–1490S
6. Andrews NC (1999) Disorders of iron metabolism. New England J Med 341(26): 1896–1995
7. Javier EQ (2004) Let's promote brown rice to combat hidden hunger. Rice Today 3(1):38
8. Butardo V Jr, Sreenivasulu N (2016) Tailoring grain storage reserves for a healthier rice diet and its comparative status with other cereals. Inter Rev Cell Mol Bio 323:31–70
9. Tuaño APP, Regalado MJC, Juliano BO (2016) Grain quality of rice in selected retail stores and supermarkets in the Philippines. Inter J Phi Sci Tech 9(1):15–22
10. Kennedy G, Burlingame B (2003) Analysis of food composition data on rice from a plant genetic resources perspective. Food Chem 80(4):589–596
11. Stangoulis J (2010) Technical aspects of zinc and iron analysis in biofortification of the staple food crops, wheat and rice, in 19th World Congress of Soil Science. Soil Solutions for a Changing World, Brisbane, pp 42–44
12. Guild GE, Stangoulis J (2016) Non-matrix matched glass disk calibration standards improve XRF micronutrient analysis of wheat grain across five laboratories in India. Front Plant Sci 7:784
13. Trijatmiko KR, Dueñas C, Tsakirpaloglou N, Torrizo L, Arines FM, Adeva C, Balindong J, Oliva N, Sapasap MV, Borrero J, Rey J, Francisco P, Nelson A, Nakanishi H, Lombi E, Tako E, Glahn RP, Stangoulis J, Chadha-Mohanty P, Johnson AAT, Tohme J, Barry G, Slamet-Loedin IH (2016) Biofortified indica rice attains iron and zinc nutrition dietary targets in the field. Sci Rep 6:19792
14. Wu S, Feng X, Wittmeier A (1997) Microwave digestion of plant and grain reference materials in nitric acid or a mixture of nitric acid and hydrogen peroxide for the determination of multi-elements by inductively coupled plasma mass spectrometry. J Anal At Spectrom 12:797–806
15. Wheal MS, Fowles TO, Palmer LT (2011) A cost-effective acid digestion method using

closed polypropylene tubes for inductively coupled plasma optical emission spectrometry (ICP-OES) analysis of plant essential elements. Anal Methods 3:2854–2863

16. Schilt AA (1979) Perchloric Acid and Perchlorates. G. F. Smith Chemical Co, Columbus, OH
17. Gorsuch TT (1970) The Destruction of Organic Matter, 1st edn. Pergamon Press, Oxford
18. Hansen TH, Laursen KH, Persson DP, Pedas P, Husted S, Schjoerring JK (2009) Micro-scaled high-throughput digestion of plant tissue samples for multi-elemental analysis. Plant Methods 5:12
19. Kim G, Kwak J, Choi J, Park K (2012) Detection of nutrient elements and contamination by pesticides in spinach and rice samples using Laser-Induced Breakdown Spectroscopy (LIBS). J Agric Food Chem 60:718–724
20. Stangoulis J, Guild GE (2014) Measuring trace micronutrient levels in crops. In: The 2nd global conference on biofortification: getting nutritious foods to people. Kigali, Rwanda
21. Paltridge NG, Palmer LJ, Milham PJ, Guild GE, JCR S (2012) Energy-dispersive X-ray fluorescence analysis of zinc and iron concentration in rice and pearl millet grain. Plant Soil 361(1):251–260
22. Perkin-Elmer Corporation (2006) Perkin Elmer WinLab32 Instrument Control Software Guide. Perkin-Elmer Corporation, Massachusetts
23. Kebbekus BB (2003) Preparation of Samples for Metals Analysis. In: Mitra S (ed) Sample Preparation Techniques in Analytical Chemistry. John Wiley & Sons, Inc., Hoboken, NJ
24. Gaines PR (2011) ICP Operations Guide: A Guide for using ICP-OES and ICP-MS. Inorganic Ventures, Inc, Christiansburg, VA, p 37

Chapter 15

Determination of Cadmium Concentration in Milled and Brown Rice Grains Using Graphite Furnace Atomic Absorption Spectrometry

Lilia Molina, Jennine Rose Lapis, Nese Sreenivasulu, and Rosa Paula O. Cuevas

Abstract

Heavy metal pollution is a growing public health concern since it poses a food risk to public health via metal transfer. Cadmium is of particular concern because it is a potential carcinogen if exceed tolerable limits in the grain. Hence, it is important to monitor the cadmium content of rice before it reaches the market to ensure public healthy safety, especially in areas known to have high cadmium levels in soil. In this chapter, the method used to determine the concentration of cadmium in milled and brown rice grain samples is described. This method involves sample digestion with concentrated nitric acid and hydrogen peroxide and analysis of cadmium by Graphite Furnace Atomic Absorption Spectrometry (GF-AAS). Because cadmium concentrations are low in rice grains, quantification of cadmium content requires the use of a more sensitive instrument, such as GF-AAS.

Key words Cadmium in rice, Heavy metals, GF-AAS

1 Introduction

Heavy metals (like cadmium, lead, mercury, and arsenic) in the environment have potential deleterious effects on humans and animals because heavy metals can be accumulated in the edible parts of food crops, which are then ingested by humans and by animals (reviewed in [1]). Cadmium, a heavy metal, can be taken up by rice plants from the soil and can accumulate in different parts of the plant, including the grain [2]. However, cadmium is mostly accumulated in the stalk. In rice samples sourced from various heavy metal-contaminated sites in China, the concentrations of cadmium in the hull and in the milled grain are reported to be mostly similar and much lower than the cadmium concentration in the stalk [1]. On the other hand, rice samples from cadmium-contaminated fields in Thailand showed that the stalk had a wider

Nese Sreenivasulu (ed.), *Rice Grain Quality: Methods and Protocols*, Methods in Molecular Biology, vol. 1892,
https://doi.org/10.1007/978-1-4939-8914-0_15,

range of cadmium concentrations than the leaf and the unpolished grain, with leaf and grain having similar concentrations [3].

Although the bioaccumulation of cadmium is lower in the rice grain than in the stalk, cadmium is still a public health concern because of its toxicity and because of the high risk of exposure through continuous consumption. The critical level for cadmium in rice grains is 0.4 mg/kg [4] and numerous studies have reported grain cadmium concentrations higher than this (e.g., [1, 3, 5–7]), indicating that consumption of rice cultivated from areas contaminated with cadmium may pose increased health risk. For instance, elevated urinary cadmium levels are reportedly associated with renal and urinary tract disorders, bone toxicities, and elevated blood pressure (reviewed in [8]) while meta-analysis of available data suggests that cadmium exposure is associated with risk of breast cancer, though more studies are needed to confirm this [9]. The risk is further exacerbated by the per capita consumption rates of rice; i.e., high per capita rice consumption in countries such as Sri Lanka, India, and Bangladesh can cause cadmium bioaccumulation to exceed the threshold [10]. Research into cadmium uptake becomes even more important in the development of biofortified rice. Cadmium has been reported to mimic zinc, such that cadmium can be taken up by the plant and compete with zinc (reviewed in [11]). Current efforts to boost zinc content in rice must ensure that cadmium is not co-transported to avoid the risk of increase of the heavy metal concentration. Therefore, it is important to ensure that biofortified rice does not accumulate cadmium, which can be done by quantitatively monitoring the cadmium content of the rice grain.

Several methods for trace cadmium analysis are available such as Flame Atomic Absorption Spectrometry (FAAS) [3], Graphite Furnace Atomic Absorption Spectrometry (GF-AAS) [12], Differential Pulse Anodic Stripping Voltammetry [13], Inductively Coupled Plasma Optical Emission Spectrometry (ICP-OES) [14], and Inductively Coupled Plasma Mass Spectrometry (ICP-MS) [15]. However, existing methodologies have not become more widely used because of several limitations, the methodologies may be too complex and expensive to perform, not sensitive enough to detect extremely low cadmium levels in rice grains and straw, or require tedious pre-concentration steps to quantify trace cadmium levels. Sample preparation procedures are also associated with sample spitting during digestion of agricultural sample matrices.

The method routinely used at the International Rice Research Institute's (IRRI) Grain Quality and Nutrition Services Laboratory (GQNSL) to measure cadmium content is via closed vessel acid digestion and measure it through graphite furnace-atomic absorption spectrometry (GF-AAS) [16]. The closed vessel digestion technique is preferred over the open vessel digestion since it reduces risk of contamination and loss of volatile analyte [17], while its complex organic matrix gets oxidized to release cadmium in solution (during closed vessel digestion) for further quantification via GF-AAS [18].

2 Materials

2.1 Glassware and Apparatus

1. 50 mL screw-capped polypropylene tube.
2. DigiPrep LS 72-tube (50 mL) block digestor (SCP Science, USA) (Fig. 1).
3. HGA 900 Graphite Furnace with AS 800 Autosampler and AAnalyst 400 atomic absorption spectrometer equipped with cadmium Electrodeless Discharge Lamp (EDL, All from Perkin Elmer, Singapore) (Fig. 2).
4. Glass volumetric flasks (1 L, 100 mL, 25 mL).
5. Horizontal shaker, Model E59100.25, Eberbach Corp., USA.
6. Calibrated dispenser, 10 mL L/I Repipet® Dispensers, Cole Parmer, Canada (*see* **Note 1**).
7. Desiccator cabinet, Boekel, Philadelphia, USA.
8. Weighing balance.

2.2 Reagents and Standards

1. Deionized water.
2. Concentrated Nitric Acid, HNO_3 (65%, CAS No. 7697-37-2).
3. Hydrogen Peroxide, H_2O_2, 30% (v/v) in H_2O (CAS No. 7722-84-1).
4. 4% (v/v) HNO_3: Dilute concentrated nitric acid (40 mL) in deionized water (1 L). This solution must be used within 1 year. This also serves as the calibration blank.
5. Pd stock solution (10,000 ppm) in 15% HNO_3 (High Purity Standards, USA).
6. $Mg(NO_3)_2$ (10,000 ppm, Perkin Elmer, Singapore).
7. Matrix modifier (Pd-$Mg(NO_3)_2$): Dilute an aliquot of the Pd stock solution (3.00 g) in 15% HNO_3 (prepared from 23.10 g

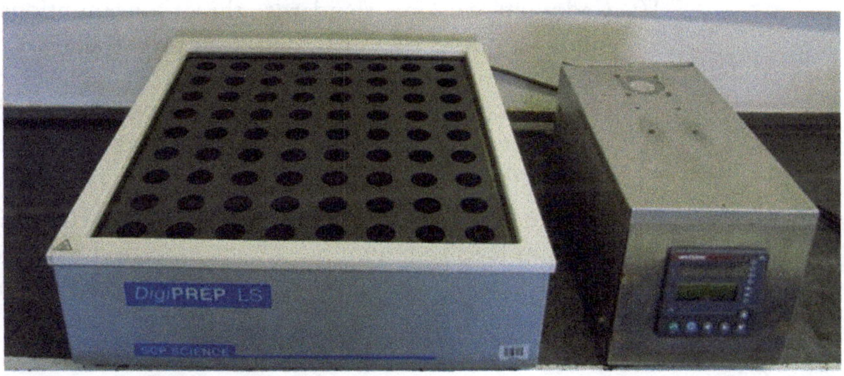

Fig. 1 The DigiPrep block digestor

Fig. 2 The GF-AAS setup

of the conc. HNO_3 and diluted to 100.00 g with deionized water) then add $Mg(NO_3)_2$ (2.00 g) in a polyethylene bottle. Make up to 10.00 g with 4% HNO_3.

8. Cadmium stock solution A (1000 ppm cadmium): Single-element cadmium stock solution (in 2% HNO_3 matrix) (High Purity Standards, Charleston, SC, USA). This solution must be used before the expiration date specified by the manufacturer (*see* **Note 2**).
9. Cadmium intermediate standard solution B (1 ppm cadmium): Weigh 0.100 g cadmium stock solution A into a polyethylene bottle and make up to 100.00 g with 4% HNO_3. This solution must be used within 6 months (*see* **Note 3**).
10. Cadmium working standards: three concentrations of cadmium solutions (0.5, 1.0, and 2.0 ppb). Weigh the required amount of cadmium standard solution B (Table 1) into polyethylene bottles and dilute to 50 g with 4% HNO_3. Mix well. The solutions must be used within 2 weeks (*see* **Note 4**).
11. National Institute for Environmental Studies (NIES) Certified Reference Materials (CRM) #10-a and #10-b, low- and medium-level cadmium rice flour, respectively; both of which are unpolished. The CRM are used for method validation and optimization.

Table 1
Components of cadmium working standards

Conc. (ppb Cd)	Weight of Cd int. standard solution B (g)	Total weight (g)
0.5	0.025	50.00
1.0	0.050	50.00
2.0	0.100	50.00

12. In-house rice flour reference material (99G, i.e., milled rice prepared from mixed paddy rice grain collected from ZES in 1999) used for routine cadmium analysis.
13. Continuing calibration verification (CCV) standard or QC insert (0.5 ppb cadmium in 4% HNO_3): In a polyethylene bottle, dilute an aliquot of cadmium standard solution B (0.025 g) to 50 g with 4% HNO_3. Mix well. The solution must be used within 2 weeks.

3 Procedure

3.1 Oven-Drying the Samples

Oven-dry rice flour prepared from milled or brown rice grain samples for 2 h at 80 °C and store the dried samples packed in paper envelope inside the desiccator cabinet prior to weighing.

3.2 Closed Tube Digestion

1. Weigh 0.30 g of oven-dried sample in a polypropylene tube and add 2.0 mL concentrated HNO_3 and 0.5 mL H_2O_2.
2. Repeat **step 1** for the in-house rice reference material.
3. For the blank sample, dispense 2.0 mL concentrated HNO_3 and 0.5 mL H_2O_2 in a polypropylene tube.
4. Cap the tubes and allow to stand overnight (pre-digestion) at room temperature (22–26 °C).
5. Digest all samples in the block digestor using the settings listed in Table 2 (*see* **Note 5**).
6. After the digestion, remove the tubes from the digestor and allow them to cool to room temperature.
7. Add deionized water (22 mL) into the tube to allow cooling of the diluted acid. Add the volume up to 25 mL with deionized water (Fig. 3).
8. Re-seal the caps and agitate in a shaker (180 rpm, 5 min).
9. Allow the undissolved materials to settle (1 h). After the settling time, decant into sample cups for GF-AAS run.

Table 2
Digestor settings for the digestion of rice grain to release cadmium

Stage	Temperature (°C)	Time (min)
Warming	80	30
Digestion	125	120

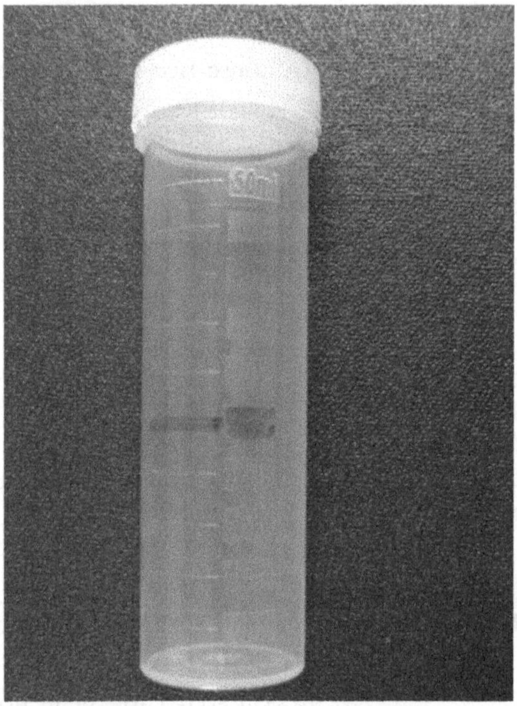

Fig. 3 Sample of polypropylene tube with mark

3.3 Analytical Determination of Cadmium Content Using Graphite Furnace Atomic Absorption Spectrometer

1. Start-up Perkin Elmer GF-AAS for cadmium.
2. Clean and check furnace windows and the graphite tube. Replace the graphite tube as needed.
3. Perform tip alignment, furnace alignment, and autosampler alignment.
4. Warm up the cadmium EDL and conduct background correction for at least 15 min.
5. Run a sensitivity check by measuring the characteristic mass, m_o of Solution B. In the instrument readings, the measured m_o should be within ±20% of the cookbook value or of the comparison m_o (0.5 pg/0.0044A) (*see* **Note 6**).

6. Perform a precision check by running Solution B five times. Relative Standard Deviation (RSD) values obtained must be <5%.
7. Prepare the sample information sheet for the sequence of the analysis. The sequence of materials for analysis includes the calibration standards (Table 1), the matrix modifier, the diluents, the samples, and the reference materials. Place sample cups containing these materials in the autosampler based on the encoded sequence.
8. Quantify the cadmium contents of the materials in the sequence using the optimized furnace program (Table 3). Check from time-to-time if the sample is being dispensed properly to the graphite tube.
9. Shut down the GF-AAS after analysis.

4 Quality Control

1. Method validation for cadmium in rice grains is done using NIES CRM #10 prepared via closed vessel digestion (Subheading 3.2) and analyzed using the GF-AAS optimized setting (Table 3).
2. The calibration curve must be linear thru zero with a correlation coefficient ≥0.995; re-calibrate if this correlation coefficient is not achieved.
3. Each sequence run or batch of samples must have at least one method blank, one reference material, and one duplicated sample run for every ten samples. The result for the in-house rice reference material, 99G, must be within ±2 s.d. of the mean in the control chart (*see* **Note 10**).
4. The relative percent difference (% RPD) for sample repeats must be ≤20%. However, owing to the heterogeneity of some samples (e.g., milled grain samples), it is not always the case. The cadmium concentrations of standards included in sequence runs as unknowns must be within ±10% of the certified or theoretical value. % RSD for ICV and CCV insert must be ≤10% RSD (*see* **Note 11**).
5. If the results are not acceptable, take corrective action or qualify the data accordingly (*see* **Note 12**).
6. Unusual situations such as presence of undigested sample after sample preparation, high method blank readings, poor sensitivity signal, and poor precision between replicated samples occurring during analysis must be documented on the work sheet, and non-conformity/corrective action reports must be issued if necessary (*see* **Note 12**).

Table 3
Optimized furnace program[a] used for the analysis of cadmium in rice grains

Step no.	Temp. (°C)	Ramp time (s)	Hold time (s)	Internal Ffow
1	110	5	20	250
2	120	10	20	250
3	600	10	30	250
4	20	1	10	250
5	1600	0	5	0
6	2450	1	3	250

[a]Refer to Perkin Elmer AAnalyst 400 atomic absorption spectrometer equipped with HGA 900 Graphite Furnace, with AS 800 Autosampler Protocol or the Perkin Elmer AA WinLab Software Guide Manual [14] for the complete guide to the operation of GF-AAS (*see* **Note 7**). The matrix modifier is Pd-Mg$(NO_3)_2$ (*see* **Notes 8** and **9**)

5 Calculations and Reporting Results

1. Results of cadmium content must be reported in μg/kg. If dilution was used, appropriate dilution factor must be incorporated in the result.
2. Calculate results for rice grain as follows:

$$\text{cd}\left(\frac{\mu g}{\text{kg}}\right)=\left(R_s-R_b\right)\times\frac{\text{vol. of digest }(\text{mL})}{\text{wt. of oven}-\text{dried sample}(\text{g})}.$$

where

R_s = μg/L Cd in sample digest obtained from standard curve

R_b = μg/L Cd in blank obtained from standard curve.

6 Notes

1. To calibrate the dispenser bottle, dispense 10 mL of the digesting solution from the dispenser into a graduated cylinder. Obtain the weight of the dispensed solution. Repeat the procedure five times. Calculate mean and RSD. The mean must be 10 ± 0.1 g and the RSD must fall within ±2%. Otherwise, adjust the dispenser and repeat the calibration.
2. The stock solution may also be prepared from ultra-high-purity-grade chemical or metal (99.99% pure or greater).

3. The weight of the cadmium stock solution may not be exactly 0.1 g, depending on the density of the solution. Take into account the density and recalculate the weight of stock solution needed.
4. The solution containing 1.0 ppb in 4% (v/v) HNO_3 both can be used as a calibration working standard and as a standards sensitivity check.
5. Pressure may build up during the warming stage. Loosen the caps to equalize the pressure and immediately re-tighten firmly and place back to the digestor.
6. The characteristic mass (m_o) is defined as the mass of analyte that gives 1% absorption or 0.0044 absorbance [18]. If calculated m_o < 20% of the comparison m_o, troubleshoot the GF-AAS for low absorbance due to possible contamination and inaccuracy in the preparation of the calibration standards. If calculated m_o > 20% of the comparison m_o, check for the integrity of the EDL; the integration time may be too short, the internal gas flow during atomization was not set to zero, or the wavelength/slit width were not corrected.
7. For cadmium analysis, use 228.8 nm wavelength and argon gas for GF-AAS.
8. For standard and sample injections, a working volume of 20 μL is used, with an additional 5 μL aliquot of matrix modifier added in a combined injection. The mixing of the reagents and the samples is performed by the autosampler. A continuous light source (deuterium lamp) is used for background correction.
9. The rice grain contains elements such as sodium, potassium, calcium, and magnesium [19] and will require a matrix modifier to suppress matrix interferences, reduce cadmium volatility, and stabilize the cadmium signal [20, 21].
10. The control chart is used to monitor how an analytical process changes over time. Data from the reference materials used in a batch of analyzed samples are plotted in time order. Measurement process is within statistical control if the results for reference materials are within ±2 s.d. in the control chart. If the data shows out of control signal (i.e., above ±3 s.d. in the control chart or observed QC trending), mark it on the chart and investigate the cause. Proper documentation on the cause and how it was corrected is required.
11. The initial calibration verification (ICV) and the continuing calibration verification (CCV) are quality control standards inserted after the calibration standards (ICV) to check the validity of the calibration standards' concentrations. The

changes in concentrations of the calibration standards may be attributed to changing conditions of the graphite tube such as corrosion or formation of cracks. These calibration verification standards are from independent standard sources and are placed at regular intervals in the sequence run (CCV).

12. These are some corrective actions to do if the results are not acceptable. Verify the quality and the homogeneity of the sample and the reference materials used by visual inspection, check the quality of the calibration standards used (its concentration and its expiry date), check for possible contamination of labware, and inspect the graphite tube if it is already old and corroded and requires replacement.

Acknowledgments

The authors thank the following laboratory staff involved in the sample preparation, in the method development and optimization, and in the preparation of the initial documentation: Edgar Amoloza, Marnol Santos, Karen Serdeña, and Eric Jhon Cruz. This work has been supported under the CGIAR thematic area Global Rice Agri-Food System CRP, RICE, The Indian Council of Agricultural Research, Taiwan support from Council of Agriculture (COA), Ministry of Foreign Affairs (MOFA) and BBSRC Newton grant funding Reference number BB/N013808/1.

References

1. Zhuang P, McBride MB, Xia H, Li N, Li Z (2009) Health risk from heavy metals via consumption of food crops in the vicinity of Dabaoshan mine, South China. Sci Total Environ 407:1551–1561
2. Chaney RL, Ryan JA, Li YM, Welch RM, Reeves PG (1996) Phyto-availability and bioavailability in risk assessment for Cd in agricultural environments. In: Sources of cadmium in the environment. Organization for Economic Cooperation and Development, Paris, pp 49–78
3. Simmons RW, Noble AD, Pongsakul P, Sukreeyapongse O, Chinabut N (2008) Analysis of field-moist Cd contaminated paddy soils during rice grain fill allows reliable prediction of grain Cd levels. Plant Soil 302(1):125–137
4. Codex Alimentarius Commission (2005) Report of the 37th session of the codex committee on food additives and contaminants, Codex Alimentarius Commission, The Hague. p 201
5. Fang Y, Sun X, Yang W, Ma N, Xin Z, Fu J, Liu X, Liu M, Mariga AM, Zhu X, Hu Q (2014) Concentrations and health risks of lead, cadmium, arsenic, and mercury in rice and edible mushrooms in China. Food Chem 147:147–151
6. Simmons RW, Pongsakul P, Chaney RL, Saiyasitpanich D, Klinphoklap S, Nobuntou W (2003) The relative exclusion of zinc and iron from rice grain in relation to rice grain cadmium as compared to soybean: Implications for human health. Plant Soil 257(1):163–170
7. Sun L, Xu X, Jiang Y, Zhu Q, Yang F, Zhou J, Yang Y, Huang Z, Li A, Chen L, Tang W, Zhang G, Wang J, Xiao G, Huang D, Chen C (2016) Genetic diversity, rather than cultivar type, determines relative grain Cd accumulation in hybrid rice. Front Plant Sci 7:1407
8. Swaddiwudhipong W, Limpatanachote P, Mahasakpan P, Krintratun S, Punta B, Funkhiew T (2012) Progress in cadmium-related health effects in persons with high envi-

ronmental exposure in northwestern Thailand: a five-year follow-up. Environ Res 112: 194–198

9. Larsson SC, Orsini N, Wolk A (2015) Urinary cadmium concentration and risk of breast cancer: a systematic review and dose-response meta-analysis. Epidemiol Rev 182(5):375–380
10. Meharg AA, Norton G, Deacon C, Williams P, Adomako EE, Price A, Zhu Y, Li G, Zhao F-J, McGrath S, Villada A, Sommelia A, De Silva PM, Brammer H, Dasgupta T, Islam MR (2013) Variation in rice cadmium related to human exposure. Environ Sci Technol 47(11):5613–5618
11. Palmgren MG, Clemens S, Williams LE, Krämer U, Borg S, Schjørring JK, Sanders D (2008) Zinc biofortification of cereals: Problems and solutions. Trends Plant Sci 13(9):464–473
12. Shi J, Li L, Pan G (2009) Variation of grain Cd and Zn concentrations of 110 hybrid rice cultivars grown in a low-Cd paddy soil. J Environ Sci 21(2):168–172
13. Kjuus BE (1985) Determination of cadmium, copper, lead, and zinc in inorganic fertilizers by anodic stripping voltammetry. Fresenius J Anal Chem 322:577–580
14. Simmons RW, Pongsakul P, Saiyasitpanich D, Klinphoklap S (2005) Elevated levels of cadmium and zinc in paddy soils and elevated lelvels of cadmium in rice grain downstream of a zinc mineralized area in Thailand: implications for public health. Environ Geochem Health 27:501
15. Shindoh K, Yasui A (2003) Changes in cadmium concentration in rice during cooking. Food Sci Tech Res 9(2):193–196
16. Gunduz S, Akman S (2013) Determination of lead in rice grains by solid sampling HR-CS GFAAS. Food Chem 141:2634–2638
17. Hansen TH, Laursen KH, Persson DP, Pedas P, Husted S, Schjoerring JK (2009) Micro-scaled high-throughput digestion of plant tissue samples for multi-elemental analysis. Plant Methods 5:12
18. Perkin-Elmer Corporation (1996) Analytical methods for atomic absorption spectroscopy. Perkin-Elmer Corporation, Massachusetts
19. Juliano BO (1993) Rice in human nutrition. In: FAO food and nutrition series. Food and Agriculture Organization, Rome, pp 4–59
20. Bettinelli M, Baroni U, Pastorelli N (1988) Determination of arsenic, cadmium, lead, antimony, selenium and thallium in coal fly ash using the stabilised temperature platform furnace and Zeeman-effect background correction. J Anal Atom Spectro 3:1005–1011
21. Schlemmer G, Radziuk B (1999) Analytical graphite furnace atomic absorption spectrometry: a laboratory guide. Birkhäuser, Germany, p 286

Chapter 16

Analysis of Developing Rice Grain Transcriptome Using the Agilent Microarray Platform

Mandy Püffeld, Christiane Seiler, Markus Kuhlmann, Nese Sreenivasulu, and Vito M. Butardo Jr.

Abstract

Transcriptome analysis reflects the status quo of transcribed genetic code present in the form of mRNA, which helps to infer biological processes and unravel metabolic status. Despite the increasing adoption of RNA-Seq technique in recent years, transcriptome analysis using the microarray platform remains the gold standard technique, which offers a simpler, more cost-effective, and efficient method for high-throughput gene expression profiling. In this chapter, we described a streamlined transcriptomic analyses pipeline employed to study developing rice grains that can also be applied to other tissue samples and species. We described a novel RNA extraction method that obviates the problem introduced by high-starch content during rice grain development that usually leads to reduction in RNA yield and quality. The detailed procedure of microarray analysis involved in cDNA synthesis, cRNA labeling, microarray hybridization, slide scanning, feature extraction to QC validation has been described. The description of a newly developed *Indica*- and *Japonica*-specific microarray slides developed from the genome information of subpopulation to study gene expression of 60,000 genes has been highlighted. The downstream bioinformatics analyses including expression QTL mapping and gene regulatory network analyses were mentioned.

Key words cDNA synthesis, cRNA labeling, Microarray, Rice grain, RNA extraction, Hybridization, Slide scanning

1 Introduction

Transcriptome is the sum of all ribonucleic acid (RNA) molecules present in a cell, which are generated from the genome through transcription. The composition of the transcriptome is dependent on the active transcription of genes, as well as processing and stability of the messenger RNA. The composition of a transcriptome reflects the response of a cell or tissue and thus the status of each individual cell can be monitored by analysis of its transcriptome. The presence, absence, or abundance of transcripts derived from specific marker genes can be used to estimate the developmental status or response of an organism to environmental changes.

Nese Sreenivasulu (ed.), *Rice Grain Quality: Methods and Protocols*, Methods in Molecular Biology, vol. 1892, https://doi.org/10.1007/978-1-4939-8914-0_16, © Springer Science+Business Media, LLC, part of Springer Nature 2019

Transcripts can be detected by various methods after the extraction of the total RNA from a selected tissue at a specific time point under a defined growth condition. Two broadly applied methods currently exist that allow the quantification of transcripts from their respective transcriptomes, which include RNA sequencing (RNA-Seq) and whole-genome microarray. The RNA-Seq technology is more recent and offers a number of advantages compared to the older microarray hybridization technology. RNA-Seq allows unbiased detection of novel transcripts because it is not limited to species- or transcript-specific probes, novel transcripts, gene fusions, single-nucleotide variants such as edited nucleotides, indels (small insertions and deletions), and other previously unknown changes which otherwise cannot be detected by microarrays. RNA-Seq also provides a broader dynamic range of detection, which depends on the depth of sequencing coverage. In addition, RNA-Seq technology is able to quantify discrete, unique sequencing read counts representing the quantity of transcripts. However, the data analysis of RNA-Seq data previously required extensive bioinformatics expertise, presenting a major hurdle to the adoption of RNA-Seq technology by biologists. Currently, several user-friendly tools are available to analyze microarray-based transcriptome analysis that can significantly simplify the analysis process, providing accessible solutions for any researcher without requiring significant learning curve for data analyses [1–6]. In contrast, array hybridization technology measurement is limited by oligos represented on the microarray which is customized and in addition signal saturation issues need to be dealt.

The comparability of results derived from RNA-Seq and array hybridization are still under discussion. Several parallel comparisons between RNA-Seq and microarray hybridization technology have been performed which revealed a 68% overlap of commonly detectable genes [7–13]. The comparable reliability of detectable differential expressed genes is approximately 90% in both technologies validated by quantitative PCR. However, comparisons like this can only be performed if sufficient genomic information of the species under investigation is available. Because sufficient sequence information is the prerequisite for reliable microarray hybridization experiments, most of the well-characterized model species such as rice have predesigned microarrays available.

In terms of cost-efficiency, microarray technology is significantly cheaper than RNA-Seq technology based on the current price lists from various providers. Also, the amount of data stored from the microarray technology is much smaller and therefore easier to handle than the RNA-Seq datasets. While the data derived from the microarray hybridization is approximately 30 MB, the RNA-Seq data set of one sample ranges from 30 to 65 GB. This will not only affect time of uploading and downloading of data, but it will also significantly slow down the time needed for analysis and increase the cost of data storage. Taken together, although RNA-Seq is now widely used, gene expression analysis by whole-genome microarray is still an applicable technology especially when several samples and multiple replications should be performed such as in the screening of a breeding population or diversity panel.

Several studies have already utilized the power of genome-wide microarray to characterize gene expression profile in developing rice grains. Microarray technology was successfully utilized in determining genes controlling nutrient partitioning [14], analyzing the effect of elevated temperature on grain filling [15], identifying multiple regulatory pathways involved in the formation of chalkiness trait [16], and associating genes for grain number [17]. Because of the economic advantage of microarray technology, it finds its application in transcriptome mapping which involves genome-wide genetic analysis of gene expression data to identify expression QTLs (eQTLs) to further unravel metabolic, regulatory, and developmental pathways [18–21]. In this method, the microarray method newly developed in our lab are thoroughly described to aide researchers in generating 60 K transcriptomic data from *indica* or *japonica* subpopulation of rice.

2 Materials

2.1 Kits, Chemicals, and Reagents

1. RNA Extraction Buffer (100 mM Tris–HCl pH 8.0, 150 mM LiCl, 50 mM EDTA, 1.5% SDS).
2. β-Mercaptoethanol (freshly add 300 μL to 20 mL RNA Extraction Buffer immediately before use).
3. Phenol:chloroform:isoamyl (25:24:1).
4. Isopropanol.
5. Absolute ethanol.
6. 1.2 M NaCl.
7. Nuclease-Free Water.
8. Qiagen RNeasy Plant Mini Kit (50).
9. Qiagen RNase-Free DNase Set (50).
10. Qiagen RNeasy Mini Kit (50).
11. Agilent Low Input Quick Amp Labeling Kit, one-color.
12. Agilent One-Color RNA Spike-In Kit.
13. Agilent Hi-RPM Gene Expression Hybridization Kit, Large Volume.
14. Agilent Gene Expression Wash Buffer Kit.

2.2 Disposable and Consumable Items

1. RNA-free pipette filter tips (maximum capacity 2, 10, 20, 100, 200, and 1000 μL volumes).
2. RNAse-free microfuge tubes (1.5 mL and 2.0 mL capacity).
3. Crushed ice.
4. Agilent gasket slides.
5. Agilent ozone barrier slides.
6. Agilent glass trays.
7. Agilent slide holder.

2.3 Instruments

1. Pipettor set (for 2, 10, 20, 100, 200, and 1000 μL tips).
2. Dewar flask for liquid nitrogen.
3. Liquid nitrogen storage tank.
4. Vortex mixer.
5. Crushed ice maker.
6. Heat blocks for 1.5 mL microfuge tubes (set at 50 °C).
7. Heat block for 2.0 mL microfuge tubes.
8. Ultralow temperature freezer (set at −80 °C).
9. Refrigerated microcentrifuge with rotor for 24 microcentrifuges (set at 4 °C).
10. Agilent DNA Microarray Scanner.
11. Hybridization Oven.
12. Mini-incubator.

3 Methods

This method describes in detail the optimized steps for One-Color Microarray-Based Gene Expression Analysis using the Low Input Quick Amp Labeling Protocol from Agilent. The protocol is conveniently subdivided into four sections: (1) RNA extraction, (2) cDNA synthesis and cRNA labeling, (3) microarray hybridization, and (4) microarray reading and data quality validation. Each section can be conveniently done within 1 day but it is suggested that all samples should all go through the same process before moving on to the next section.

3.1 RNA Extraction

This newly developed RNA extraction procedure takes into account that developing rice grains have significantly elevated starch content that can reduce RNA quality and quantity. Interference due to high amounts of starch is addressed in the first part of the protocol by selective fractionation of starch pellets while ensuring optimal solubility of RNA in the aqueous phase. Cryogenic grinding is the most time-consuming part of this phase. It is suggested that all samples be ground before moving on to RNA extraction proper. A total of 24 ground samples can be conveniently subjected to RNA extraction per day. As this step involves the use of harmful organic solvents, all steps should be conducted inside the fume hood.

1. Grind developing rice grains to fine powder under strictly cryogenic conditions (*see* **Note 1**). Collect the powdered samples using stainless steel spatula scoops precooled by dipping in liquid nitrogen. Place samples in precooled 2.0 mL microfuge tubes. Place tubes with samples in liquid nitrogen and store immediately in ultralow temperature freezer at −80 °C until use.
2. On the day of RNA extraction, place 750 μL RNA Extraction Buffer and 500 μL phenol:chloroform:isoamyl (25:24:1) in 2 mL RNAse-free microfuge tubes. Vortex mix briefly.
3. Using a small metal spatula scoops previously dipped in liquid nitrogen, add approximately 200 mg of powdered rice grain sample in each tube. Immediately mix by vortexing. Keep on ice until all samples in 2.0 mL microfuge tubes are thoroughly resuspended in the extraction buffer.
4. Mix all samples simultaneously in a multi-tube vortex mixer at room temperature for 5 min at 13,394 × *g*. Keep all tubes on ice for 2 min.
5. Separate the supernatant by centrifugation at 14,000 × *g* for 10 min. Do all centrifugation steps for this section at 4 °C.
6. While spinning the samples by centrifugation, put 700 μL isopropanol and 400 μL 1.2 M NaCl in a fresh 2 mL microfuge tube. The number of tubes to be prepared will depend on the number of samples being centrifuged.
7. After centrifugation, collect and transfer supernatant (approximately 800 μL) into new 2.0 mL microfuge tube containing isopropanol and NaCl solution.
8. Mix by gentle inversion 5 times. Incubate in ultralow freezer (−80 °C) for 1 h or in regular freezer (−20 °C) overnight (*see* **Notes 2** and **3**).
9. Pellet the RNA by centrifugation at 18,800 × *g* for 15 min.
10. While spinning, prepare RLT buffer (100 μL β-mercaptoethanol in 10 mL RLT buffer) and DNAse 1 solution (50 μL frozen DNAse stock +350 μL RDD using the Qiagen DNAse set).
11. Discard the supernatant and dissolve the pellet in freshly prepared 450 μL RLT buffer. Dissolve pellet in freshly prepared 450 μL RLT buffer by simultaneous mixing in the multi-tube vortexer at room temperature (*see* **Note 4**).
12. Transfer the solution in QIAshredder mini spin column (purple) with 2 mL collection tube. Clean the lysate by passing

through the purple spin column by centrifugation for 1 min at 9000 × *g*.

13. Discard the purple spin column and add 0.5 volume of absolute ethanol (225 μL) to the cleared lysate in the 2 mL collection tube and mix by pipetting up and down several times.
14. Transfer the solution immediately to a new RNeasy mini spin column (pink) with 2 mL collection tube (*see* **Note 5**). Collect the RNA in the pink column with RNAeasy silica-gel membrane by centrifugation at 9000 × *g* for 15 s. Discard flow-through.
15. Wash the RNA sample in the membrane with 350 μL buffer RW1 by centrifugation at 9000 × *g* for 15 s.
16. Directly add 80 μL diluted DNAse 1 onto the membrane and incubate on the benchtop for at least 15 min.
17. Remove the DNAse 1 and clean the RNA extract further by a series of washing steps. Add 350 μL buffer RW1 and centrifuge at 9000 × *g* for 15 s. Discard flow-through. Repeat the procedure but this time, use 500 μL buffer RPE. Wash again with the same amount of buffer RPE and centrifuge for 2 min.
18. Dry the membrane by transferring to a new 2 mL collection tube and spinning at 9000 × *g*.
19. Transfer the pink spin column to a new 1.5 mL collection tube. Add 50 μL RNAse-free milliQ water pre-warmed at 50 °C. Incubate for 3 min at 50 °C heat block.
20. Elute the RNA samples by centrifugation at 9000 × *g* for 1 min. Immediately place the samples on ice.
21. Determine the concentration by measuring the absorbance of each sample using NanoDrop. Dilute the sample accordingly (usually 1:10) and determine the quality using BioAnalyzer. RNA extracts with RIN value of 8.0 or higher are amenable for microarray experiment.

3.2 cDNA Synthesis Followed by cRNA Amplification, Labeling and Purification

One staff can conveniently process a maximum of 24 samples per day for this section of the experiment. The steps below are designed to be conducted continuously within one whole day but optional stop points are identified whenever possible. Three heat blocks (or small water baths) are needed for this part of the protocol. Before commencing the protocol below, turn on each heat block at 80 °C, 65 °C and 37 °C. The 80 °C heat block should be dedicated to 2.0 mL microfuges while the other two heat blocks should accommodate 1.5 mL microfuges. The 37 °C heat block is set to 40 °C instead if the Spike Mix will not be prepared in **step 2**. The initial temperature settings will be changed later based on the steps below.

Table 1
Sample spreadsheet for dilution of samples for cDNA synthesis

Sample ID	Stock RNA (ng/ μL)	RIN value	First dilution	Concentration of first dilution(ng/ μL)	Volume of RNA needed for 50 ng (μL)	DEPC-water needed for 50 ng (μL)	Total volume (μL)
RNA_001	1596.4	8.3	1:15	56.8	0.9	0.6	1.5
RNA_002	1709.8	8.2	1:17	57.0	0.9	0.6	1.5
RNA_003	1134.7	8.2	1:11	58.9	0.8	0.7	1.5

Prepare one-color spike mix, T7 promoter primer mix, and cDNA master mix based on Agilent protocol. The prepared mix will depend on the amount of starting material, which can range from 10 to 200 ng total RNA or 5 ng PolyA RNA. In our laboratory, we routinely use 50 ng total RNA as starting material for microarray hybridization. This method is described below. To prepare master mixes for 24 samples, add 2 extra reactions to compensate for pipetting variations.

1. Dilute each total RNA sample to 50 ng in a 1.5 mL microfuge tube to make a final volume of 1.5 μL. A preliminary 100-fold dilution of the original total RNA extract is usually needed to fit within the 50–100 ng concentration range. Measure the actual concentration of the First Dilution using Nanodrop. Perform a second dilution of each sample using the First Dilution to make 1.5 μL 50 ng RNA sample. To facilitate tracking of sample and ease of dilution, the sample spreadsheet in Table 1 can be used. Keep all diluted samples on ice.
2. Pre-warm the Spike Mix stock at 37 °C in a 2 mL heat block for 5 min prior to serial dilution. Change the temperature to 80 °C after use. The actual serial dilution is done in 1.5 mL microfuge tubes using the Dilution Buffer from the Agilent Spike-In kit as described in Table 2. After each dilution, the diluted spike mix is resuspended thoroughly by vortex mixing. Store the First and Second Dilution of the Agilent One-Color Spike mix positive controls at −80 °C, taking note of date when they were prepared. They can only go through freeze/ thaw cycles up to eight times. On the other hand, the Third and Fourth Dilution should be prepared fresh daily and stored on ice prior to use.
3. Prepare the T7 promoter mix in a 1.5 mL microfuge tube. The volume needed depends on the number of samples for labeling. The volume required for 24 samples is summarized in Table 3. Store the mix on ice prior to use.

Table 2
Serial dilution of Agilent One-Color Spike Mix for Cyanine-3 labeling using 50 ng total RNA as starting material

Dilution	Dilution buffer (μL)	Amount (μL)	Stability
First dilution (1:20)	38	2 μL spike mix stock	2 months
Second dilution (1:25)	48	2 μL first dilution	1 week
Third dilution (1:20)	38	2 μL second dilution	Prepare fresh
Fourth dilution (1:2)	30	30 μL third dilution	Prepare fresh

Note that dilutions vary when different RNA amount is used as starting material, please refer to the **Low Input Quick Amp Labeling Kit** manual for details (Agilent)

Table 3
T7 Promoter Primer mix for 24 labeling reactions

Component	Volume Per Reaction (μL)	Volume Per 24 + 2 Reactions (μL)
T7 promoter primer (green cap)	0.8	20.8
Nuclease-free water (white cap)	1.0	26.0
Total volume	**1.8**	**46.8**

4. Prepare the primer and the template by adding each of the components in Table 1 to Table 3, using the microfuge tube in Table 1 containing 1.5 μL of 50 ng RNA sample. Mix the components as specified in Table 4 by pipetting several times. Denature the mixture by incubating at 65 °C heat block for 10 min.
5. While denaturing, pre-warm the 5× First Strand Buffer at the first heat block at 80 °C for at least 4 min, in preparation for **step 7**.
6. Immediately place the denatured primer-template mix from **step 4** on ice for 5 min.
7. While incubating on ice in **step 6**, quickly prepare the cDNA master mix as detailed in Table 5. Mix each component by gentle pipetting and keep the master mix at room temperature prior to use. However, the AffinityScript RNAse Block Mix should be kept in the freezer and immediately placed back after use.

Table 4
Preparation of cDNA synthesis reaction mix

Components	Volume Per Reaction (μL)
50 ng total RNA (from Table 1)	1.5
Fourth dilution of spike mix (from Table 2)	2.0
T7 primer mix (from Table 3)	1.8
Total mix per tube	**5.3**

Table 5
Components of cDNA master mix

Components	Volume Per Reaction (μL)	Volume Per 24 + 2 Reactions (μL)
Pre-warmed 5× First Strand Buffer (green cap)	2.0	52.0
0.1 M DTT (white cap)	1.0	26.0
10 mM dNTP mix (green cap)	0.5	13.0
AffinityScript RNAse Block Mix (violet cap)	1.2	31.2
Total volume	**4.7**	**122.2**

8. Remove the samples from ice (**step 6**) and centrifuge the tubes briefly at room temperature to spin down the contents to the bottom in preparation for the next step.
9. Add 4.7 μL of the cDNA Master Mix from Table 5 to each tube and mix gently by pipetting.
10. Synthesize cDNA by incubating the samples for 2 h at 40 °C heat block. During the incubation, the 65 °C heat block can be changed to 70 °C in preparation for the heat inactivation (**step 11**).
11. Inactivate the AffinityScript enzyme by incubating at 70 °C for 15 min (**Optional stopping point**: The samples can be stored at −80 °C overnight. Continue the next day after heat inactivation).
12. Immediately transfer the tubes on ice and incubate for 5 min.
13. While incubating on ice, prepare Transcription Master Mix as detailed in Table 6. Just keep the mix at room temperature prior to dispensing. However, the T7 RNA polymerase blend should be kept in the freezer and immediately placed back after use. As the last component is light sensitive, mixing and

Table 6
Components of Transcription Master Mix

Component	Volume Per Reaction (μL)	Volume Per 24 + 2 Reactions (μL)
Nuclease-free water (white cap)	0.75	19.5
5× Transcription Buffer (blue cap)	3.2	83.2
0.1 M DTT (white cap)	0.6	15.6
NTP mix (blue cap)	0.1	26.0
T7 RNA Polymerase Blend (red cap)	0.21	5.5
Cyanine 3-CTP (Cy3)	0.24	6.2
Total volume	**6.0**	**156.0**

dispensing of the mix (**step 15**) should be done in low light conditions. The light directly above the work bench should be turned off. Instead, indirect lights from nearby lab benches or windows can be used.

14. Centrifuge the samples from **step 12** briefly in a microcentrifuge to drive down tube contents to the bottom.
15. Add 6 μL Transcription Master Mix to each sample tube, gently mixing by pipetting. The total volume of the tube is now 16 μL.
16. Transcribe and label the cDNA to produce Cy3-labeled cRNA by incubating samples at 40 °C for 2 h. (**Optional stopping point**: The samples can be stored at −80 °C after transcription).
17. In preparation for the next step, convert the 70 °C heat block to 55 °C and add nuclease-free water in preparation for the elution of purified cRNA (**step 27**).
18. After transcription and labeling (**step 16**), purify the labeled cRNA using a modified method for Qiagen RNeasy Mini Kit as detailed below (**step 19–28**).
19. Adjust the volume to 100 μL of each sample by adding 84 μL nuclease-free water.
20. Add 350 μL Buffer RLT.
21. Add 250 μL absolute ethanol.
22. Mix thoroughly by pipetting. Transfer 700 μL of each cRNA sample to an RNAeasy mini column with 2 mL collection tube.

23. Centrifuge samples at 4 °C for 30 s at 7,534 × *g*. Discard flow-through.
24. Wash the samples by adding 500 μL Buffer RPE to each column (do not forget to add ethanol to this buffer prior to first use). Centrifuge samples at 4 °C for 30 s at 7,534 × *g*. Discard flow-through. Repeat this step once.
25. Transfer each RNAeasy column to a new collection tube. Remove traces of RPE buffer and dry the samples by centrifugation at 4 °C for 30 s at 7,534 × *g*.
26. Discard the collection tube and transfer the mini spin column into a new 1.5 mL microfuge tube provided in the kit.
27. Elute the cRNA sample by adding 30 μL RNAse-free water pre-warmed at 55 °C (**step 17**) directly onto the RNAeasy filter membrane. Incubate for 60 s at the 55 °C heat block. Centrifuge at room temperature for 30 s at 7,534 × *g*.
28. Remove the RNAeasy column and close each microfuge tube. Immediately place all tubes on ice (**Optional stopping point**: The samples can be stored at −80 °C after elution).
29. Quantify the cRNA using the microarray measurement feature of Nanodrop based on the manufacturer's method. The sample type should be set to RNA-40. The following values are needed: Cyanine 3 dye concentration (pmol/μL), RNA absorbance ratio (260 nm/280 nm), cRNA concentration (ng/μL).
30. Immediately store the cRNA samples at −80 °C after Nanodrop readings.
31. Compute for the cRNA yield using the following formula:

$$\text{cRNA yield}\ (\mu g) = \frac{(\text{cRNA concentration in ng} / \mu L) \times (30\ \mu L\ \text{elution volume})}{1000}$$

32. Determine the specific activity using the following formula:

$$\text{Specific activity}\ (\text{pmol Cy3} / \mu g\ \text{cRNA}) = \frac{\text{concentration of Cy3}\ (\text{pmol} / \mu L)}{\text{concentration of cRNA}\ (\text{ng} / \mu L)} \times 1000$$

33. The computations in **steps 31** and **32** can easily be done using a spreadsheet. The recommended yield for 8-pack microarray format is at least 0.825 μg, a 260/280 ratio of greater than 2.0 and a specific activity of 6 pmol Cy3/μg cRNA. Consult the Agilent manual for the recommended yield and specific activity values for the other formats. A summary spreadsheet should be prepared similar to Table 7.

Table 7
Summary spreadsheet for cRNA quantification

Sample ID	Cy3 (pmols/μL)	260/280 ratio	cRNA (ng/μL)	cRNA yield (μg)	Specific activity (pmols Cy3/μg cRNA)
RNA_001	0.95	2.18	73.49	2.20	12.9
RNA_002	1.20	2.22	82.94	2.49	14.5
RNA_003	0.81	2.19	68.79	2.06	11.8

3.3 Microarray Hybridization and Scanning

A new custom 8x60K microarray slide for *japonica* (Order Code: 054269) and *indica* (Order Code: 054270) rice transcriptomes were designed. These are available for purchase from Agilent. The microarray hybridization protocol can be started in the afternoon. It will only take around 3 to 4 h to complete the hybridization of 24 samples. Up to 8 slides (64 samples) can be processed in the washing step but one staff can optimally process only up to 4 slides (32 samples) per day. The following morning is devoted to washing of and scanning of the microarray slides. Additional microarray hybridizations in the afternoon of the second day can be started. The process is repeated until all samples are hybridized and scanned. Use color-free latex gloves to avoid contaminating the hybridization mix and the microarray slides with colored pigments that can interfere with the signals during the microarray scanning.

1. Prior to the actual hybridization, prepare the 10× Blocking Agent based on the manufacturer's instructions. The large volume 10× Blocking Agent is resuspended in 1250 μL nuclease-free water. Heating the mixture for up to 5 min at 37 °C usually helps in pellet dissolution. The solution is gently vortexed until the pellet is dissolved. Centrifuge the solution for 5 s to drive down everything to the bottom. A 200 μL aliquot of the 10× Blocking Agent are dispensed in 1.5 mL microfuge tubes and stored at −20 °C freezer. Each aliquot is stable for up to 2 months and is good for up to 40 hybridizations.
2. During the actual day for hybridization, get one tube aliquot of 10× Blocking Agent and thaw on ice. Turn on the hybridization oven and set the temperature at 65 °C. Also equilibrate a heat block (or a small water bath) to 60 °C for cRNA fragmentation. If the lab is equipped with an air purification system like IQAir, one can also turn this on prior to starting the hybridization protocol.
3. Prepare the fragmentation mix based on Agilent protocol. The volume of the components will depend on the microarray format being used. For example, one can routinely hybridize 600 ng of linearly amplified Cy3-labeled cRNA to an 8-pack

Table 8
Fragmentation mix for 8-pack microarray format using 600 ng of cRNA

Components	Volume/mass 8-pack microarrays
Cyanine 3-labeled, linearly amplified cRNA	600 ng
10× Blocking Agent	5 μL
Nuclease-free water	Bring volume to 24 μL
25× Fragmentation Buffer	1 μL
Total volume	**25 μL**

microarray format. The components for this format are enumerated in Table 8. Each fragmentation mix is prepared individually in 1.5 mL microfuge tubes on ice. When all of the components for each sample are dispensed by pipetting, each fragmentation reaction is then mixed gently and thoroughly using a vortex mixer. Spin the samples briefly in the microcentrifuge.

4. Incubate the samples for **exactly** 30 min in the pre-equilibrated 60 °C heatblock.
5. Immediately stop the fragmentation reaction by cooling on ice for 1 min and proceeding to the next step.
6. Add 2× GEx Hybridization Buffer HI-RPM to each of the tube based on the manufacturer's instructions. The volume depends on the microarray format. For 8-pack microarray format, add 25 μL of Hybridization Buffer to stop the fragmentation reaction. Mix well by gently pipetting up and down. Be careful not to introduce bubbles during mixing.
7. Centrifuge the tubes for 1 min at room temperature at 15,720 × *g* to drive all solution to the bottom and to remove any bubbles formed during pipetting.
8. Immediately place all samples on ice and load onto the microarray slide as soon as possible.
9. Prepare the hybridization assembly. Details of the procedure can be found on Agilent Microarray Chamber User Guide enclosed in the Agilent Microarray Chamber Kit and available for download at the Agilent website. Open the packet and load a clean gasket slide into the Agilent SureHyb chamber base (Fig. 1a). Ensure that the label is facing up and aligned with the rectangular section of the chamber base. Also see to it that the gasket slide is flushed with the chamber base and is not misaligned.

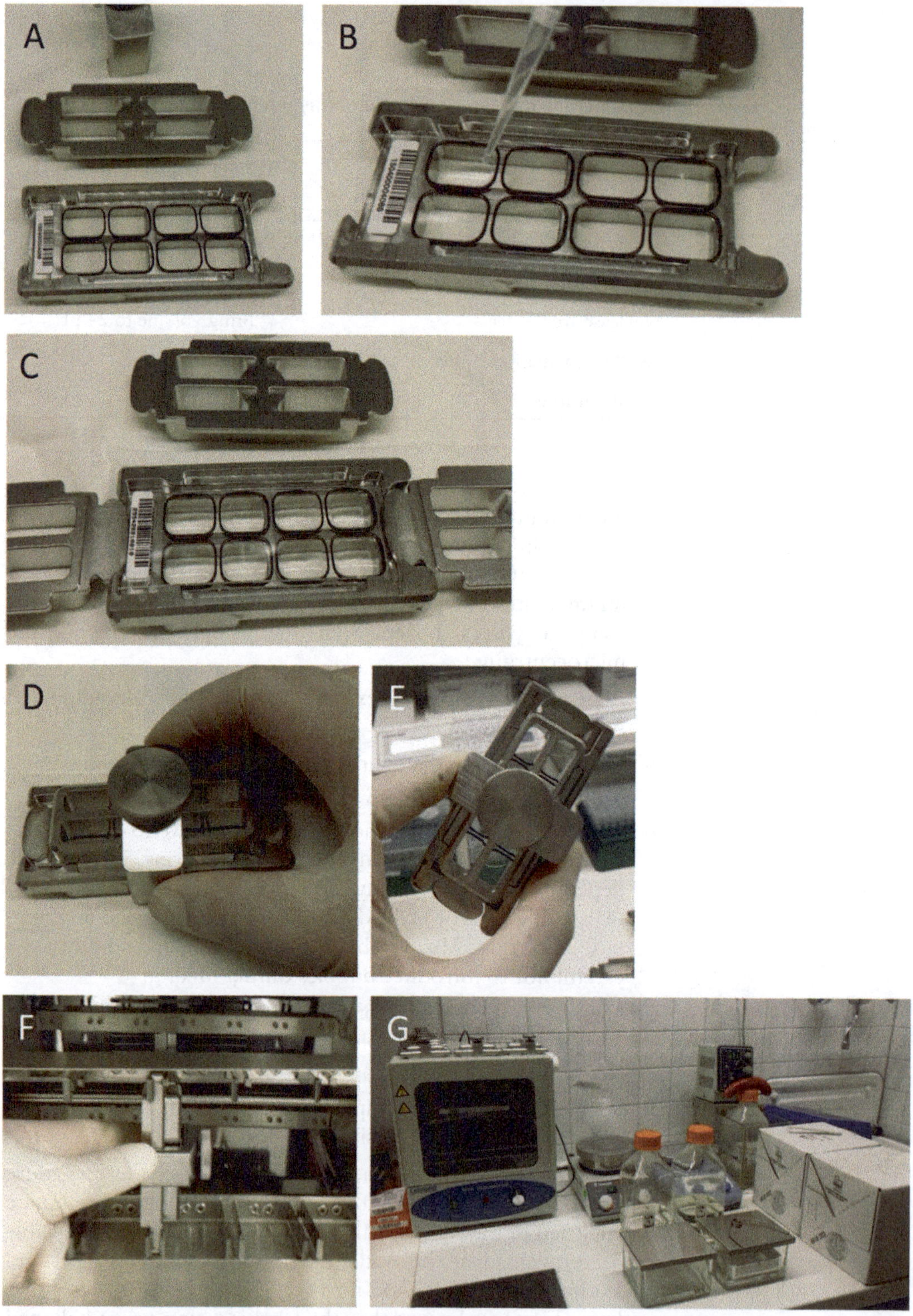

Fig. 1 Setting up overnight hybridization: (**a**) setup for the Agilent SureHyb chamber, (**b**) dispensing of samples in the middle of each gasket well, (**c**) simplified technique of placing the microarray slide on top of the gasket slide, (**d**) securing the SureHyb chamber assembly, (**e**) ensuring sample distribution and mobility along each hybridization chamber, (**f**) securing the SureHyb assembly in the hybridization oven, and (**g**) slide wash setup before leaving the lab for overnight hybridization

10. Pipette out hybridization mix and dispense slowly in the middle of each gasket well (Fig. 1b). The volume needed to hybridize will depend on the microarray format. Usually, 44 μL for the 8-pack format is required. Do not allow the pipette tip or the hybridization mix to touch the gasket walls to avoid leakage. Dispense until most of the mix is placed onto the slide but do not press the pipette until the second stop to avoid introducing bubbles. Add 44 μL of 1× Hybridization Buffer as blank for unused wells.
11. Place a microarray "active side" down onto the SureHyb gasket slide so that the "Agilent"-labeled barcode is facing down and the numeric barcode is facing up. Make sure that the microarray slide is parallel to the gasket slide and both are properly aligned. Be sure not to haphazardly drop the array slide onto the gasket to prevent the samples from mixing between gasket wells. To ensure that this crucial step is done seamlessly, two extra top covers of a SureHyb chamber can be used by placing bottom up on the left and on the right slots of the SureHyb chamber base in use. The microarray slide is placed on top, with the two extra top covers preventing the two slides from touching each other. Slowly but firmly pull out the two extra top covers on opposite direction parallel to the lab bench to place and properly align the microarray slide onto the gasket slide (Fig. 1c). Practice this step with used microarray slide to gain confidence before actually performing this critical step.
12. Immediately place the SureHyb chamber cover onto the sandwiched glass slides. Slide the clamp assembly. Hand-tighten the clamp onto the chamber by turning the thumbscrew clockwise to secure the assembly (Fig. 1d).
13. Rotate the hybridization chamber assembly to distribute the samples evenly and assess the mobility of the big bubble formed (Fig. 1e). Stationary bubbles can be mobilized by tapping the assembly onto the lab bench surface.
14. Place the hybridization chamber assembly in the rotisserie in a hybridization oven pre-equilibrated at 65 °C (Fig. 1f). When using the 2× GEx Hybridization Buffer, set the rotator speed at 10 rpm. Hybridize at 65 °C for 17 h. Ensure proper balance of weight across the rotisserie by adding empty hybridization chamber assembly whenever necessary.
15. Before leaving the lab for the overnight 17-h hybridization, prepare Gene Expression Wash Buffers 1 and 2 based on the manufacturer's protocols. Also prepare three wash chamber assemblies as detailed in **steps 16** and **17**. The addition of 0.005% Triton X-102 to both buffers is purely optional but highly recommended to reduce the incidence of microarray

wash artifacts. Open the cubitainer cardboard box and remove the outer and inner caps. Directly add 2 mL 10% Triton X-102 to the buffer and replace the original inner and outer caps. Thoroughly mix the buffer by inverting and shaking the container 5 to 6 times. Discard the outer and inner caps and install the spigot provided with the wash buffer. Do this for both Wash Buffers 1 and 2. Label each cubitainer as "Wash Buffer 1" and "Wash Buffer 2" accordingly and indicate the date when the Triton X-102 was added.

16. Prepare two wash chamber assemblies. Each assembly should be washed thoroughly with copious amounts of Milli-Q water at least five times and then dried inverted in paper towel before use. Dish 1 should be empty, without any microarray slide rack. This will be filled with Wash Buffer 1 which will be used to disassemble the sandwiched microarray slides (**step 22**). Dish 2 should have microarray slide rack and a small magnetic stirrer inside. It should be labeled with "Wash Buffer 1" as it will be used to wash the microarray slides with the first buffer. Also keep a rack holder nearby as it will be used to transfer the rack from Wash Buffer 1 to Wash Buffer 2. Keep these two wash chamber assemblies in laboratory bench.

17. Pre-warm a mini incubator to 37 °C and place a third wash chamber assembly. Dish 3 should be empty like the first wash chamber but should contain a small magnetic stirrer like Dish 2. Label this as "Wash Buffer 2" as it will be used to wash the slides using the second buffer.

18. Obtain 500 mL of Wash Buffer 1 and Wash Buffer 2 directly from the spigot and place individually in two sterile 1 L reagent bottles. Leave Wash Buffer 1 at room temperature but place Wash Buffer 2 in a 37 °C water bath overnight. Prepare an extra bottle of Wash Buffer 1 and leave at room temperature similar to the other Wash Buffer 1. Label it as "Wash Buffer 1 Reuse" as this will be used later as a reagent bottle to save the Wash Buffer 1 that was used in the first wash chamber (**steps 22–24**). The final setup from **steps 15–18** is illustrated in Fig. 1g. This washing setup should be accomplished before leaving the lab in preparation for the next day to ensure that no bubbles are introduced in the wash buffers.

19. The following day, prepare the wash buffers. The summary of wash conditions is listed in Table 9. Fill Dish 1 to three-fourths volume with Gene Expression Buffer 1. Similarly fill Dish 2 with Gene Expression Buffer 1 enough to submerge the metal slide rack. Dish 1 will be used to soak the microarray slide sandwich while disassembling (**step 22**). Wash Buffer 1 for hybridization chamber disassembly should be used for maximum 10–15 slides. Each microarray slide will then be trans-

Table 9
Summary of wash conditions for microarray slides

Steps	Dish	Wash buffer	Temperature	Time
Disassembly	1	GE wash buffer 1	Room temperature	As fast as possible
First wash	2	GE wash buffer 1	Room temperature	1 min
Second wash	3	GE wash buffer 2	37 °C	1 min

ferred to Dish 2 and placed on the microarray slide rack as detailed in the next step. Always use fresh Wash Buffer 1 and Wash Buffer 2 for each batch of washing, for up to 4–5 slides.

20. After the 17 h hybridization at 65 °C (**step 14**), get one hybridization chamber from the incubator. Optionally, the start time can be recorded but as experience is gained, one can just conduct the succeeding steps very quickly. Small bubbles formed during hybridization can also be noted and whether they are all rotating freely or not.
21. Disassemble the hybridization assembly on a flat surface (Fig. 2a). This is usually done on a lab bench with lint-free paper underneath. Loosen the thumbscrew by turning counter clockwise. Slide off the clamp and remove the chamber cover. Remove the array-gasket sandwich from the chamber base, ensuring that the two glass slides are still firmly adhering to each other.
22. Transfer the sandwich to Dish 1. Keep the microarray numeric barcode facing up as the slide sandwich is submerged onto Wash Buffer 1 in a slanted position, ensuring that the barcode label is not submerged in the buffer to avoid contaminating the buffer with ink from the barcode label. Handle the slides from their ends only and avoid touching especially the active side of the microarray slide.
23. While maintaining the sandwich in completely submerged position in Wash Buffer 1, separate the two glass slides from the barcode end by inserting one of the blunt ends of a plastic forceps between the two slides (Fig. 2b). Gently twist the forceps upward or downward to separate the slides. Allow the gasket slide to drop to the bottom of Dish 1 while ensuring that the microarray slide is held firmly for the next step.
24. Gently lift the microarray slide side sideways and immediately transfer the glass slide from Dish 1 to into the rack in Dish 2 containing Wash Buffer 1 (Fig. 2c). As one does this crucial step, minimize exposure of the slide to air. Avoid touching the active side by ensuring that only the barcode portion of the microarray slide and its edges are touched. This step can be

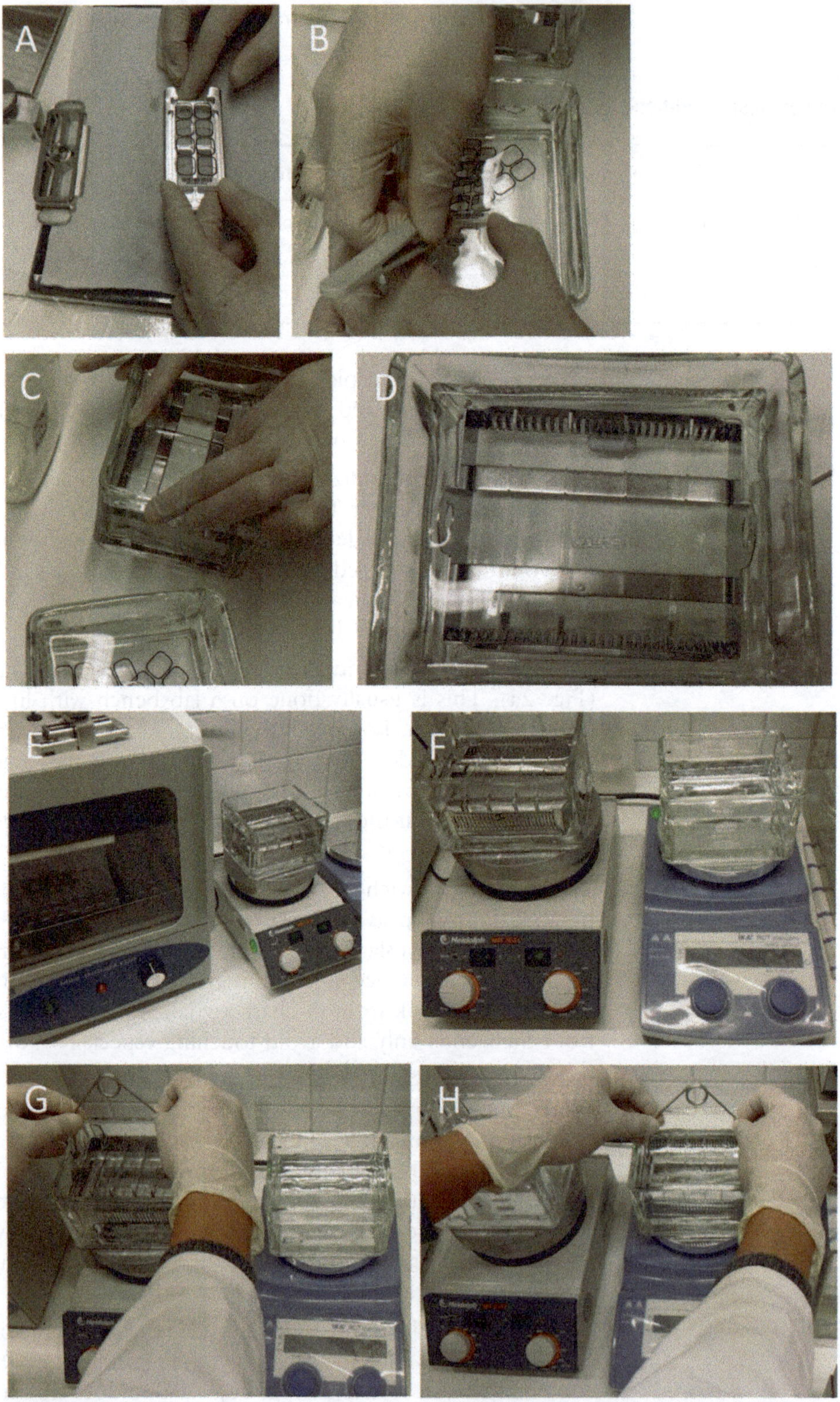

Fig. 2 Washing of microarray slides: (**a**) disassembling hybridization chamber on a flat surface, (**b**) dislodging of the microarray slide from the gasket slide while immersed in Wash Buffer 1, (**c**) transferring of the microarray slide into the metal rack containing Wash Buffer 1, (**d**) distributing the microarray slides evenly along the metal rack, (**e**) washing of the microarray slides in Wash Buffer 1, (**f**) side-by-side washing setup for Wash Buffer 1 and Wash Buffer 2, and (**g**) transferring of rack from Wash Buffer 1 to (**h**) Wash Buffer 2

practiced using the used gasket and microarray slides to gain confidence before actually performing the procedure.

25. Repeat **steps 21** through to **24** for up to seven additional slides to ensure uniformity of washing. Distribute the microarray slides evenly along the rack (Fig. 2d). Each staff can optimally do these steps unassisted for up to 4 slides.
26. Once all the slides are placed in the rack, attach the rack holder and transfer the entire setup to the first magnetic stirrer. Stir gently using setting 4 for exactly 1 min (Fig. 2e).
27. While stirring for 1 min, remove Dish 3 from the mini 37 °C incubator and place on top of the second magnetic stirrer. Also remove Wash Buffer 2 from the 37 °C water bath and gently pour the buffer in Dish 3 placed on the top of the second magnetic stirrer. Ensure that no bubbles are formed while pouring. The two magnetic stirrers should be positioned side by side, in close proximity with each other to ensure faster transfer of slide rack from Dish 2 to Dish 3 (Fig. 2f).
28. Once the 1 min stirring is finished, gently and very slowly lift the slide rack from Dish 2 and transfer it gently and slowly to Dish 3 (Fig. 2g, h). Remove the rack holder and stir using setting 4 for exactly 1 min.
29. Slowly remove the slide rack from Dish 3 ensuring that the formation of buffer droplets is minimized (Fig. 3a). This step should be done slowly and should take 5–10 s. Dry the sides of each slide by touching gently in lint-free paper (Fig. 3b). Place the slide in a used slide box and do the same for the remaining slides. Dry for approximately 15 min in the dark (**Stopping point:** short break).
30. Put each microarray slide in a slide holder with the Agilent barcode facing up (Fig. 3c). The ozone levels inside the laboratory should be 50 ppb (100 μg/m^3) or less. In our laboratory, an Agilent Ozone Barrier Slide Cover is routinely used to prevent ozone-induced degradation of cyanine dyes (Fig. 3d). Place the Ozone Barrier Slide Cover on the top of the array.
31. Once the microarray slide is securely in place inside a slide holder with ozone barrier, scan the slides immediately to minimize the possibility of reduction of signal intensities due to environmental oxidation. A SureScan or Agilent C microarray scanner is needed for SurePrint G3 format microarray slides. A feature extraction compatibility matrix for non-Agilent scanners is available online.
32. Place the assembled slide holders into the scanner carousel. Load the samples in sequence, starting from slot 1 onward (Fig. 3e).

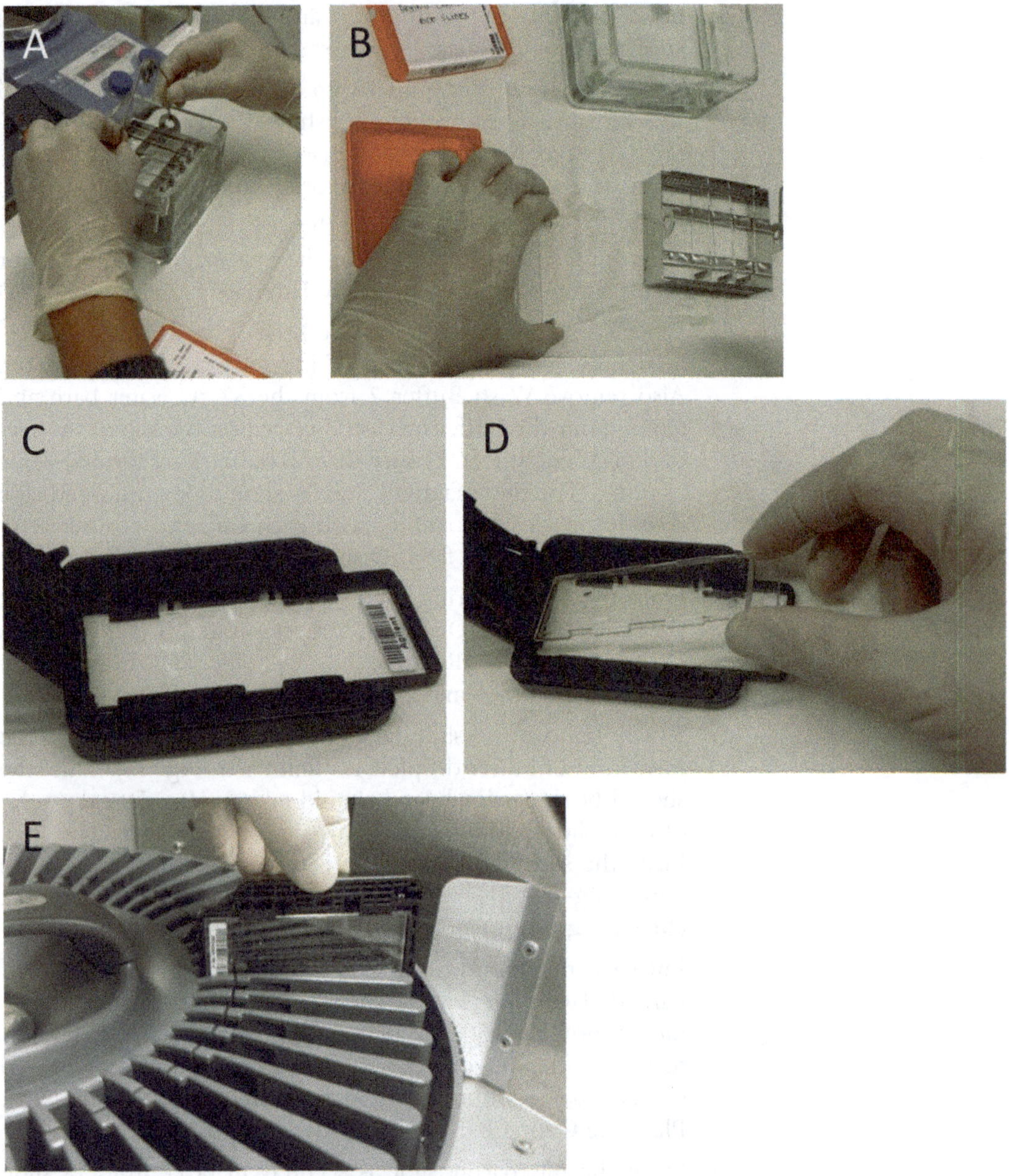

Fig. 3 Scanning microarray slides: (**a**) preparing microarray slide for microarray scanning, (**b**) drying of microarray slide, (**c**) placing microarray slide in slide holder, (**d**) placing of ozone barrier slide on top of the microarray slide, and (**e**) loading of the microarray slide assembly in the microarray scanner

33. Open the Scan Control software. In the main window, choose the slot number of the first slide for **Start Slot** and the slot number for the last slide as the **End Slot**.

34. Read the slides with the prescribed scan settings based on the manufacturer's specifications. The scan profile depends on the manufacturer's specifications. In our laboratory, we use the

Table 10
Recommended scan settings

Parameters	Settings for 8x60K G3 microarray format
Profile	AgilentG3_GX_1Color
Dye channel	Green
Scan region	Scan area (61 × 21.6 mm)
Scan resolution (μm)	3 μm double pass
Tiff size	20 bit

profile **AgilentG3_GX_1Color** which can be downloaded in the Agilent website. The scan settings are summarized in Table 10.

35. Scan the samples when the scanner status in the main window says **Scanner Ready**. Click **Start Scan** to commence the scanning process. This procedure usually takes 30 min per slide.
36. After scanning, disassemble the glass slides and store them in the orange slide boxes in a nitrogen purge box. Alternatively, they can be stored sealed in aluminum foil under vacuum and protected from light.

3.4 Microarray Reading, Feature Extraction, and QC Validation

Once all the samples are scanned, the process of feature extraction can commence. During this step, the information from the probe features is extracted from microarray scan data to facilitate the measurement of gene expression in the experiment. The software can read and processed up to 100 raw microarray image files in TIFF format. It can automatically find and place microarray grids, reject outlier pixels, accurately determine feature intensities and ratios, flag outlier pixels, and calculate statistical confidences. This software is available from the Agilent website.

1. Open the Agilent Feature Extraction (FE) Program from the PC by double-clicking on the icon. All Feature Extraction Software from version 10.7 and above can automatically download Grid Templates, protocols and QC metrics. This can be set up under **Advanced Options** and clicking **Use eArray server during extraction**.
2. Add the .tif images to be extracted to the FE Project. This is done either by clicking **Add New Extraction Set(s)** icon found on the toolbar or by right-clicking the **Project Explorer** and selecting **Add Extraction**. The default grid template and the protocol are usually automatically assigned by the FE program for each extraction set.

3. Set the FE Project properties. The **Operator** can be set in the **General** section. Verify that the default settings in the **Input** section are specified. MAGE and JPEG can be selected if the data will be imported into Rosetta Resolver. Project Properties for MAGE and JPEG can be set as **local file only**. The **Overwrite Previous Results** feature can be set to **False** to prevent accidental loss of files. The customization of other optional settings can be accomplished by consulting the user manual.
4. Check the **Extraction Set Configuration** tab. Verify that the program assigned the correct grid template from the **Grid Name** column. A different grid template can be assigned to an extraction set by selecting from the pull down menu. To update a grid template to the latest version, right-click **Grid Template Browser** and select online update. Grid templates will not be automatically available for each new microarray slide design but the information can be downloaded from the Agilent website. Go to **Grid Template Browser** and select **Add** and specify the grid template code. The design file provided in .xml format can also be browsed. Click **Open** to load grid template into the FE database. Every time a new grid template is added, specify the default protocol to ensure that the FE software will be able to automatically assign an FE protocol to an extraction set.
5. In addition, verify that the correct protocol is assigned to each extraction set under the **Protocol Name** column. A different protocol can be specified from the pull down menu. The correct protocol always begins with "GE1" for one-color analysis. FE Protocol Browser can also import a protocol if it is not available. This is done by browsing for an appropriate FE protocol in .xml format and loading the protocol into the FE database by clicking **Open**.
6. Save the FE Project in .fep format by clicking **File > Save As** and specifying the correct file directory position. Ensure that the icons for the image files in the FE Project Window has no more red "X" which indicates that an extraction protocol was not selected. Click **Project > Start Extracting** to commence the feature extraction procedure. This takes around 30 min per microarray slide to finish.
7. After feature extraction, QC report for each microarray slide extraction is available for viewing in the **Summary Report** tab. The proper placement of the grid can be inspected using Spot Finding at the four corners of the array. If a QC Metric Set has been assigned to the FE Project, the QC results can be viewed via (1) Project Run Summary, (2) QC Report, and (3) QC Chart Tool. An application note for specific for this step can be obtained from the Agilent website.

8. For downstream analysis of Agilent microarray data, use GeneSpring GX 9.0 or later. To compare data across a set of one-color microarray data, the 75th percentile scaling is employed by the GeneSpring software for signal intensity normalization. The detailed bioinformatic approaches involved in normalization of microarray data, application of dimension reduction techniques to detect outliers, implementation of clustering methods to identify clusters to define coexpressed genes, implementing statistical methods to identify differentially expressed genes (DEGs), and annotating functional pathway of DEGs were described earlier [22]. Hence, this will not be covered as this is beyond the scope of this protocol.

4 Notes

1. In our laboratory, we use Retsch mill MM400 with cryo kit because it can grind two samples at a time. We soak the 50 mL stainless steel grinding jar (Retsch P/N 01.462.0216) and 25 MM grinding ball (05.368.0105) under liquid nitrogen before placing the rice grain samples. Grinding is done for 1 min at speed of 25/s.
2. The experiment can be stopped at this point. This can be continued the following day.
3. The samples may freeze at this stage but one can proceed to centrifugation in the next step without thawing. The samples will thaw during the next step.
4. Some pellets might not dissolve completely within the 5 min allocated time. Just extend the vortex mixing until the pellet is dissolved.
5. Up to 750 μL clarified lysate can be transferred. Discard flow-through and repeat until all lysates are transferred.

Acknowledgments

This work has been supported under the CGIAR thematic area Global Rice Agri-Food System CRP, RICE, Stress-Tolerant Rice for Africa and South Asia (STRASA) Phase III funding.

References

1. Trapnell C, Roberts A, Goff L, Pertea G, Kim D, Kelley DR, Pimentel H, Salzberg SL, Rinn JL, Pachter L (2012) Differential gene and transcript expression analysis of RNA-seq experiments with TopHat and cufflinks. Nat Protoc 7(3):562–578
2. Trapnell C, Pachter L, Salzberg SL (2009) TopHat: discovering splice junctions with RNA-Seq. Bioinformatics 25(9):1105–1111
3. Law CW, Chen YS, Shi W, Smyth GK (2014) Voom: precision weights unlock linear model

analysis tools for RNA-seq read counts. Genome Biol 15(2):R29

4. Robertson G, Schein J, Chiu R, Corbett R, Field M, Jackman SD, Mungall K, Lee S, Okada HM, Qian JQ, Griffith M, Raymond A, Thiessen N, Cezard T, Butterfield YS, Newsome R, Chan SK, She R, Varhol R, Kamoh B, Prabhu AL, Tam A, Zhao YJ, Moore RA, Hirst M, Marra MA, Jones SJM, Hoodless PA, Birol I (2010) De novo assembly and analysis of RNA-seq data. Nat Methods 7(11):U909–U962
5. Trapnell C, Roberts A, Goff L, Pertea G, Kim D, Kelley DR, Pimentel H, Salzberg SL, Rinn JL, Pachter L (2014) Differential gene and transcript expression analysis of RNA-seq experiments with TopHat and Cufflinks (vol 7, p 562, 2012). Nat Protoc 9(10):2513
6. Grabherr MG, Haas BJ, Yassour M, Levin JZ, Thompson DA, Amit I, Adiconis X, Fan L, Raychowdhury R, Zeng QD, Chen ZH, Mauceli E, Hacohen N, Gnirke A, Rhind N, di Palma F, Birren BW, Nusbaum C, Lindblad-Toh K, Friedman N, Regev A (2011) Full-length transcriptome assembly from RNA-Seq data without a reference genome. Nat Biotechnol 29(7):644–U130
7. Bottomly D, Walter NAR, Hunter JE, Darakjian P, Kawane S, Buck KJ, Searles RP, Mooney M, McWeeney SK, Hitzemann R (2011) Evaluating gene expression in C57BL/6J and DBA/2J mouse striatum using RNA-Seq and microarrays. PLoS One 6(3):e17820
8. Fu X, Fu N, Guo S, Yan Z, Xu Y, Hu H, Menzel C, Chen W, Li YX, Zeng R, Khaitovich P (2009) Estimating accuracy of RNA-Seq and microarrays with proteomics. BMC Genomics 10:161
9. Marioni JC, Mason CE, Mane SM, Stephens M, Gilad Y (2008) RNA-seq: an assessment of technical reproducibility and comparison with gene expression arrays. Genome Res 18(9): 1509–1517
10. Sirbu A, Kerr G, Crane M, Ruskin HJ (2012) RNA-Seq vs dual- and single-channel microarray data: sensitivity analysis for differential expression and clustering. PLoS One 7(12):e50986
11. Xu X, Zhang YH, Williams J, Antoniou E, McCombie WR, Wu S, Zhu W, Davidson NO, Denoya P, Li E (2013) Parallel comparison of Illumina RNA-Seq and Affymetrix microarray platforms on transcriptomic profiles generated from 5-aza-deoxy-cytidine treated HT-29 colon cancer cells and simulated datasets. Bmc Bioinformatics 14:S1
12. Zhang W, Ferguson J, Ng SM, Hui K, Goh G, Lin AP, Esplugues E, Flavell RA, Abraham C, Zhao HY, Cho JH (2012) Effector CD4+T cell expression signatures and immune-mediated disease associated genes. PLoS One 7(6):e38510
13. Zhao SR, Fung-Leung WP, Bittner A, Ngo K, Liu XJ (2014) Comparison of RNA-Seq and microarray in Transcriptome profiling of activated T cells. PLoS One 9(1):e78644
14. Zhu T, Budworth P, Chen W, Provart N, Chang H-S, Guimil S, Su W, Estes B, Zou G, Wang X (2003) Transcriptional control of nutrient partitioning during rice grain filling. Plant Biotechnol J 1(1):59–70
15. Yamakawa H, Hakata M (2010) Atlas of rice grain filling-related metabolism under high temperature: joint analysis of metabolome and transcriptome demonstrated inhibition of starch accumulation and induction of amino acid accumulation. Plant Cell Physiol 51(5):795–809. https://doi.org/10.1093/pcp/pcq034
16. Liu X, Guo T, Wan X, Wang H, Zhu M, Li A, Su N, Shen Y, Mao B, Zhai H, Mao L, Wan J (2010) Transcriptome analysis of grain-filling caryopses reveals involvement of multiple regulatory pathways in chalky grain formation in rice. BMC Genomics 11:730. https://doi.org/10.1186/1471-2164-11-730
17. Deshmukh R, Singh A, Jain N, Anand S, Gacche R, Singh A, Gaikwad K, Sharma T, Mohapatra T, Singh N (2010) Identification of candidate genes for grain number in rice (Oryza sativa L.). Funct Integr Genomics 10(3):339–347
18. Li H, Deng HW (2010) Systems genetics, bioinformatics and eQTL mapping. Genetica 138(9–10):915–924
19. Jansen RC, Nap JP (2001) Genetical genomics: the added value from segregation. Trends Genet 17(7):388–391
20. Jansen RC (2003) Studying complex biological systems using multifactorial perturbation. Nat Rev Genet 4(2):145–151
21. Butardo VM Jr, Anacleto R, Parween S, Samson I, de Guzman K, Alhambra CM, Misra G, Sreenivasulu N (2017) Systems genetics identifies a novel regulatory domain of amylose synthesis. Plant Physiol 173(1):887–906. https://doi.org/10.1104/pp.16.01248
22. Sreenivasulu N, Sunkar R, Wobus U, Strickert M (2010) Array platforms and bioinformatics tools for the analysis of plant transcriptome in response to abiotic stress. Methods Mol Biol 639:71–93

Chapter 17

Quantification of DNA Methylation as Biomarker for Grain Quality

Christiane Seiler and Markus Kuhlmann

Abstract

DNA methylation is an important biomarker for gene activity. It contributes to gene silencing and is involved in regulating various seed developmental processes in plants. Many of these processes are involved in important traits associated with aspects of grain quality. A reliable, fast, and cheap method is the estimation of DNA methylation utilizing methylation sensitive restriction enzymes (MSRE) and quantitative real-time PCR (qPCR) for selected candidate regions. The presented method can be used to confirm an effect of RNAi constructs on their target genes or trans-activity. Analysis of promoter regions can contribute to estimation of gene activity and related traits.

Key words DNA methylation, MSRE, qPCR, Grain quality, Epigenetic

1 Introduction

High amounts of methylated 5-methylcytosine (5-mC) can be found in plant genomic DNA. The presence of a methyl-group at position 5 of cytosine is associated with inactivation of genes, formation of heterochromatin and displays a hallmark of epigenetic modification per se [1, 2]. In contrast to other eukaryotes, DNA cytosine methylation in plants appears in three sequence contexts: CpG, CpHpG, and CpHpH, where H stands for A, C, or T. DNA methylation in plants is species-, tissue-, organelle- and age-specific. It is controlled by phytohormones and changes on seed germination, flowering and under the influence of various pathogens (viral, bacterial, fungal). DNA methylation of respective regions influences plant growth and development, with particular involvement in regulation of gene expression and DNA replication.

Electronic supplementary material: The online version of this chapter (https://doi.org/10.1007/978-1-4939-8914-0_17) contains supplementary material, which is available to authorized users.

Nese Sreenivasulu (ed.), *Rice Grain Quality: Methods and Protocols*, Methods in Molecular Biology, vol. 1892, https://doi.org/10.1007/978-1-4939-8914-0_17, © Springer Science+Business Media, LLC, part of Springer Nature 2019

DNA methylation is controlled by three main pathways: de novo methylation, maintenance of methylation, and active removal of methylation marks on the genomic DNA [3]. Sequence specificity of modification is conferred by short interfering RNAs (siRNA) in a mechanism termed RNA-directed DNA methylation (RdDM) that was first observed in tobacco plants infected with the RNA pathogen potato spindle tuber viroid [4].

The RdDM mechanism is the main pathway contributing to de novo DNA methylation [5, 6]. This mechanism is plant-specific and involved in various regulatory processes such as basal heat tolerance [7]. DNA methylation acquired by RdDM can be associated with transcriptional gene silencing—TGS [8], genome stability by transposon inactivation [9], hybrid vigor [10], and DNA virus defence [11]. When the region subjected to RdDM contains promoter elements, the transcription of the corresponding gene can be supressed correlating with the amount of DNA methylation present [12]. The term grain quality can be separated in various aspects from different perspectives.

The quality depends on the intended use of the grain and is manifested in the biological features and chemical composition. Main aspects are high germination percentage (viability of seed) and chemical composition of the grain such as starch content, flavor, appearance, and healthy compounds. During fertilization and grain development massive transcriptional changes occur, which are controlled by epigenetic regulators and chromatin remodeling factors [13–15]. These events of remodeling are accompanied by histone and DNA modification and/or Polycomb-complexes as mechanisms affecting the gene expression.

As several physiological parameters are changed during the early phase of grain development one of the most important compounds with respect to grain quality is the starch content. During embryo development a decline of starch level was detected in *Arabidopsis thaliana* seeds [16]. A similar regulation was proposed for the monocot species *Brachypodium dystachion* [17] followed by a strong accumulation of starch in the germinating embryo. Within the same study it is concluded that epigenetic modifications may play an important role in the organs transmitting stimuli to the embryo during seed maturation, desiccation, and germination. Nevertheless up to now no target gene regulated via DNA methylation of its promoter was identified in a crop plant. The reason for this is most probably the lack of knowledge on the genome sequence for several species and methods estimating DNA methylation. Using MSAP (methylation-sensitive amplified polymorphism) markers an endosperm-specific decrease of CG and CHG methylation could be identified in Sorghum [18]. This indicates that processes involving DNA methylation and demethylation are contributing to gene regulation at this phase of seed development in the endosperm but also contributing to the proper germination of the embryo.

The described method can be used to validate a methylation pattern of a putative target gene and correlate the level of methylation with an evaluated quality trait. The presence of DNA methylation in a promoter region [19] is generally related to transcriptional suppression of a target gene.

Another valuable approach is the detection of DNA methylation as footprint of a functional RNAi-construct at the target gene level. The identification of DNA methylation within the target region is much faster and simpler than detection of small RNAs as hallmark of functional RNAi: After cleavage (dicing) of the transcribed hairpin RNA the produced small RNAs will result in DNA methylation as a consequence of the RdDM mechanism. This DNA methylation can be detected in all genomic regions homologous to the 24mers siRNAs derived from the RNAi-construct. If the selected RNAi target gene is related to grain quality, DNA methylation is an easy detectable and reliable marker. But also for several crop genes related to grain development and germination an epigenetic regulation was proposed [20, 21].

1.1 Experimental Approaches to Quantify DNA Cytosine Methylation

For the analysis of DNA methylation within a defined region various methods based on different principles are available:

*Me*thylated *D*NA *I*mmuno*P*recipitation (MeDIP) is based on the immunological detection of the 5-mC-Epitope by a specific antibody. The bound DNA can be precipitated and based on the assay quantified by PCR, displayed by slot blotting procedure or used for microarray hybridization and next-generation sequencing.

The second method: bisulfite sequencing is based on the chemical property of the methyl-group attached to the cytosine's to protect them against chemical modification with sodium bisulfite. Based on this technique sodium bisulfite is used to reduce unmodified cytosines to uracil. After PCR amplification all uracils are converted into thymines. The products can be sequenced or analyzed by other methods (e.g., *CO*mbined *B*isulfite *R*estriction *A*nalysis, COBRA [22]). With technical methods available this technique can be applied genome wide. The classical method of methylation analysis is based on the property of restriction enzymes to be unable to cleave methylated DNA. This property is based on the evolutionary use of restriction enzymes developed in bacteria and archea. In these prokaryotes restriction enzymes protect the organism from foreign DNA and virus particles by recognizing DNA sequences and cleaving them into nonfunctional pieces. Endogenous sequences are masked by a modification enzyme (a methyltransferase) that modifies the host DNA and blocks cleavage of the corresponding cleaving enzyme. Such two component system is referred to as restriction modification system [23].

The quantification of the proportion of cleaved to not cleaved DNA can be performed in several ways. Cleavage can be displayed directly by Southern blotting and hybridization of a probe related to the target region [12] or quantified by qPCR. The advantage of using methylation sensitive restriction enzymes in combination with quantitative real-time PCR is relatively simple handling and low analysis costs. This will allow testing a high number of individual plants. With this, the methylation level of a particular target region in a small population, cultivar or breeding line can be estimated and used as biomarker.

One drawback of the method is that only the selected restriction site can be monitored.

2 Materials

2.1 Extraction of Genomic DNA

1. Any DNA extraction kit.
2. Water bath or heating block for heating at 65 °C and 80 °C.
3. Vortexer.
4. Ethanol (96–100%).
5. Microcentrifuge with rotor for 2 ml tubes.

2.2 Methylation Sensitive Restriction Digestion

1. Water bath or heating block for heating at 37 °C and 85 °C.
2. Methylation sensitive restriction enzymes (e.g., *Hpa*II) with reaction buffer.
   (a) A list of methylation sensitive restriction enzymes and specificities of their methylation sensitivity are described at REBASE.NEB.COM [24] and can be found in Supplementary Table 1.

2.3 Real-Time PCR

1. Real-time PCR cycler.
2. 96- or 384-Well plates.
3. Optical seal.
4. SyBr-green-based PCR mix.
5. Specific forward and reverse primer for control and target region.

3 Method

DNA methylation assay based on cytosine methylation-sensitive restriction cleavage and quantitative PCR

An exemplary quantification of DNA cytosine methylation present at a particular sequence by PCR-based quantification of that respective DNA part that is resistant to cleavage by cytosine methylation-sensitive restriction enzymes will be shown as an example. The analyzed target region that will be analyzed is a genic, coding region, which is not methylated under any environmental condition in a wild-type barley plant (Fig. 1). After transformation of a hairpin-RNAi-construct the post transcriptional gene silencing of the corresponding target gene was monitored. As by-product of the PTGS related small RNAs also 24mer small RNAs are generated. This class of small RNAs is targeting the RdDM machinery to homologous sequences. As a consequence, the genomic DNA encoding the hairpin sequence as well as the homologous target gene region in the nucleus is subjected to DNA methylation. For the exemplary analysis two cytosine methylation sensitive enzymes, *Hpa*II (CCGG) and *Nhe*I (CGTACG), with single-recognition sites in this region (Fig. 1) are selected. It is generally suggested to include an enzyme that is insensitive to cytosine methylation to confirm completeness of the cleavage reaction if a suited restriction site is included in the region analyzed. The level of methylation will be displayed as amplification relative to the uncut control, respective control cleavage.

3.1 Cleavage of Genomic DNA

Genomic DNA preparations from individual plants should be isolated according to the manufacturer's protocols. Uniform sampling of selected tissue is recommended. Plants should be of similar age, developmental stage, and grown under similar controlled environmental conditions without pathogens (all these factors have influences on DNA methylation) unless specific effects should be tested. To get statistical meaningful data it is recommended to test 5 biological replicates.

1. For each DNA preparation:
   0.1 μg genomic DNA is diluted in 500 μl H_2O
   separate aliquots of 87 μl reaction tubes (~17.5 ng), e.g.,

Tube I	Tube II	Tube III
No enzyme	*Hpa*II	*Nhe*I

2. Add 10 μl buffer (10x).
3. Add 3 μl enzyme (30 U, 10 U/μl).
4. Incubate over night at 37 °C.
5. Inactivate restriction enzymes for 5 min at 85 °C (do also for "no enzyme" control).
6. Add 900 μl bidistilled water to achieve a final volume of 1000 μl.
   Standards that can be used for quantitative PCR calibration

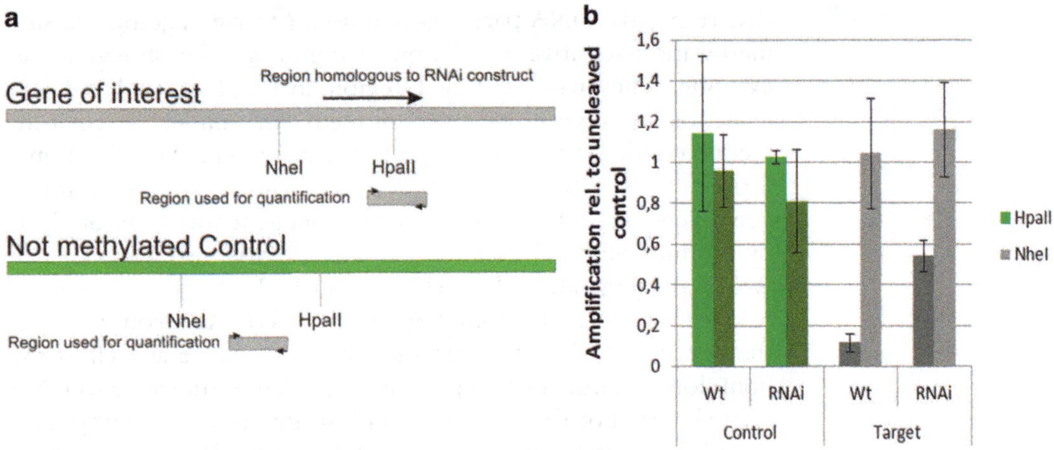

Fig. 1 Principle of methylation analysis of target (Gene of interest) and Control region by MSRE-qPCR. (**a**) Genomic DNA is cleaved by methylation sensitive restriction enzymes and selected regions are amplified by qPCR (gray: gene of interest/green: not methylated control). The region homologous to the RNAi construct is methylated by RNA directed DNA methylation. The arrows indicate the forward and reverse primer defining the amplicon used for quantitative PCR. (**b**) The relative level of amplification indicates the level of protection of the restriction sites by methylation. DNA was extracted from barley leaves. The level of amplification is given relative to the uncleaved control (1 equals 100%) as mean with standard deviation. Green bars show the negative cleavage of the control region (*Hpa*II: dark green, *Nhe*I: light green). Gray bars indicate the relative amplification after analytical cleavage of the target region (*Hpa*II: dark gray, *Nhe*I: light gray). $N = 5$ biological replicates. The level of estimated DNA methylation in the region homologous to the RNAi construct is ~50%

gDNA serial dilution:	0.01 μg/10 μl
	0.001 μg/10 μl
	0.0001 μg/10 μl
	0.00001 μg/10 μl
	Water-control

3.2 Setup of Quantitative "Real-Time" PCR

1. Use 10 μl (~0.000225 μg) of cleaved or control DNA preparation per 25 μl PCR setup:

   10 μl template.

   12.5 μl SyBr green Supermix.

   1.25 μl 100 pmol/μl forward primer.

   1.25 μl 100 pmol/μl reverse primer.

   25 μl total volume per reaction.
2. Prepare a master-mix for all samples (SyBr green Supermix, for. & rev. primers).
3. Set up samples for PCR in 3 technical replicates of a 96-well plate.
4. Transfer into real-time PCR device.

3.3 PCR Program

1. 5 min 95 °C.
2. 15 s 95 °C.
3. 30 s 62 °C.
4. 30 s 72 °C Detection.
5. Repeat **steps 2–4** 40×.
   Melting curve
6. 65 °C Detection.
7. +0.5 °C.
8. Repeat **step** 7 60×.

3.4 Evaluation of Data

Evaluation of real-time PCR data should always be performed according to the MIQE (*M*inimum *I*nformation for publication of *Q*uantitative real-time PCR *E*xperiments) guidelines as a minimal standard [25]. The calculation of the relative amount of amplified target is done according to [26] with the $2^{-\Delta\Delta CT}$ formula. After the PCR run, the melting curve of the PCR products provides an important control for the homogeneity of the amplification products (Fig. 2). A sharp, narrow peak is desired. Shoulders or signals beside the peak at higher temperatures are hints for unspecific amplification "background." As these unspecific products also contribute to the fluorescence signal on which quantification is based, PCR conditions should be carefully optimized for each primer combination. The values for the correlation coefficient of the PCR should be around 0.9 and the PCR efficiency around 100%. In routine operation, values between 80% and 110% are acceptable and might vary according to pipetting errors or sample quality. The CT (cycle threshold) is the value at which the fluorescence signal curve of a sample crosses the preset threshold of

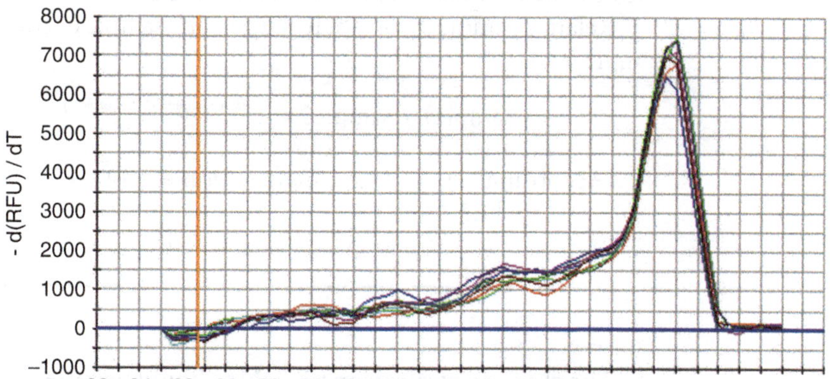

Fig. 2 Melting-cure of the PCR product obtained for a fragment of 460 bp length and a melting point at 88.5–89 °C. The melting curve analysis should result in one uniform peak. Additional peaks indicate additional products and the reaction should not be used for further evaluation. Peaks in the lower temperature range (60–70 °C) are usually results of primer dimer formation

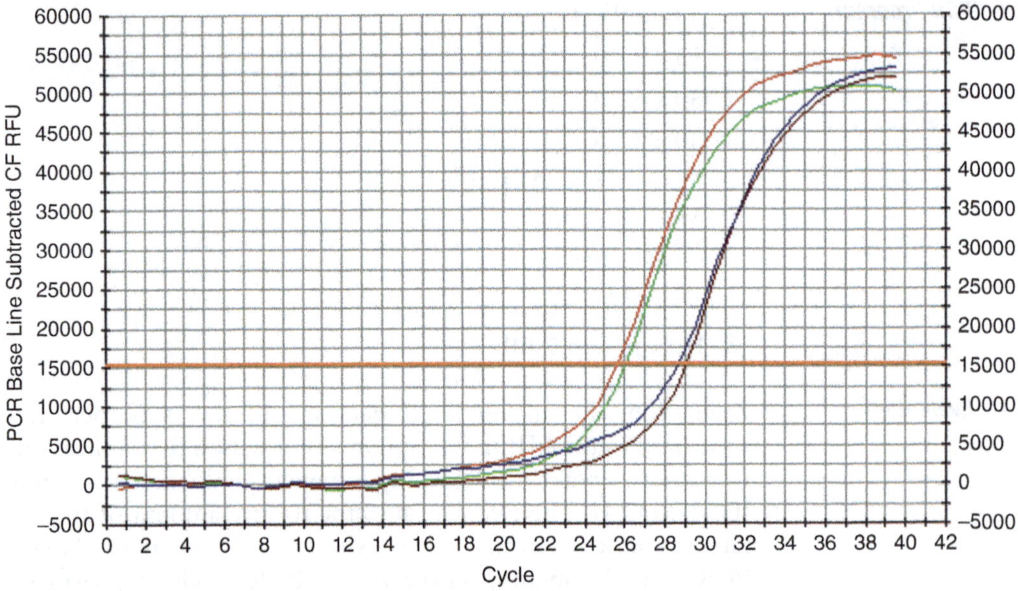

Fig. 3 "Real-time" fluorescence signals during PCR amplification. The graph indicates two technical replicates of cleaved (blue/brown) to uncleaved reactions (red/green). The line indicates the Cycle threshold level (CT) that should be positioned in the exponential amplification phase of the PCR reaction

amplification (orange line ~15,000) in a particular analysis (Fig. 3). The CT should be set in the range of the exponential amplification phase of the PCR for all samples.

The "CT value" of a PCR reaction is defined as the number of amplification cycles until the fluorescence signal passes the CT line. Measurements are usually done in triplicate and the mean value of CT values for each sample will be used for further calculation. The CT values generated for the different samples can be compared with the help of the "delta CT method" [26].

Relative amount of PCR target in sample1 compared to sample2:

$$\text{sample1} / \text{sample2} = 2^{-(\text{CT sample1} - \text{CT sample2})}$$

[Accordingly, percentage methylation = 100% × $2^{-(\text{CT sample cleaved} - \text{CT sample uncleaved})}$].

4 Notes

1. The cleavage of the genomic DNA should be tested. The entire genome will never be cleaved completely. The presence of residual uncut DNA can be tested by adding non-methylated PCR products or plasmid DNA as tester. Alternatively, a non-methylated region containing a restriction site of the selected enzyme can be used as reference.

2. It is recommended to analyze reference regions with described DNA methylation levels. Meanwhile, several regions of crop plants are analyzed by bisulfite sequencing and available in public databases [27].
3. To make general statements about the level of methylation it has to be considered that CG methylation is in general higher abundant than asymmetric DNA methylation. This is caused by the maintenance mechanism acting in the methylated region based on symmetric sequences. Using a restriction enzyme such as *Hpa*II (CCGG) methylation on one of the four cytosines present in the cleavage recognition motif will be detected.
4. Using enzymes with a longer recognition motif will increase specificity and completeness of the genome cleavage.
5. If DNA methylation levels of high repetitive sequence should be analyzed a higher dilution of genomic DNA might be required.

References

1. Cokus SJ et al (2008) Shotgun bisulphite sequencing of the Arabidopsis genome reveals DNA methylation patterning. Nature 452(7184):215–219
2. Lister R et al (2008) Highly integrated single-base resolution maps of the epigenome in Arabidopsis. Cell 133(3):523–536
3. Matzke M et al (2007) Targets of RNA-directed DNA methylation. Curr Opin Plant Biol 10(5):512–519
4. Wassenegger M et al (1994) RNA-directed de novo methylation of genomic sequences in plants. Cell 76(3):567–576
5. Chan SW, Henderson IR, Jacobsen SE (2005) Gardening the genome: DNA methylation in *Arabidopsis thaliana*. Nat Rev Genet 6(5): 351–360
6. Matzke MA, Birchler JA (2005) RNAi-mediated pathways in the nucleus. Nat Rev Genet 6(1):24–35
7. Popova OV et al (2013) The RdDM pathway is required for basal heat tolerance in Arabidopsis. Mol Plant 6(2):396–410
8. Castel SE, Martienssen RA (2013) RNA interference in the nucleus: roles for small RNAs in transcription, epigenetics and beyond. Nat Rev Genet 14(2):100–112
9. Bucher E, Reinders J, Mirouze M (2012) Epigenetic control of transposon transcription and mobility in Arabidopsis. Curr Opin Plant Biol 15(5):503–510
10. Shen H et al (2012) Genome-wide analysis of DNA methylation and gene expression changes in two Arabidopsis ecotypes and their reciprocal hybrids. Plant Cell 24(3):875–892
11. Wang MB et al (2012) RNA silencing and plant viral diseases. Mol Plant-Microbe Interact 25(10):1275–1285
12. Fischer U et al (2008) Local DNA features affect RNA-directed transcriptional gene silencing and DNA methylation. Plant J 53(1):1–10
13. Shi J, Dong A, Shen WH (2014) Epigenetic regulation of rice flowering and reproduction. Front Plant Sci 5:803
14. Zhang H, Ogas J (2009) An epigenetic perspective on developmental regulation of seed genes. Mol Plant 2(4):610–627
15. Zhang L et al (2011) ABA treatment of germinating maize seeds induces VP1 gene expression and selective promoter-associated histone acetylation. Physiol Plant 143(3):287–296
16. Andriotis VM et al (2010) Starch turnover in developing oilseed embryos. New Phytol 187(3):791–804
17. Wolny E, Braszewska-Zalewska A, Hasterok R (2014) Spatial distribution of epigenetic modifications in *Brachypodium distachyon* embryos during seed maturation and germination. PLoS One 9(7):e101246
18. Zhang M et al (2011) Tissue-specific differences in cytosine methylation and their associa-

tion with differential gene expression in sorghum. Plant Physiol 156(4):1955–1966

19. Kuhlmann M et al (2014) DNA methylation maintenance consolidates RNA-directed DNA methylation and transcriptional gene silencing over generations in *Arabidopsis thaliana*. Plant J 80(2):269–281
20. Bai F, Settles AM (2014) Imprinting in plants as a mechanism to generate seed phenotypic diversity. Front Plant Sci 5:780
21. Li J et al (2013) Integration of epigenetic and genetic controls of seed size by cytokinin in Arabidopsis. Proc Natl Acad Sci U S A 110(38):15479–15484
22. Xiong Z, Laird PW (1997) COBRA: a sensitive and quantitative DNA methylation assay. Nucleic Acids Res 25(12):2532–2534
23. Kobayashi I (2001) Behavior of restriction-modification systems as selfish mobile elements and their impact on genome evolution. Nucleic Acids Res 29(18):3742–3756
24. Roberts RJ et al (2015) REBASE--a database for DNA restriction and modification: enzymes, genes and genomes. Nucleic Acids Res 43(Database issue):D298–D299
25. Bustin SA et al (2009) The MIQE guidelines: minimum information for publication of quantitative real-time PCR experiments. Clin Chem 55(4):611–622
26. Livak KJ, Schmittgen TD (2001) Analysis of relative gene expression data using real-time quantitative PCR and the 2(−Delta Delta C(T)) method. Methods 25(4):402–408
27. Li X et al (2012) Single-base resolution maps of cultivated and wild rice methylomes and regulatory roles of DNA methylation in plant gene expression. BMC Genomics 13:300

Chapter 18

CRISPR-Cas9-Mediated Genome Editing of Rice Towards Better Grain Quality

Anindya Bandyopadhyay, Xiaojia Yin, Akshaya Biswal, Robert Coe, and William Paul Quick

Abstract

With continued economic development in Asia the demand for high yielding varieties with premium grain quality traits is set to increase. This presents a significant challenge to plant breeders because varieties must be tailored to meet regional preferences. It is already apparent that traditional breeding techniques cannot meet this challenge and so emerging genomics technologies will have to be utilized. Genome editing tools afford the ability to efficiently and precisely manipulate the genome. Among these, the bacterial clustered, regularly interspaced, short palindromic repeat (CRISPR) associated protein 9 (Cas9) or CRISPR-Cas9 has emerged as the easiest, most economic, and efficient technology to undertake genome editing in rice. This technique allows precise site-specific gene modification or integration. In this chapter we present a method for utilizing CRISPR-Cas9 for improving grain quality traits in rice; this should enable molecular breeders to quickly and efficiently produce high yielding rice varieties tailored to meet specific cultural and regional requirements for grain quality.

Key words Genome, Rice, Grain quality

1 Introduction

The market value of rice and the adoption of new varieties by farmers is strongly driven by grain quality traits [1, 2]. Underlying these are consumer preferences for particular physical, nutritional, cooking, and eating characteristics [3–5]. Desirable physical properties include uniform grain length, width, weight, color (whiteness and translucence), and chalk content. Cooking and eating (organoleptic) characteristics, such as cooking time (gelatinization temperature and viscosity [6]), the ability of rice to remain soft after cooking (gel consistency [7]), textural properties of the cooked rice (amylose content [8–10] and aroma [11]) are also very important. Aroma is considered to be the premium quality trait that brings the highest economic gain for the farmer [11–15].

Nese Sreenivasulu (ed.), *Rice Grain Quality: Methods and Protocols*, Methods in Molecular Biology, vol. 1892,
https://doi.org/10.1007/978-1-4939-8914-0_18,

The current benchmark varieties with premium grain quality traits were developed during the 1960–1980s [16]. These varieties, although low yielding and susceptible to stress, continue to be grown today because of their superior grain quality and high acceptance among consumers. The release of new varieties has been limited by a focus on increasing yield and the difficulty associated with combining this with superior grain quality [17]. This presents a challenge for breeders as continued economic growth in Asia [18, 19] will only increase the demand for premium quality rice [1, 20]. Meeting this demand is complicated by the fact that there are distinctive cultural differences in the traits that are preferred [21], meaning that breeders must tailor rice varieties to meet regional preferences. The genomics era brings many opportunities to help breeders to address this challenge.

While not complete, our understanding of the genetic basis underlying many grain quality traits is growing [3, 22]. Quantitative trait loci (QTL) controlling for chalkiness (reviewer in [22]), appearance (reviewed [3]), protein content [23], weight, and length [24–26] have been reported. Regions on chromosome 3, 6, and 8 have been identified as hotspots for QTLs controlling a variety of grain quality traits [22, 27, 28], indicating that there remain numerous novel genes to be identified. The genes associated with these QTLs are starting to be identified and characterized, such as BETADINE ALDEHYDE DEHYDROGENASE 2 (*BADH2*) controlling the flavor and aroma of Jasmine and Basmati rice (reviewed in [3, 29]), the *Waxy* locus associated with amylose content and grain texture of japonica and indica rice [30] and GRAIN INCOMPLETE FILLING 1 (*GIF1*) controlling grain size [31]. As a deeper understanding of the genotype underlying varieties with diverse grain quality traits becomes available, the potential to use biotechnology to engineer these traits into high-yielding varieties will become a reality [32–35]. The discovery and fine-mapping of QTLs can be used for genomic selection of varieties with desirable traits. Genes can be cloned to unravel the allelic variation in the polymorphisms or transformed into high-yielding lines to satisfy regional preferences in grain quality. However, as it is now possible to sequence the entire genomes of a large number of rice accessions [36], breeders will soon be able to identify the genes underlying grain quality traits and to mine for variation in alleles. This affords an unprecedented opportunity to understand how allelic variation relates to phenotypic variation. The challenges for molecular breeders then become how best to use this information. The major drawbacks with conventional transformation technologies are consumer acceptance and the fact they cannot eliminate the endogenous target gene, thus there is an emerging role for modern tools such as targeted genome editing technologies.

Genome editing allows precise, targeted changes to be made in the genome by inserting, replacing, or removing DNA to induce a

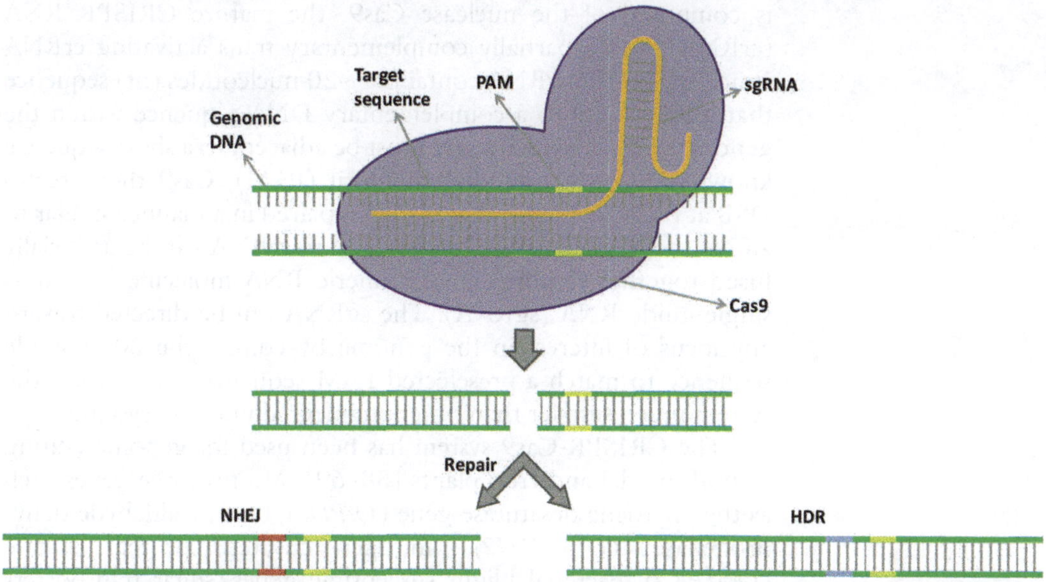

Fig. 1 Illustration of CRISPR-Cas9 and its mode of action. CRISPR-Cas9 is composed of Cas9 nuclease and a sgRNA. The 20 nt guide sequence binds to the target sequence upstream to the PAM (in yellow) and the Cas9 nuclease creates a DSB. The DSB is then repaired by either NHEJ that leads to indels (in red), or by HDR that could confer precise genome editing (in blue)

desired trait. Several artificially engineered nucleases have emerged in the last decade, including customized homing endonucleases (mega-nulceases) [37], zinc finger nucleases [ZFNs; [38–40], and transcription activator like effector nucleases [TALENs; [40–47]. These techniques generate double strand breaks (DSBs) at specific loci in the genome, which are then repaired by either nonhomologous end joining (NHEJ) or homology-directed repair (HDR; *see* Fig. 1). NHEJ is a DSB repair pathway that ligates two broken ends together leading to insertions or deletions (indels) at the site of the break. These often induce mutations that alter, or lead to frame-shifts that knock out gene function. Although uncommon, HDR is a template-dependent pathway for DSB repair which can be exploited to insert gene(s) of interest (GOIs) or single-nucleotide substitutions at a target locus. A site-specific nuclease is used to cleave the DNA, a homology-containing donor template is then provided that is faithfully inserted by HDR.

While these technologies offer an unprecedented ability to efficiently and precisely modify the genome they are technically challenging, time consuming, and costly. The emergence of the bacterial clustered, regularly interspaced, short palindromic repeat (CRISPR) associated protein 9 (Cas9) or CRISPR-Cas9 [48–57] from *Streptococcus pyogenes* offers a much easier, more economic, and efficient technology for genome editing. CRISPR-Cas9 is a microbial adaptive immune mechanism that utilizes RNA-guided nucleases to cleave foreign genetic elements [48–51]. The system is composed of the nuclease Cas9, the mature CRISPR RNA (crRNA), and a partially complementary trans activating crRNA (tracrRNA). The crRNA contains a ~20 nucleotides (nt) sequence that guides Cas9 to a complementary DNA sequence within the genome. This DNA sequence must be adjacent to a short sequence known as protospacer-adjacent motif (PAM). Cas9 then creates DSB at the target locus that is then repaired in a manner similar to ZFNs and TALENs. The crRNA and tracrRNA can be artificially fused together resulting in a chimeric RNA molecule known as single-guide RNA (sgRNA). The sgRNA can be directed toward any locus of interest in the genome by editing the 20-nt guide sequence to match a preselected PAM sequence. This makes the system much simpler than other engineered nuclease systems.

The CRISPR-Cas9 system has been used for genome editing of both model and crop plants [58–69]. Multiple rice genes such as the phytoene desaturase gene (*OsPDS*), betaine aldehyde dehydrogenase (*OsBADH2*), mitogen-activated protein kinase (*OsMPK2*), bacterial blight susceptible genes/sugar transporters (*OsSWEETs*), Myb family transcription factor (*OsMYB1*), rice outermost cell-specific gene 5 (*ROC5*), stromal processing peptidase (*SPP*), young seedling albino gene (*YSA*), stress-responsive ricemitogen-activated protein kinase (*OsMPK5*), chlorophyll A oxygenase-1 (*CAO1*), LA1/rice tiller angle controlling gene (*LAZY01*), and rice stomatal developmental gene Epidermal Patterning Factor Like-9 (*EPFL9*) have been modified [60, 62–65, 69–71]. There remains an opportunity to utilize this technology to enhance grain quality traits. An example of the potential is to modify grain texture by manipulating the amylose content of the endosperm [72]. The lower the amylose content the stickier the rice. Then enzyme granule bound starch synthase (*GBSS*), encoded by the *waxy* locus on chromosome 6, is required for amylose synthesis. Two wild-type alleles are found in rice, Wx^a is associated with high amylose content of indica rice, and Wx^b associated with lower amylose content of japonica rice [30]. In Wx^b a single-nucleotide polymorphism (SNP) change of G to T at +1 position of the intron 1 reduces the ability of the enzyme to excise the intron 1 from the *Waxy* pre-mRNA, reducing the abundance of the mature *Waxy* transcript, GBSS protein, and amylose content. The CRISPR-Cas9 system could offer breeders the potential to engineer high yielding varieties to meet local preferences for stickiness by artificially mimicking natural variation and inducing a SNP. The system could also be used to introduce non-native genes into the rice genome. Homology directed insertion allows these to be introduced into a safe loci, such as intergenic regions, where there is minimal effect on the adjacent genes. This should reduce the pleiotropic effects associated with random gene insertion associated with *Agrobacterium* or biolistic mediated transformation techniques that are currently widely used.

2 Materials

2.1 Cloning

1. PCR thermocycler (Q-CyclerII, Quanta Biotech, UK).
2. Water bath (SW22, Julabo, Germany).
3. Liquid nitrogen flask.
4. NanoDrop spectrophotometer (Thermo Fisher Scientific, USA).
5. Digital Gel Documentation Systems (Uvitec, Cambridge, UK).
6. Laminar flow hood (ELITE LH-48, Elite Scientific & Diagnostic International Supplies Corp, Philippines).
7. Horizontal gel electrophoresis systems (Thermo Fisher Scientific, USA).
8. Plating spreader (glass or plastic).
9. Microcentrifuge tubes (2 mL, 1.5 mL, and 0.6 mL, Axygen, Corning, USA).
10. PCR tubes (0.2 mL, Axygen, Corning, USA).
11. 96-Well PCR plates (Axygen, Corning, USA).
12. Pipette tips (1000, 200, and 10 μL, Axygen, Corning, USA).
13. 94 mm × 16 mm petri dishes (Biologix Plastics Co., Ltd., China).
14. Liquid nitrogen.
15. SOB Medium (Pronadisa, 1541).
16. Bacto agar (BD, 214010).
17. SOC Medium (Pronadisa, 2019).
18. Miller's LB Broth (Pronadisa, 1551).
19. Miller's LB Agar (Pronadisa, 1552).
20. YEB *Agrobacterium* Growth Medium (bioT, 30627031-1).
21. Bacto agar (BD, 214010).
22. Kanamycin sulfate (Sigma Aldrich, K1377).
23. Rifampicin (Sigma Aldrich, R3501).
24. Agarose, low EEO (ASIAGEL, 8010).
25. Goodview DNA dye (SBS Bio, China).
26. 1 kb Plus DNA ladder (Invitrogen, USA).
27. AarI (Thermo Fisher Scientific, ER1581 or ER1582).
28. HindIII-HF® (NEB, R3104S).
29. EcoRI-HF® (NEB, R3101S).
30. KpnI-HF® (NEB, R3142S).
31. SpeI-HF® (NEB, R3133S).

32. AarI (Thermo Fisher Scientific, USA).
33. T4 DNA ligase (NEB, M0202S).
34. DNA Polymerase I, Large (Klenow) Fragment (NEB, M0210S).
35. PCR reagent (U-taq Kit, SBS Bio, China).
36. dNTP set, 100 mM × 0.4 mL each (SBS Bio, China).
37. GenUP™ Plasmid Kit (Biotechrabbit, BR0700203).
38. GenUP™ Gel Extraction Kit (Biotechrabbit, BR0700403).
39. Plasmid: IRS154a (modified pCAMBIA1300).
40. Plasmid: pOsU3-sgRNA.
41. Plasmid: pJIT163-2NLSCas9.
42. *E. coli* DH5α chemically competent cells.
43. *Agrobacterium tumefaciens* LBA4404 competent cells.

2.2 Rice Transformation

1. Environmentally controlled growth chamber (Conviron, Winnipeg, Canada).
2. Laminar flow hood (NUAIRE, Plymouth, USA).
3. Stereo microscope (Olympus, Tokyo, Japan).
4. 28 °C and 37 °C incubator (LM-570RD, Yihder Co., Ltd).
5. Cell density meter (Ultrospec10, Biochrom Ltd., UK).
6. Micropipettors (EPPEDORF, Hamburg, Germany).
7. pH meter(BP3001, Trans Instrument, Singapore).
8. Rotator AG (FINE PCR, South Korea).
9. Water bath (JULABO TW20, Julabo GmbH, Germany).
10. Centrifuge (Thermo Fisher Scientific, USA).
11. Weighing balance (Ohaus, USA)
12. Steri 350 (Inotech, USA).
13. Graduated cylinder (glass, Corning, USA).
14. Beaker (glass, Corning, USA).
15. Forceps and scalpels (KFI K-14 stainless steel).
16. Nicrome Loop.
17. 50 mL Falcon tubes (Thermo Fisher Scientific, USA).
18. Tween20 (M147, AMRESCO).
19. Ethanol (E193, AMRESCO).
20. 94 mm × 16 mm petri dishes (Biologix Plastics Co., Ltd., China).
21. Filter paper (GE Healthcare Life Sciences, 3030–917).
22. 1.5 mL microfuge tubes (Axygen, Corning, USA).
23. 1 mL pipette tips (Axygen, Corning, USA).

24. 3 M Micropore tape (3 M, USA) (Tables 1 and 2).
25. Reagents for transformation are listed in Tables 1 and 2.

2.3 Plant Verification

1. PCR thermocycler (Q-CyclerII, Quanta Biotech, UK).
2. Water bath (SW22, Julabo, Germany).
3. NanoDrop spectrophotometer (Thermo Fisher Scientific, USA).
4. Digital Gel Documentation Systems (Uvitec, Cambridge, UK).
5. Horizontal gel electrophoresis systems (Thermo Fisher Scientific, USA).
6. Standard microcentrifuge tubes (2 mL, 1.5 mL, and 0.6 mL, Axygen, Corning, USA).
7. PCR tubes (0.2 mL, Axygen, Corning, USA).
8. Pipette tips (1000, 200, 10 μL).
9. Agarose, low EEO (ASIAGEL, 8010).
10. Goodview (SBS Bio, China).
11. 1 kb plus DNA ladder (Invitrogen, USA).
12. Surveyor® Mutation Detection Kit for Standard Gel Electrophoresis (IDT, 706020).
13. Phusion® High-Fidelity DNA Polymerase (NEB, M0530S).
14. Zero Blunt® pCR Cloning Kit (Invitrogen, K2700).
15. GenUP™ Plasmid Kit (Biotechrabbit, BR0700203).
16. GenUP™ Gel Extraction Kit (Biotechrabbit, BR0700403).
17. 100 mM dNTP set (Invitrogen, 10,297,018).
18. *E. coli* DH5α chemically competent cells.
19. SOB Medium (Pronadisa, 1541).
20. Bacto agar (BD, 214010).

3 Methods

The work flow for the CRISPR-Cas9 system is illustrated in Fig. 2.

1. Identify the target sequence in rice genome. Perform in silico evaluation of the target sequence to eliminate possible off target effects. Synthesize the guide sequence.
2. Clone the guide sequence into a binary vector harboring sgRNA scaffold, Cas9, and a hygromycin resistant gene that acts as a selectable marker.
3. Transform the sgRNA-Cas9 binary construct into immature rice embryos using *Agrobacterium*-mediated transformation.

Table 1
Stock solutions 1 (store at 4 °C)

Stock solution	Ingredients	Amount for 1 L solution	
AA SALT MACRO	$CaCl_2 \cdot 2H_2O$	1.47 g	(10 mM)
	$MgSO_4 \cdot 7H_2O$	2.46 g	(10 mM)
	$NaH_2PO_4 \cdot 2H_2O$	1.7 g	(10.87 mM)
	KCl	29.82 g	(400 mM)
AA SALTS MICRO	$CoCl_2 \cdot 6H_2O$	23.79 mg	(0.1 mM)
	$CuSO_4 \cdot 5H_2O$	24.97 mg	(0.1 mM)
	H_3BO_3	3 g	(48.5 mM)
	KI	750 mg	(4.5 mM)
	$MnSO_4 \cdot H_2O$	1 g	(5.9 mM)
	$Na_2MoO_4 \cdot 2H_2O$	250 mg	(1 mM)
	$ZnSO_4 \cdot 7H_2O$	2 g	(7 mM)
B5 MINOR-1	$FeSO_4 \cdot 7H_2O$	2.78 g	(10 mM)
	$Na_2EDTA \cdot 2H_2O$	3.72 g	(10 mM)
B5 MINOR-2	$MnSO_4 \cdot 4H_2O$	1 g	(4.5 mM)
	$ZnSO_4 \cdot 7H_2O$	200 mg	(0.7 mM)
	H_3BO_3	200 mg	(4.85 mM)
B5 MINOR-3	KI	75 mg	(0.45 mM)
B5 MINOR-4	$CuSO_4 \cdot 5H_2O$	2.5 mg	(0.01 mM)
	$Na_2MoO_4 \cdot 2H_2O$	25 mg	(0.1 mM)
	$CoCl_2 \cdot 6H_2O$	2.5 mg	(0.01 mM)
B5 VITAMINS	Thiamine HCl	20 mg	(0.06 mM)
	Pyridoxine HCl	2 mg	(0.01 mM)
	Nicotinic acid	2 mg	(16 μM)
	i-Inositol	200 mg	(1.1 mM)
GLYCINE	Glycine	7.5 g	(100 mM)
MS-1	KNO_3	95 g	(940 mM)
	NH_4NO_3	82.5 g	(1.03 M)
MS-2	$MgSO_4 \cdot 7H_2O$	37 g	(150 mM)
	$MnSO_4 \cdot 4H_2O$	2.23 g	(10 mM)
	$ZnSO_4 \cdot 7H_2O$	860 mg	(3 mM)

(continued)

Table 1
(continued)

Stock solution	Ingredients	Amount for 1 L solution	
	$CuSO_4 \cdot 5H_2O$	2.5 mg	(0.01 mM)
MS-3	$CaCl_2 \cdot 2H_2O$	44.1 g	(300 mM)
	KI	83 mg	(0.5 mM)
	$CoCl_2 \cdot 6H_2O$	2.5 mg	(0.01 mM)
MS-4	KH_2PO_4	17 g	(125 mM)
	H_3BO_3	620 mg	(10 mM)
	$Na_2MoO_4 \cdot 2H_2O$	25 mg	(0.1 mM)
MS VITAMINS	Nicotinic acid	1 mg	(8.1 μM)
	Pyridoxine HCl	1 mg	(5 μM)
	Thiamine HCl	200 μg	(0.6 μM)
	Glycine	4 mg	(53 μM)
	Myoinositol	200 mg	(1.07 mM)
N6 MAJOR-1	KNO_3	141.5 g	(1.4 M)
N6 MAJOR-2	$MgSO_4 \cdot 7H_2O$	18.5 g	(75 mM)
	$(NH_4)_2SO_4$	46.3 g	(350 mM)
N6 MAJOR-3	KH_2PO_4	40 g	(294 mM)
N6 MAJOR-4	$CaCl_2 \cdot 2H_2O$	16.6 g	(113 mM)
YCS MAJOR-1	NH_4NO_3	914 g	(11.4 M)
YCS MAJOR-2	$NaH_2PO_4 \cdot 2H_2O$	403 g	(2.58 M)
YCS MAJOR-3	K_2SO_4	714 g	(4.1 M)
YCS MAJOR-4	$CaCl_2$	886 g	(8 M)
YCS MAJOR-5	$MgSO_4 \cdot 7H_2O$	3240 g	(13.1 M)
YCS MINOR	$MnC_2 \cdot 4H_2O$	15 g	(75.8 mM)
	$(NH_4)_6Mo_7O_{24} \cdot 4H_2O$	740 mg	(0.6 mM)
	H_2BO_3	9.34 g	(153.5 mM)
	$ZnSO_4 \cdot 7H_2O$	350 mg	(1.22 mM)
	$CuSO_4 \cdot 5H_2O$	310 mg	(1.24 mM)
	$FeCl_3 \cdot 6H_2O$	77 g	(285 mM)
	Citric acid (monohydrate)	119 g	(566 mM)

Table 2
Stock solutions 2

Stock solution name	Amount for 50 mL solution	Preparation notes
2,4-D	50 mg	Dissolve in 2 mL of 1 M KOH, and dilute with Nanopure H_2O to a final volume of 50 mL, filter, sterilize and store at 4 °C
NAA	50 mg	Dissolve in 2 mL of 1 M NaOH, and dilute with Nanopure H_2O to a final volume of 50 mL, filter-sterilize and store at 4 °C
6BA	50 mg	Dissolve in 1 mL of 1 M NaOH, and dilute with Nanopure H_2O to final 50 mL, filter-sterilize and store at 4 °C
Kinetin	50 mg	Dissolve in 1 mL of 1 M NaOH, and dilute with Nanopure H_2O to final 50 mL, filter-sterilize and store at −20 °C
Cafotaxime	10 g	Dissolve in sterilized H_2O, filter-sterilize and store at −20 °C
Cabernicillin	5 g	Dissolve in autoclaved Nanopure H_2O, filter-sterilize and store at −20 °C
Hygromycin B	Supplied as solution	Supplied as sterilized 1 mg/mL solution, store at 4 °C

4. Analyze the regenerated plants to verify that the targeted genome sequence has been edited.

If homologous recombination HDR was used in the introduction-mediated gene insertion to be performed, an extra donor vector containing the gene of interest, flanked by a long homology arm, is required. Homology arms of the donor vector are homologues to the sequences flanking both sides of the target site.

3.1 Generation of the CRISPR-Cas9 Binary Vector

In order to establish a robust CRISPR-Cas9 rice genome editing system that is optimized for rice transformation, the CRISPR-Cas9 binary vector is inserted into the pCAMBIA background containing a hygromycin selectable marker (pCAMBIA1300, http://www.cambia.org/daisy/cambia/585.html). The pCAMBIA backbone is well tested and provides for efficient transformation of rice [75, 76]. Scaffold-sgRNA can be acquired from any of the available scaffold vectors. Similarly, the Cas9 cassette can be acquired

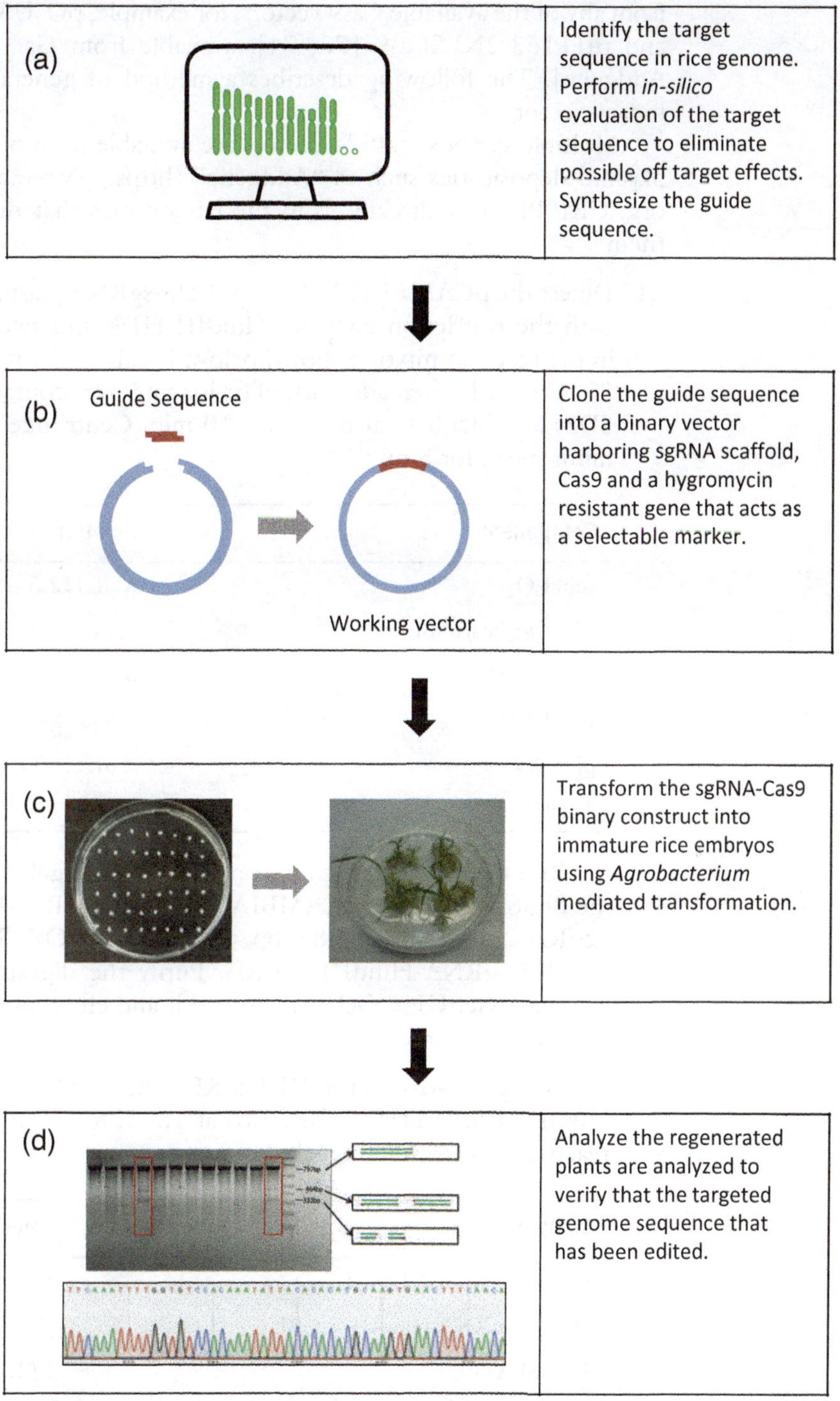

Fig. 2 Work flow for plant genome editing using CRISPR-Cas9

from any of the available Cas9 vectors, for example, pOsU3-sgRNA and pJIT163-2NLSCas9 [73, 74], available from Gao lab and Addgene]. The following describes a method of generating the binary vector:

Multiple sgRNA scaffold vectors are available from non-profit plasmid depositories such as "Addgene" (https://www.addgene.org/CRISPR/) or directly from the laboratories that developed them.

1. Digest the pCAMBIA1300 and pOsU3-sgRNA plasmid DNA with the restriction enzymes HindIII-HF® and EcoRI-HF® in the reaction mixture shown below. Incubate the mixture at 37 °C for 1 h, then add 3 μL of 6x loading dye (containing 1% SDS) and incubate at 65 °C for 10 min. Centrifuge at maximum speed for 5 min.

Component	Amount
snpH_2O	X μL (12.5 – X) μL
10× CutSmart® Buffer	1.5 μL
HindIII-HF®	0.5 μL
Eco RI-HF®	0.5 μL
pDNA	X μL (1000 ng)
Total	15 μL

Run the digested mixture on 0.8% agarose gel and excise the 8858 bp band of pCAMBIA 1300 (pCAMBIA-HindIII-EcoRI) and the 558 bp band of pOsU3-sgRNA (pOsU3-sgRNA-HindIII-EcoRI). Purify the digested product using GenUP™ Gel Extraction Kit and elute into a maximum total volume of 20 μL.

2. Ligate pCAMBIA-HindIII-EcoRI and pOsU3-sgRNA-HindIII-EcoRI (1:1 molar ratio) at 16 °C for 30 min in the reaction mixture shown below.

Component	Amount
IRS154a-HindIII-EcoRI	X μL
pOsU3-sgRNA-HindIII-EcoRI	(7.5 – X) μL
10 mM ATP	1 μL
10× T4 DNA ligase reaction buffer	1 μL
T4 DNA ligase	0.5 μL
Total	10 μL

3. Transform the ligated product into *E. coli* using the heat shock method (5 μL per transformation). Also transform 3 μL of linearized pCAMBIA-HindIII-EcoRI plasmid as a negative control.
4. Screen colonies using OsU3-F + sgRNA-R primers, the PCR product should be 455 bp. Select 2 positive colonies and culture in SOB medium containing 50 mg/L kanamycin overnight.
5. Isolate plasmid DNA (pCAMBIA_sgRNA) from the overnight culture and digest the plasmid DNA with HindIII-HF® and EcoRI-HF® at 37 °C for 45 min for verification. Successfully ligated pCAMBIA_sgRNA should be digested, producing one DNA fragment of 8858 bp and 558 bp.

Component	Amount
snpH_2O	(12.5 – X) μL
10× CutSmart® Buffer	1.5 μL
HindIII-HF®	0.5 μL
EcoRI-HF®	0.5 μL
pDNA	X μL (200 ng)
Total	15 μL

6. Linearize pCAMBIA_sgRNA by digesting with HindIII-HF® at 37 °C for 1 h.

Component	Amount
snpH_2O	(13.0 – X) μL
10× CutSmart® Buffer	1.5 μL
HindIII-HF®	0.5 μL
pDNA	X μL (1000 ng)
Total	15 μL

7. Digest pJIT163-2NLSCas9 with KpnI-HF® and SpeI-HF® at 37 °C for 1 h.

Component	Amount
snpH_2O	(12.5 – X) μL
10× CutSmart® Buffer	1.5 μL
KpnI-HF®	0.5 μL
SpeI-HF®	0.5 μL
pDNA	X μL (1000 ng)
Total	15 μL

8. Run the digested mixture on 0.8% agarose gel and excise the 9418 bp band of pCAMBIA_sgRNA (pCAMBIA_sgRNA-HindIII) and the 5720 bp band of pJIT163-2NLSCas9 (pJIT163-2NLSCas9-KpnI-SpeI). Purify the two digested products using GenUP™ Gel Extraction Kit and elute each into a maximum total volume of 20 μL.
9. Blunt both fragments in the reaction mixture shown below. Incubate the mixture at 25 °C for 20 min, then add 1 μL of 150 mM EDTA and incubate at 70 °C for 20 min to inactivate the enzyme.

Component	Amount
DNA fragment	12 μL
10× T4 DNA ligase reaction buffer	1.4 μL
10 mM dNTP mix	0.1 μL
DNA Polymerase I, Large (Klenow) Fragment	0.5 μL
Total	14 μL

10. Ligate the two blunted fragments (1:1 molar ratio) in the reaction mixture below at 16 °C for 3 h.

Component	Amount
snpH_2O	1.5 μL
Blunt-pC_sgRNA-HindIII	X μL
Blunt-pJIT163-2NLSCas9-KpnI-SpeI	(10 − X) μL
10 mM ATP	1.5 μL
10× T4 DNA ligase reaction buffer	0.5 μL
T4 DNA ligase	1.5 μL
Total	15 μL

11. Transform 5 μL of the ligated product into *E. coli* using the heat shock method.
12. Screen colonies using Cas9-F + Cas9-R primers, the PCR product should be 390 bp. Select 2 positive colonies and culture in SOB medium containing 50 mg/L kanamycin overnight.
13. Isolate the plasmid DNA. Successful cloning should be verified by sequencing. The Cas9 cassette could have been inserted in either orientation as it was cloned with a blunt end.

14. Store the verified CRISPR-Cas9 backbone vector (pCRISPR_UniDir) as the *E. coli* glycerol stock (500 μL of *E. coli* culture and 250 μL of 50% glycerol, deep in liquid nitrogen before keeping at −80 °C).

3.2 Cloning the Working Vector with Desired sgRNA

Cas9 is directed to the targeted loci in plant cells by the help of the 20 nt guide sequence of the sgRNA. Target sequence is identified according to experimental purpose. In silico evaluation needs to be performed to analyze the on-target score, as well as the possible off-target effects.

1. The 20 nt guide sequence should be the same sequence as the 20 nt at the 5′ upstream of the NGG PAM.
2. Digest the plasmid DNA (pCRISPR_UniDir) with restriction enzyme AarI in the following reaction mixture, incubated at 37 °C for 5 h.

Component	Amount
snpH_2O	(17.1 − X) μL
10× Buffer AarI	2 μL
pCRISPR_UniDir	X μL (1000 ng)
50× oligonucleotide	0.4 μL
AarI	0.5 μL
Total	20 μL

3. Run the digested mixture on a 0.7% agarose gel and excise the 15,089 bp band. Purify the digested product using GenUP™ Gel Extraction Kit. Elute purified product (pCRISPR_UniDir-AarI) in a maximum total volume of 20 μL.
4. The guide sequence can be synthesized as single-strand oligos. Forward oligo (FP) sequence is GGCA$(N)_{20}$, of which the $(N)_{20}$ sequence is the 20 nt identified target sequence at the 5′ upstream of the PAM. Reverse oligo (RP) sequence is AAAC$(N)_{20}$, of which the $(N)_{20}$ sequence is complementary to the 20 nt identified target sequence. The two oligos are diluted to 10 μM. Mix 10 μL of each oligos, and heat to 95 °C for 3 min on a heating block and then turn it off to cool down to room temperature naturally. The annealed oligos have sticky ends that are compatible to the AarI digested pCRISPR_UniDir, as shown in Fig. 3.
5. Ligate the linearized (AarI digested) backbone plasmid (pCRISPR_UniDir-AarI) and the annealed oligos in the reaction mixture shown below, for 30 min at 16 °C.

```
FP    5′-GGCANNNNNNNNNNNNNNNNNNNN-3′
                ||||||||||||||||||||
RP          3′-NNNNNNNNNNNNNNNNNNNNCAAA-3′
```

Fig. 3 Annealed oligos that have compatible sticky ends to AarI digested pCRISPR_UniDir

Component	Amount
pCRISPR_UniDir-AarI	3 μL
annealed oligos (FP + RP)	4.5 μL
10 mM ATP	1 μL
10× T4 DNA ligase reaction buffer	1 μL
T4 DNA ligase	0.5 μL
Total	10 μL

6. Transform 5 μL of the ligated product into *E. coli* using the heat shock method. Also transform 3 μL of linearized pCRISPR_UniDir-AarI plasmid as a negative control.
7. To identify the successful insertion of the guide sequence, use OsU3-F and RP primers for colony PCR. The band size should be 315 bp. Select 2 positive colonies and culture in SOB medium containing 50 mg/L kanamycin overnight.
8. Isolate plasmid DNA from the overnight culture and validate the vector by sequencing using the OsU3-F primer.
9. Transform the sequence-verified plasmid into *Agraobactrium* using the freeze thaw method. Two days after transformation, perform colony PCR with the OsU3-F and RP primer.
10. Inoculate 1 positive colony and extract plasmid DNA from the *Agrobacterium* and then transform it back into *E. coli*. Verify the back-transformed plasmid by PCR (OsU3-F and RP primer) and sequencing.
11. Store the verified working CRISPR-Cas9-sgRNA vector as the *E. coli* and *Agrobacterium* glycerol stock (500 μL of bacterium culture and 250 μL of 50% glycerol, deep in liquid nitrogen before keeping at −80 °C) respectively.

3.3 Preparation of Medium for Transformation of Rice

1. *Infection medium*: For 1 L total volume, add 100 mL of AA salt Macro, 1 mL of AA SALTS MICRO, 10 mL of B5 MINOR-1, 1 mL of B5 VITAMIN, 1 mL of 100 mM glycine, 876 mg of L-glutamine, 266 mg of aspartic acid, 174 mg of Arginine, 500 mg of Casamino Acid, 20 g of Sucrose, and 10 g of D-glucose. Adjust pH to 5.2 and filter-sterilize the solution. Store the solution at 4 °C for a maximum of 2 weeks.

For transformation of 300 rice immature embryos (IEs), 25 mL Infection medium should be prepared with 25 μL of 100 mM of fresh Acetosyringone (490.5 μg dissolved in 25 μL of DMSO).

2. *Co-cultivation medium*: For 1 L total volume, add 20 mL of N6 MAJOR-1, 10 mL of N6 MAJOR-2, 10 mL of N6 MAJOR-3, 10 mL of N6 MAJOR-4, 10 mL of B5 MINOR-1, 10 mL of B5 MINOR-2, 10 mL of B5 MINOR-3, 10 mL of B5 MINOR-4, 5 mL of B5 VITAMINS, 500 mg of casamino acid, 500 mg of L-proline, 20 g of sucrose, and 10 g of D-glucose. Adjust to a pH of 5.2 and add 5.5 g of ttype I agarose. Autoclave the mixture for 15 min, cool to 50 °C, and add 2 mL of 2,4-D, 1 mL of NAA, 1 mL of 6BA and 1 mL of fresh 100 mM acetosyringone prior to dispensing in to the petri dishes (25 mL per 94 mm × 16 mm petri dish). Seal petri dishes with micropore tape and store at 4 °C for a maximum of 1 week.
3. *Resting medium*: For 1 L total volume, add 20 mL of N6 MAJOR-1, 10 mL of N6 MAJOR-2, 10 mL of N6 MAJOR-3, 10 mL of N6 MAJOR-4, 10 mL of B5 MINOR-1, 10 mL of B5 MINOR-2, 10 mL of B5 MINOR-3, 10 mL of B5 MINOR-4, 5 mL of B5 VITAMINS, 500 mg of casamino acid, 500 mg of L-proline, 300 mg of L-glutamine, 36 g of mannitol, 20 g of maltose. Adjust pH to 5.8 then add 5 g of gelrite. Autoclave the mixture for 15 min, cool to 50 °C and add 1 mL of 2,4-D, 1 mL of NAA, 200 μL of 6BA, 1 mL of cafotaxime, and 1 mL of cabernicillin right prior to dispensing in to the petri dishes (25 mL per 94 mm × 16 mm petri dish). Seal petri dishes with micropore tape and store at 4 °C for a maximum of 2 weeks.
4. *Selection medium*: For 1 L total volume, add 20 mL of N6 MAJOR-1, 10 mL of N6 MAJOR-2, 10 mL of N6 MAJOR-3, 10 mL of N6 MAJOR-4, 10 mL of B5 MINOR-1, 10 mL of B5 MINOR-2, 10 mL of B5 MINOR-3, 10 mL of B5 MINOR-4, 5 mL of B5 VITAMINS, 500 mg of casamino acid, 500 mg of L-proline, 300 mg of L-glutamine, 36 g of mannitol, and 20 g of maltose. Adjust pH to 5.8 then add 5 g of gelrite. Autoclave the mixture for 15 min, cool to 50 °C and add 1 mL of 2,4-D, 1 mL of NAA, 200 μL of 6BA, 1 mL of cafotaxime, 1 mL of cabernicillin, and 0.6 mL of hygromycin prior to dispensing into the petri dishes (25 mL per 94 mm × 16 mm petri dish). Seal petri dishes with micropore tape and store at 4 °C for a maximum of 2 weeks.
5. *Pre-regeneration medium*: For 1 L total volume, add 20 mL of MS-1, 10 mL of MS-2, 10 mL of MS-3, 10 mL of MS-4, 5 mL of MS VITAMINS, 10 mL of B5 MINOR-1, 30 g of maltose, and 20 g of sorbitol. Adjust pH to 5.8 and then add 10 g of ttype I agarose. Autoclave the mixture for 15 min, cool

to 50 °C, add 2 mL of kinetin, 500 μL of NAA, 1 mL of cafotaxime, and 1 mL of hygromycin prior to dispensing into the petri dishes (25 mL per 94 mm × 16 mm petri dish). Seal petri dishes with micropore tape and store at 4 °C for a maximum of 2 weeks.

6. *Regeneration medium*: For 1 L medium, add 20 mL of MS-1, 10 mL of MS-2, 10 mL of MS-3, 10 mL of MS-4, 5 mL of MS VITAMINS, 10 mL of B5 MINOR-1, and 30 g of sucrose. Adjust pH to 5.8 and then add 3 g of Gelrite. Autoclave the mixture for 15 min, cool to 50 °C, and add 2 mL of kinetin, 1 mL of NAA, 1 mL of cafotaxime and 1 mL of hygromycin prior to dispensing into the petri dishes (25 mL per 94 mm × 16 mm petri dish). Seal petri dishes with micropore tape and store at 4 °C for a maximum of 2 weeks.
7. *YCS solution*: For 10 L solution, add 12.5 mL of YCS MAJOR-1, 12.5 mL of YCS MAJOR-2, 12.5 mL of YCS MAJOR-3, 12.5 mL of YCS MAJOR-4, 12.5 mL of YCS MAJOR-5, and 12.5 mL of YCS MONOR. Adjust pH to 5.0. Store at room temperature.

3.4 Rice Transformation

Transform the CRISPR-Cas9-sgRNA vector into immature embryos (IEs) of *Oryza sativa* IR64 using *Agrobacterium tumefaciens*-mediated transformation [75, 76].

1. *Agrobacterium preparation*: Streak *Agrobacterium* glycerol stock onto a YEB plate containing 50 mg/L kanamycin and incubate at 28 °C for 2 days. One hour before infecting the IEs, take the *Agrobacterium* from the YEB plate using a loop and mix well with 5 mL of infection medium. Adjust the *Agrobacterium* suspension to an OD of 0.3 using a cell density meter. Incubate the *Agrobacterium* suspension at 25 °C for 1 h in the dark without shaking.
2. *Preparation of immature embryos*: Collect immature seeds (milk to soft dough stage) from panicles harvested 12 days after anthesis. Soak de-hulled immature seeds in 70% ethanol for 1 min and then rinse with sterilized distilled water. Sterilize immature seeds with 1% sodium hypochlorite solution (containing one drop of Tween20) in a 50 mL falcon tube for 10 min. Rinse the seeds with sterilized distilled water until all the sodium hypochlorite has been removed. Isolate IEs and place 50 on a co-cultivation medium plate and air dry for 3 h in laminar flow hood to prevent the over spreading of bacterium.
3. *Infection*: Rearrange IEs on a co-cultivation medium plate with the scutellum facing up. Drop 5 μL of *Agrobacterium* suspension on top of each IE, seal the plates with micropore tape, and incubate at 25 °C for 7 days in the dark.

4. *Blotting*: After the co-cultivation, remove elongated shoots from IEs using a sterilized scalpel. Blot the IEs gently on sterile filter paper to prevent the overgrowth of bacterium. Transfer the blotted IEs to resting medium with the scutellum facing up. Seal the plates with micropore tape and incubate at 30 °C for 5 days under continuous illumination.
5. *Selection 1*: Cut each IE into 4 pieces and place them on selection medium containing 30 mg/L hygromycin with scutellum facing up (10 IEs/plate, i.e., 40 pieces/plate). Four pieces from a single IE should be grouped together on a single plate. Seal the plates with micropore tape and incubate at 30 °C for 10 days under continuous illumination.
6. *Selection 2*: Transfer all the cut IEs onto fresh selection medium containing 30 mg/L hygromycin. Seal the plates with micropore tape and incubate at 30 °C for 10 days under continuous illumination. .
7. *Selection 3*: Select hygromycin-resistant calli that are well proliferated and yellowish to white in color and transfer onto fresh selection medium containing 30 mg/L hygromycin. Label each calli corresponding to the cut IE number. Seal the plates with micropore tape and incubate at 30 °C for 10 days under continuous illumination.
8. *Regeneration*: Transfer (8 calli/plate) hygromycin-resistant calli onto pre-regeneration medium containing 50 mg/L hygromycin and incubate at 30 °C for 10 days. Transfer the proliferating calli with visible green tissue onto regeneration medium containing 50 mg/L hygromycin. Allow the calli to grow on regeneration medium for 10 to 15 days or until roots are 2 mm long.
9. *Hydroponics*: Regenerated plantlets should be transferred to Yoshida Culture Solution (YCS) until large enough to plant into soil. The YCS solution should be changed weekly.

3.5 Plant Verification Using Surveyor Assay

1. Extract and normalize the concentration of genomic DNA, from both transformed plants and tissue culture controls.
2. PCR should be performed using Phusion® High-Fidelity DNA Polymerase with primers that flank the target site. One primer should be designed 200–300 bp away from the target site, and the second primer 400–500 bp away from the opposite side. This is to ensure that the digested bands can easily be distinguished from each other (Fig. 4). The primers should have a high melting temperature (Tm) to avoid the formation of primer dimmers [77].
3. Take 6 μL of PCR product from each sample and mix with 6 μL of WT PCR product to produce heteroduplex DNA using the following protocol:

Temperature	Time	Temperature ramp
95 °C	10 min	
95 °C to 85 °C		(−2.0 °C/s)
85 °C	1 min	
85 °C to 75 °C		(−0.3 °C/s)
75 °C	1 min	
75 °C to 65 °C		(−0.3 °C/s)
65 °C	1 min	
65 °C to 55 °C		(−0.3 °C/s)
55 °C	1 min	
55 °C to 45 °C		(−0.3 °C/s)
45 °C	1 min	
45 °C to 35 °C		(−0.3 °C/s)
35 °C	1 min	
35 °C to 25 °C		(−0.3 °C/s)
25 °C	1 min	
4 °C	Hold ∞	

4. Prepare Surveyor Nuclease reaction mixture as shown below. Incubate the mixture at 42 °C for 60 min.

Component	Amount
Hybridized heteroduplex DNA	12 μL
Surveyor Enhancer S	1 μL
Surveyor Nuclease S	2 μL
Total	15 μL

5. Add 1.5 μL of Stop Solution and run the digested product on a 2% agarose gel. Herteroduplex DNA with mismatched base pairs will be digested and should render a banding pattern similar to that shown in Fig. 4, this indicates a Cas9-induced mutation.
6. Samples that showed digested bands should be amplified by Phusion® High-Fidelity DNA and PCR products should be sequenced directly. The sequencing results can be analyzed using TIDE [78] and Poly Peak Parser [79].

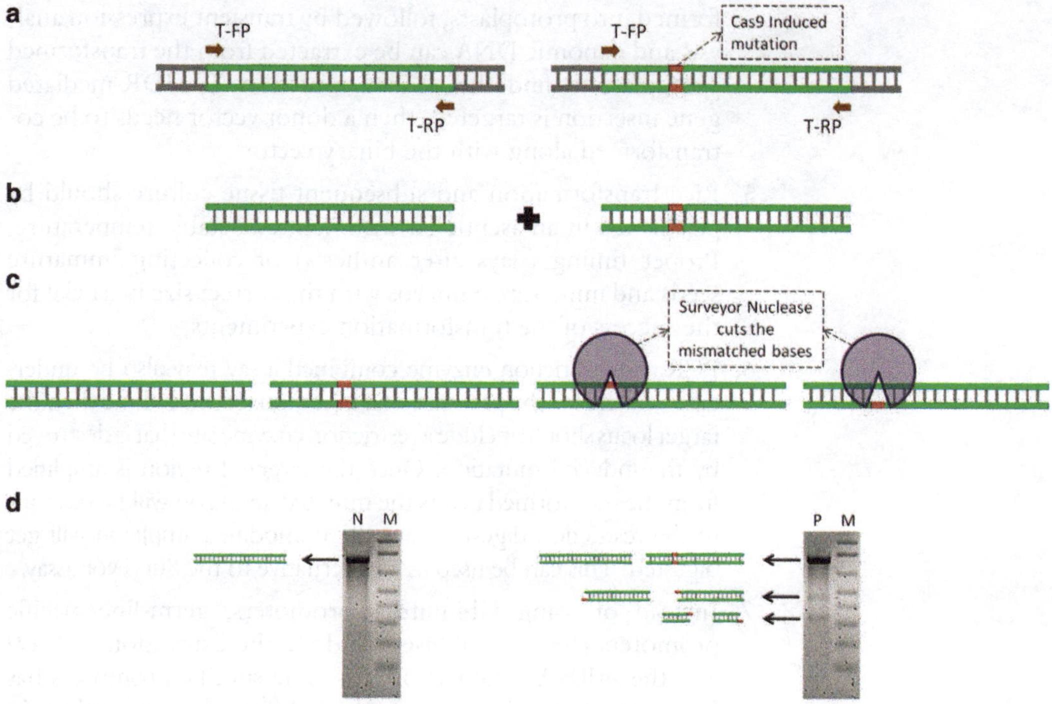

Fig. 4 The illustration of Surveyor Assay for detection of Cas9-induced mutations. PCR is performed on transformed and tissue culture controls plants (**a**) and the products are hybridized to form hetroduplex DNA (**b**). The surveyor nuclease cuts the mismatched base pairs from hetroduples DNA (**c**) yielding a distinctive banding pattern, allowing plants carrying the Cas9-induced mutation to be identified (**d**)

4 Notes

1. When designing the target sequence, introns and the region near the 3′ end of the coding sequence should be avoided.
2. Once the 20 nt target sequence has been identified, off targets should be identified using either Blast search (http://blast.ncbi.nlm.nih.gov/) or web tools such as Cas Offinder (http://www.rgenome.net/cas-offinder/, [80]) or CRISPR-P (http://cbi.hzau.edu.cn/crispr/, [81]).
3. If HDR-based gene insertion is to be performed, a double stranded DNA donor molecule containing the long homology arm is required. Single-stranded DNA oligos with a shorter homology arm can also be used [73, 82].
4. Prior to transformation, a transient expression assay of the protoplast should be performed in order to validate the activity of the guide sequence that has been designed. The binary vector containing the guide sequence and Cas9 can be trans-

formed into protoplasts, followed by transient expression analysis and genomic DNA can be extracted from the transformed protoplast to find the desired mutation. If HDR-mediated gene insertion is targeted, then a donor vector needs to be co-transformed along with the binary vector.

5. Rice transformation and subsequent tissue culture should be performed in an aseptic environment at a stable temperature. Proper timing (days after anthesis) of collecting immature seeds and immature embryos with the correct size is crucial for the success of the transformation experiments.
6. PCR and restriction enzyme combined assay may also be undertaken to test for the presence of a mutation. If this is done then the target locus should include a restriction enzyme site that is destroyed by the induced mutation. Once the targeted region is amplified from the transformed plants the mutated amplicon will be resistant to the restriction digestion but the unmodified amplicon will get digested. This can be used as an alternative to the Surveyor assay.
7. Instead of using Ubiquitous promoters, germ-line-specific promoters [83] can be used to drive the expression of Cas9 and the sgRNA. The use of germ-line-specific promoters has been reported to improve the heritability of the mutations in the T_1 and T_2 generations resulting in fewer chimeric plants.

Acknowledgments

The authors thank Dr. Tobias Kretzschmar for insightful comments and discussions. This work has been supported by a grant from International Rice Research Institute and the Bill and Melinda Gates foundation.

References

1. Wang H, Pandey S, Velarde O, Hardy B (2012) Patterns of varietal adoption and economics of rice production in Asia. International Rice Research Institute, Los Baños
2. Dalton TJ (2004) A household hedonic model of rice traits: economic values from farmers in West Africa. Agric Econ 31:149–159
3. Fitzgerald MA, McCouch SR, Hall RD (2009) Not just a grain of rice: the quest for quality. Trends Plant Sci 14:133–139
4. Juliano BO (1979) The chemical basis of rice grain quality. Chemical aspects of rice grain quality. International Rice Research Institute, Los Baños
5. Juliano BO, Villareal CP (1993) Grain quality evaluation of world rices. International Rice Research Institute, Los Baños, Laguna
6. Juliano B (2003) Rice chemistry and quality. International Rice Research Institute, Los Baños, Laguna
7. Philpot K, Martin M, Butardo V, Willoughby D, Fitzgerald M (2006) Environmental factors that affect the ability of amylose to contribute to retrogradation in gels made from rice flour. J Agric Food Chem 54:5182–5190
8. Champagne ET, Bett KL, Vinyard BT, McClung AM, Barton FE, Moldenhauer K, Linscombe S, McKenzie K (1999) Correlation between cooked rice texture and rapid visco analyser measurements. Cereal Chem 76:764–771
9. Cagampang GB, Perez CM, Juliano BO (1973) A gel consistency test for eating quality of rice. J Sci Food Agric 24:1589–1594

10. Pandey MK, Rani NS, Madhav MS, Sundaram RM, Varaprasad GS, Sivaranjani AK, Bohra A, Kumar GR, Kumar A (2012) Different isoforms of starch-synthesizing enzymes controlling amylose and amylopectin content in rice (*Oryza sativa* L.). Biotechnol Adv 30:1697–1706
11. Fitzgerald MA, Sackville Hamilton NR, Calingacion MN, Verhoeven HA, Butardo VM (2008) Is there a second fragrance gene in rice? Plant Biotechnol J 6:416–423
12. Champagne ET (2008) Rice aroma and flavor: a literature review. Cereal Chem 85:447–456
13. Chen S, Yang Y, Shi W, Ji Q, He F, Zhang Z, Cheng Z, Liu X, Xu M (2008) Badh2, encoding betaine aldehyde dehydrogenase, inhibits the biosynthesis of 2-acetyl-1-pyrroline, a major component in rice fragrance. Plant Cell 20:1850–1861
14. Huang TC, Teng CS, Chang JL, Chuang HS, Ho CT, Wu ML (2008) Biosynthetic mechanism of 2-acetyl-1-pyrroline and its relationship with Δ1-pyrroline-5-carboxylic acid and methylglyoxal in aromatic rice (*Oryza sativa* L.) callus. J Agric Food Chem 56:7399–7404
15. Kovach MJ, Calingacion MN, Fitzgerald MA, McCouch SR (2009) The origin and evolution of fragrance in rice (*Oryza sativa* L.). Proc Natl Acad Sci 106:14444–14449
16. Leung H, Hettel GP, Cantrell RP (2002) International Rice Research Institute: roles and challenges as we enter the genomics era. Trends Plant Sci 7:139–142
17. Peng S, Khush GS, Virk P, Tang Q, Zou Y (2008) Progress in ideotype breeding to increase rice yield potential. Field Crops Res 108:32–38
18. Lee JW, Hong K (2012) Economic growth in Asia: determinants and prospects. Japan World Econ 24:101–113
19. FAO, WFP, IFAD (2012) The State of Food Insecurity in the World 2012. Economic growth is necessary but not sufficient to accelerate reduction of hunger and malnutrition. FAO, Rome
20. ADB (2012) Food Security and Poverty in Asia Pacific: Key Challenges and Policy Issues. Asian Development Bank, Los Baños
21. Son JS, Do VB, Kim KO, Cho MS, Suwonsichon T, Valentin D (2014) Understanding the effect of culture on food representations using word associations: The case of 'rice' and 'good rice. Food Qual Prefer 31:38–48
22. Sreenivasulu N, Butardo VM, Misra G, Cuevas RP, Anacleto R, Kavi Kishor PB (2015) Designing climate-resilient rice with ideal grain quality suited for high-temperature stress. J Exp Bot 66:1737–1748
23. Peng B, Kong H, Li Y, Wang L, Zhong M, Sun L, Gao G, Zhang Q, Luo L, Wang G, Xie W, Chen J, Yao W, Peng Y, Lei L, Lian X, Xiao J, Xu C, Li X, He Y (2014) OsAAP6 functions as an important regulator of grain protein content and nutritional quality in rice. Nat Commun 5:4847
24. Fan C, Xing Y, Mao H, Lu T, Han B, Xu C, Li X, Zhang Q (2006) GS3, a major QTL for grain length and weight and minor QTL for grain width and thickness in rice, encodes a putative transmembrane protein. Theor Appl Genet 112:1164–1171
25. Song XJ, Huang W, Shi M, Zhu MZ, Lin HX (2007) A QTL for rice grain width and weight encodes a previously unknown RING-type E3 ubiquitin ligase. Nat Genet 39:623–630
26. Weng J, Gu S, Wan X, Gao H, Guo T, Su N, Lei C, Zhang X, Cheng Z, Guo X, Wang J, Jiang L, Zhai H, Wan J (2008) Isolation and initial characterization of GW5, a major QTL associated with rice grain width and weight. Cell Res 18:1199–1209
27. Kwon SW, Cho YC, Lee JH, Suh JP, Kim JJ, Kim MK, Choi IS, Hwang HG, Koh HJ, Kim YG (2011) Identification of quantitative trait loci associated with rice eating quality traits using a population of recombinant inbred lines derived from a cross between two temperate japonica cultivars. Mol Cells 31:437–445
28. Liu X, Wan X, Ma X, Wan J (2011) Dissecting the genetic basis for the effect of rice chalkiness, amylose content, protein content, and rapid viscosity analyzer profile characteristics on the eating quality of cooked rice using the chromosome segment substitution line population across eight environments. Genome 54:64–80
29. Bradbury LMT, Fitzgerald TL, Henry RJ, Jin Q, Waters DLE (2005) The gene for fragrance in rice. Plant Biotechnol J 3:363–370
30. Isshiki M, Morino K, Nakajima M, Okagaki RJ, Wessler SR, Izawa T, Shimamoto K (1998) A naturally occurring functional allele of the rice waxy locus has a GT to TT mutation at the 5′ splice site of the first intron. Plant J 15:133–138
31. Wang E, Wang J, Zhu X, Hao W, Wang L, Li Q, Zhang L, He W, Lu B, Lin H, Ma H, Zhang G, He Z (2008) Control of rice grain-filling and yield by a gene with a potential signature of domestication. Nat Genet 40:1370–1374
32. Lucca P, Hurrell R, Potrykus I (2001) Genetic engineering approaches to improve the bioavailability and the level of iron in rice grains. Theor Appl Genet 102:392–397

33. Paine JA, Shipton CA, Chaggar S, Howells RM, Kennedy MJ, Vernon G, Wright SY, Hinchliffe E, Adams JL, Silverstone AL, Drake R (2005) Improving the nutritional value of Golden Rice through increased pro-vitamin A content. Nat Biotechnol 23:482–487
34. Lucca P, Poletti S, Sautter C (2006) Genetic engineering approaches to enrich rice with iron and vitamin A. Physiol Plant 126:291–303
35. Bhullar NK, Gruissem W (2013) Nutritional enhancement of rice for human health: The contribution of biotechnology. Biotechnol Adv 31:50–57
36. The 3,000 rice genomes project (2014) The 3,000 rice genomes project. Gigascience 3:7
37. Gao H, Smith J, Yang M, Jones S, Djukanovic V, Nicholson MG, West A, Bidney D, Falco SC, Jantz D, Lyznik LA (2010) Heritable targeted mutagenesis in maize using a designed endonuclease. Plant J 61:176–187
38. Miller JC, Holmes MC, Wang J, Guschin DY, Lee YL, Rupniewski I, Beausejour CM, Waite AJ, Wang NS, Kim KA, Gregory PD, Pabo CO, Rebar EJ (2007) An improved zinc-finger nuclease architecture for highly specific genome editing. Nat Biotechnol 25:778–785
39. Sander JD, Dahlborg EJ, Goodwin MJ, Cade L, Zhang F, Cifuentes D, Curtin SJ, Blackburn JS, Thibodeau-Beganny S, Qi Y, Pierick CJ, Hoffman E, Maeder ML, Khayter C, Reyon D, Dobbs D, Langenau DM, Stupar RM, Giraldez AJ, Voytas DF, Peterson RT, Yeh JR, Joung JK (2011) Selection-free zinc-finger-nuclease engineering by context-dependent assembly (CoDA). Nat Methods 8:67–69
40. Wood AJ, Lo T-W, Zeitler B, Pickle CS, Ralston EJ, Lee AH, Amora R, Miller JC, Leung E, Meng X, Zhang L, Rebar EJ, Gregory PD, Urnov FD, Meyer BJ (2011) Targeted genome editing across species using ZFNs and TALENs. Science 333:307
41. Christian M, Cermak T, Doyle EL, Schmidt C, Zhang F, Hummel A, Bogdanove AJ, Voytas DF (2010) Targeting DNA double-strand breaks with TAL effector nucleases. Genetics 186:757–761
42. Zhang F, Cong L, Lodato S, Kosuri S, Church GM, Arlotta P (2011) Efficient construction of sequence-specific TAL effectors for modulating mammalian transcription. Nat Biotechnol 29:149–153
43. Hockemeyer D, Wang H, Kiani S, Lai CS, Gao Q, Cassady JP, Cost GJ, Zhang L, Santiago Y, Miller JC, Zeitler B, Cherone JM, Meng X, Hinkley SJ, Rebar EJ, Gregory PD, Urnov FD, Jaenisch R (2011) Genetic engineering of human pluripotent cells using TALE nucleases. Nat Biotechnol 29:731–734
44. Reyon D, Tsai SQ, Khayter C, Foden JA, Sander JD, Joung JK (2012) FLASH assembly of TALENs for high-throughput genome editing. Nat Biotechnol 30:460–465
45. Boch J, Scholze H, Schornack S, Landgraf A, Hahn S, Kay S, Lahaye T, Nickstadt A, Bonas U (2009) Breaking the code of DNA binding specificity of TAL-type III effectors. Science 326:1509–1512
46. Moscou MJ, Bogdanove AJ (2009) A simple cipher governs DNA recognition by TAL effectors. Science 326:1501
47. Sanjana NE, Cong L, Zhou Y, Cunniff MM, Feng G, Zhang F (2012) A transcription activator-like effector toolbox for genome engineering. Nat Protoc 7:171–192
48. Deveau H, Garneau JE, Moineau S (2010) CRISPR/Cas system and its role in phage-bacteria interactions. Annu Rev Microbiol 64:475–493
49. Horvath P, Barrangou R (2010) CRISPR/Cas, the immune system of bacteria and archaea. Science 327:167–170
50. Makarova KS, Haft DH, Barrangou R, Brouns SJJ, Charpentier E, Horvath P, Moineau S, Mojica FJ, Wolf YI, Yakunin AF, van der Oost J, Koonin EV (2011) Evolution and classification of the CRISPR-Cas systems. Nat Rev Microbiol 9:467–477
51. Bhaya D, Davison M, Barrangou R (2011) CRISPR-Cas systems in bacteria and archaea: versatile small RNAs for adaptive defense and regulation. Annu Rev Genet 45:273–297
52. Cong L, Ran FA, Cox D, Lin S, Barretto R, Habib N, Hsu PD, Wu X, Jiang W, Marraffini LA, Zhang F (2013) Multiplex genome engineering using CRISPR/Cas systems. Science 339:819–823
53. Mali P, Yang L, Esvelt KM, Aach J, Guell M, DiCarlo JE, Norville JE, Church GM (2013) RNA-guided human genome engineering via Cas9. Science 339:823–826
54. Jinek M, East A, Cheng A, Lin S, Ma E, Doudna J (2013) RNA-programmed genome editing in human cells. Elife 2:1–9
55. Cho SW, Kim S, Kim JM, Kim JS (2013) Targeted genome engineering in human cells with the Cas9 RNA-guided endonuclease. Nat Biotechnol 31:230–232
56. Garneau JE, Dupuis MÈ, Villion M, Romero DA, Barrangou R, Boyaval P, Fremaux C, Horvath P, Magadán AH, Moineau S (2010) The CRISPR/Cas bacterial immune system cleaves bacteriophage and plasmid DNA. Nature 468:67–71

57. Mazumdar S, Quick WP, Bandyopadhyay A (2016) CRISPR-Cas9 mediated genome editing in rice, advancements and future possibilities. Indian J Plant Physiol 21:437–445
58. Li J, Norville JE, Aach J, McCormack M, Zhang D, Bush J, Church GM, Sheen J (2013) Multiplex and homologous recombination-mediated genome editing in *Arabidopsis* and *Nicotiana benthamiana* using guide RNA and Cas9. Nat Biotechnol 31:688–691
59. Nekrasov V, Staskawicz B, Weigel D, Jones JD, Kamoun S (2013) Targeted mutagenesis in the model plant *Nicotiana benthamiana* using Cas9-guided endonuclease. Nat Biotechnol 31:691–693
60. Shan Q, Wang Y, Li J, Zhang Y, Chen K, Liang Z, Zhang K, Liu J, Xi JJ, Qiu JL, Gao C (2013) Targeted genome modification of crop plants using a CRISPR–Cas system. Nat Biotechnol 31:686–688
61. Belhaj K, Chaparro-Garcia A, Kamoun S, Nekrasov V (2013) Plant genome editing made easy: targeted mutagenesis in model and crop plants using the CRISPR/Cas system. Plant Methods 9:39
62. Feng Z, Zhang B, Ding W, Liu X, Yang DL, Wei P, Cao F, Zhu S, Zhang F, Mao Y, Zhu JK (2013) Efficient genome editing in plants using a CRISPR/Cas system. Cell Res 23:1229–1232
63. Jiang W, Zhou H, Bi H, Fromm M, Yang B, Weeks DP (2013) Demonstration of CRISPR/Cas9/sgRNA-mediated targeted gene modification in *Arabidopsis*, tobacco, sorghum and rice. Nucleic Acids Res 41:e188
64. Mao Y, Zhang H, Xu N, Zhang B, Gou F, Zhu JK (2013) Application of the CRISPR-Cas system for efficient genome engineering in plants. Mol Plant 6:2008–2011
65. Miao J, Guo D, Zhang J, Huang Q, Qin G, Zhang X, Wan J, Gu H, Qu LJ (2013) Targeted mutagenesis in rice using CRISPR–Cas system. Cell Res 23:1233–1236
66. Sugano SS, Shirakawa M, Takagi J, Matsuda Y, Shimada T, Hara-Nishimura I, Kohchi T (2014) CRISPR/Cas9-mediated targeted mutagenesis in the liverwort *Marchantia polymorpha* L. Plant Cell Physiol 55:475–481
67. Jiang H, Wang N (2014) Targeted genome editing of sweet orange using Cas9/sgRNA. PLoS One 9:e93806
68. Wang Y, Cheng X, Shan Q, Zhang Y, Liu J, Gao C, Qiu JL (2014) Simultaneous editing of three homoeoalleles in hexaploid bread wheat confers heritable resistance to powdery mildew. Nat Biotechnol 32:947–951
69. Yin X, Biswal AK, Dionora J, Perdigon KM, Balahadia CP, Mazumdar S, Chater C, Lin HC, Coe RA, Kretzschmar T, Gray JE, Quick WP, Bandyopadhyay A (2017) CRISPR-Cas9 and CRISPR-Cpf1 mediated targeting of a stomatal developmental gene EPFL9 in rice. Plant Cell Rep 36:1–13
70. Xie K, Yang Y (2013) RNA-Guided genome editing in plants using a CRISPR-Cas system. Mol Plant 6:1975–1983
71. Kumar V, Jain M (2015) The CRISPR-Cas system for plant genome editing: advances and opportunities. J Exp Bot 66:47–57
72. Fitzgerald MA (2004) Starch. In: Champagne ET (ed) Rice Chemistry and Technology, 3rd edn. American Association of Cereal Chemists, Eagan, pp 109–133
73. Shan Q, Wang Y, Li J, Gao C (2014) Genome editing in rice and wheat using the CRISPR/Cas system. Nat Protoc 9:2395–2410
74. Chen K, Gao C (2015) Developing CRISPR technology in major crop plants. Adv New Technol Target Modif Plant Genomes 145–159
75. Hiei Y, Komari T (2008) *Agrobacterium*-mediated transformation of rice using immature embryos or calli induced from mature seed. Nat Protoc 3:824–834
76. Slamet-Loedin IH, Chadha-Mohanty P, Torrizo L (2014) *Agrobacterium*-mediated transformation: rice transformation. Methods Mol Biol 1099:261–271
77. Ran FA, Hsu PDP, Wright J, Agarwala V, Scott DA, Zhang F (2013) Genome engineering using the CRISPR-Cas9 system. Nat Protoc 8:2281–2308
78. Brinkman EK, Chen T, Amendola M, van Steensel B (2014) Easy quantitative assessment of genome editing by sequence trace decomposition. Nucleic Acids Res 42(22):e168
79. Hill JT, Demarest BL, Bisgrove BW, Su YC, Smith M, Yost HJ (2014) Poly peak parser: method and software for identification of unknown indels using sanger sequencing of polymerase chain reaction products. Dev Dyn 243(12):1632–1636
80. Bae S, Park J, Kim J (2014) Cas-OFFinder: a fast and versatile algorithm that searches for potential off-target sites of Cas9 RNA-guided endonucleases. Bioinformatics 30(10):1473–1475
81. Lei Y, Lu L, Liu H, Li S, Xing F, Chen L (2014) CRISPR-P: a web tool for synthetic single-guide RNA design of CRISPR-system in plants. Mol Plant 7(9):1494–1496
82. Strouse B, Bialk P, Niamat RA, Rivera-Torres N, Kmiec EB (2014) Combinatorial gene edit-

ing in mammalian cells using ssODNs and TALENs. Sci Rep 4:3791

83. Mao Y, Zhang Z, Feng Z, Wei P, Zhang H, Botella JR, Zhu JK (2016) Development of germ-line-specific CRISPR-Cas9 systems to improve the production of heritable gene modifications in *Arabidopsis*. Plant Biotechnol J 14:519–532. https://doi.org/10.1111/pbi.12468

Index

Nese Sreenivasulu (ed.), *Rice Grain Quality: Methods and Protocols*, Methods in Molecular Biology, vol. 1892, https://doi.org/10.1007/978-1-4939-8914-0,

MIX
Papier aus verantwortungsvollen Quellen
Paper from responsible sources
FSC® C105338

Printed by Books on Demand, Germany